14.00
1-EC

Analytical Profiles
of
Drug Substances
Volume 1

EDITORIAL BOARD

Glenn A. Brewer, Jr. Arthur F. Michaelis
Lester Chafetz   Stephen M. Olin
Jack P. Comer   Gerald J. Papariello
Klaus Florey   Carl R. Rehm
David E. Guttman   Bernard Z. Senkowski
Eric H. Jensen   Frederick Tishler

Analytical Profiles of Drug Substances
Volume 1

Edited by
Klaus Florey
The Squibb Institute for Medical Research
New Brunswick, New Jersey

Contributing Editors

Glenn A. Brewer, Jr.   Stephen M. Olin
Lester Chafetz   Gerald J. Papariello
Jack P. Comer   Bernard Z. Senkowski

*Compiled under the auspices of the
Pharmaceutical Analysis and Control Section
Academy of Pharmaceutical Sciences*

Academic Press   New York and London   1972

COPYRIGHT © 1972, BY THE AMERICAN PHARMACEUTICAL ASSOCIATION
ALL RIGHTS RESERVED
NO PART OF THIS BOOK MAY BE REPRODUCED IN ANY FORM,
BY PHOTOSTAT, MICROFILM, RETRIEVAL SYSTEM, OR ANY
OTHER MEANS, WITHOUT WRITTEN PERMISSION FROM
THE PUBLISHERS.

ACADEMIC PRESS, INC.
111 Fifth Avenue, New York, New York 10003

United Kingdom Edition published by
ACADEMIC PRESS, INC. (LONDON) LTD.
24/28 Oval Road, London NW1 7DD

LIBRARY OF CONGRESS CATALOG CARD NUMBER: 70-187259

PRINTED IN THE UNITED STATES OF AMERICA

CONTENTS

AFFILIATIONS OF EDITORS, CONTRIBUTORS, AND REVIEWERS. vii
FOREWORD. ix
PREFACE xi

Acetohexamide 1
   C. E. Shafer

Chlordiazepoxide 15
   A. MacDonald, A. F. Michaelis, and B. Z. Senkowski

Chlordiazepoxide Hydrochloride. 39
   A. MacDonald, A. F. Michaelis, and B. Z. Senkowski

Cycloserine 53
   J. W. Lamb

Cyclothiazide 65
   C. D. Wentling

Diazepam 79
   A. MacDonald, A. F. Michaelis, and B. Z. Senkowski

Erythromycin Estolate 101
   J. M. Mann

Halothane 119
   R. D. Daley

Levarterenol Bitartrate 149
   C. F. Schwender

Meperidine Hydrochloride 175
   Nancy P. Fish and Nicholas J. DeAngelis

CONTENTS

Meprobamate 207
  C. Shearer and P. Rulon

Nortriptyline Hydrochloride 233
  J. L. Hale

Potassium Phenoxymethyl Penicillin 249
  John M. Dunham

Propoxyphene Hydrochloride 301
  B. McEwan

Sodium Cephalothin 319
  R. J. Simmons

Sodium Secobarbital 343
  I. Comer

Triamcinolone 367
  K. Florey

Triamcinolone Acetonide 397
  K. Florey

Triamcinolone Diacetate 423
  K. Florey

Vinblastine Sulfate 443
  J. H. Burns

Vincristine Sulfate 463
  J. H. Burns

AFFILIATIONS OF EDITORS, CONTRIBUTORS, AND REVIEWERS

G. A. Brewer, Jr., The Squibb Institute for Medical Research, New Brunswick, New Jersey

J. H. Burns, Eli Lilly and Company, Indianapolis, Indiana

L. Chafetz, Warner-Lambert Research Institute, Morris Plains, New Jersey

I. Comer, Eli Lilly and Company, Indianapolis, Indiana

J. P. Comer, Eli Lilly and Company, Indianapolis, Indiana

R. D. Daley, Ayerst Laboratories, Rousses Point, New York

N. J. DeAngelis, Wyeth Laboratories, Philadelphia, Pennsylvania

J. M. Dunham, The Squibb Institute for Medical Research, New Brunswick, New Jersey

N. P. Fish, Wyeth Laboratories, Philadelphia, Pennsylvania

K. Florey, The Squibb Institute for Medical Research, New Brunswick, New Jersey

D. E. Guttman, Smith, Kline and French Laboratories, Philadelphia, Pennsylvania

J. L. Hale, Eli Lilly and Company, Indianapolis, Indiana

E. H. Jensen, The Upjohn Company, Kalamazoo, Michigan

J. W. Lamb, Eli Lilly and Company, Indianapolis, Indiana

A. MacDonald, Hoffmann-La Roche Inc., Nutley, New Jersey

B. McEwan, Eli Lilly and Company, Indianapolis, Indiana

AFFILIATIONS

J. M. Mann, Eli Lilly and Company, Indianapolis, Indiana

A. F. Michaelis, Sandoz Pharmaceuticals, Hanover, New Jersey

N. Neuss, Eli Lilly and Company, Indianapolis, Indiana

S. M. Olin, Ayerst Laboratories, New York, New York

G. J. Papariello, Wyeth Laboratories, Philadelphia, Pennsylvania

E. L. Pratt, The Sterling-Winthrop Research Institute, Rensselaer, New York

C. R. Rehm, Ciba-Geigy Inc., Summit, New Jersey

N. E. Rigler, Lederle Laboratories, Pearl River, New York

P. Rulon, Wyeth Laboratories, Philadelphia, Pennsylvania

C. F. Schwender, Warner-Lambert Research Institute, Morris Plains, New Jersey

B. Z. Senkowski, Hoffmann-La Roche Inc., Nutley, New Jersey

C. E. Shafer, Eli Lilly and Company, Indianapolis, Indiana

C. Shearer, Wyeth Laboratories, Philadelphia, Pennsylvania

R. J. Simmons, Eli Lilly and Company, Indianapolis, Indiana

L. H. Sternbach, Hoffmann-La Roche Inc., Nutley, New Jersey

F. Tishler, Ciba-Geigy Inc., Ardsley, New York

C. D. Wentling, Eli Lilly and Company, Indianapolis, Indiana

FOREWORD

The concept for gathering together and publishing pertinent information on the physical and chemical properties of various official and new drug substances had its origin with the members of the Section on Pharmaceutical Analysis and Quality Control of the Academy of Pharmaceutical Sciences. More than two years of consideration preceded the authorization of this ambitious project by the Executive Committee of the Academy in the Spring of 1970. The immediate and virtually spontaneous enlistment of the first group of contributors to this work attested to its importance and the wisdom of pursuing its publication.

By coincidence, the delegates to the sesquicentennial anniversary meeting of the United States Pharmacopeial Convention, Inc., in Washington, D. C. on April 8-10, 1970, adopted the following resolution:

> Whereas widespread interest has been expressed in the inclusion of additional information about physical and chemical properties of drugs recognized in the United States Pharmacopeia
>
> *Be It Resolved* that the Board of Trustees consider publishing in the Pharmacopeia, or in a companion publication, information on such attributes as solubilities, pH and pK values, spectra and spectrophotometric constants, and stability data, pertaining to pharmacopeial drugs.

The U.S.P.C. Board of Trustees unanimously approved the resolution in principle on June 4, 1970 and authorized the Director of Revision to include in the U S.P. monographs such physical-chemical information as he deemed proper and also to cooperate with the Academy of Pharmaceutical Sciences to secure the publication of other physical-chemical data.

It was my privilege to be the President of the Academy during the period when *Analytical Profiles* was under consideration. It is my unusual and unique honor as President of the Academy and Director of U.S.P Revision to assist in the institution and dedication of this first volume. I trust that it will serve immeasurably in providing the scientific community with an authoritative source of information on the properties of many of our important drug compounds.

Thomas J. Macek
Director of Revision
The United States Pharmacopeia

PREFACE

Although the official compendia define a drug substance as to identity, purity, strength, and quality, they normally do not provide other physical or chemical data, nor do they list methods of synthesis or pathways of physical or biological degradation and metabolism. At present such information is scattered through the scientific literature and the files of pharmaceutical laboratories.

For drug substances important enough to be accorded monographs in the official compendia such supplemental information should also be made readily available. To this end the Pharmaceutical Analysis Section, Academy of Pharmaceutical Sciences, has started a cooperative venture to compile and publish *Analytical Profiles of Drug Substances* in a series of volumes of which this is the first. It is also planned to revise and update these profiles at suitable intervals.

Our endeavor has been made possible through the encouragement we have received from many sources and through the enthusiasm and cooperative spirit of our contributors. For coining the term Analytical Profile we are indebted to Dr. James L. Johnson of the Upjohn Company.

We hope that this, our contribution to the better understanding of drug characteristics, will prove to be useful. We welcome new collaborators, and we invite comment and counsel to guide the infant to maturity.

Klaus Florey

ACETOHEXAMIDE

C. E. Shafer

CONTENTS

1. Description
   1.1 Name, Formula, Molecular Weight
2. Physical Properties
   2.1 Infrared Spectrum
   2.2 Nuclear Magnetic Resonance Spectrum
   2.3 Ultraviolet Spectrum
   2.4 Melting Range
   2.5 Differential Thermal Analysis
   2.6 Thermogravimetric Analysis
   2.7 Solubility
3. Synthesis
   3.1 First Example
   3.2 Second Example
4. Stability
   4.1 Infrared Analysis
   4.2 Solubility Analysis
5. Drug Metabolic Products
6. Methods of Analysis
   6.1 Titrimetric
   6.2 Ultraviolet Spectrophotometric (Alkali)
   6.3 Ultraviolet Spectrophotometric (Alcohol)
7. Pharmacokinetics
8. Identification
9. References

ACETOHEXAMIDE

1. **Description**

    1.1 **Name, Formula, Molecular Weight**
    Acetohexamide is N-(p-acetylphenylsulfonyl)-N'-cyclohexylurea[1], and is also known as 1-[(p-acetylphenyl)sulfonyl]-3-cyclohexylurea[2,4].

    $$CH_3\overset{O}{\overset{\|}{C}}-\text{\textlangle{}phenyl\textrangle{}}-SO_2-NH\overset{O}{\overset{\|}{C}}NH-\text{\textlangle{}cyclohexyl\textrangle{}}$$

    $C_{15}H_{20}N_2O_4S$ Ml. wt. = 324.40

2. **Physical Properties**

    2.1 **Infrared Spectrum**
    The infrared absorption spectrum of acetohexamide (Lilly Reference Standard, Lot No. 2KT47) is presented in Figure No. 1. The spectrum was taken in a KBr pellet with a Perkin-Elmer 221 Infrared Spectrophotometer. The band at 3300, 3200 cm.$^{-1}$ is typical of an N-H stretch; the band at 1680 cm.$^{-1}$ of a conjugated ketone, and the band at 1445 cm.$^{-1}$ of a C-CH$_3$ group.

    2.2 **Nuclear Magnetic Resonance Spectrum**
    The NMR spectrum of acetohexamide (Lilly Reference Standard, Lot No. 2KT47) is presented in Figure 2. The spectrum was produced using a Varian A60 NMR Instrument. The quartet at low field (<8δ) is an A_2B_2 pattern that is typical of para substitution and the singlet at 2.52δ is typical of a methyl group adjacent to a carbonyl function.

    2.3 **Ultraviolet Spectrum**
    The UV absorption spectrum of acetohexamide (Lilly Reference Standard, Lot No. 2KT47) is presented in Figure 3. The spectrum was produced using a Cary 14 instrument. The sample was dissolved in 95% ethanol using 8.06 mg. per 25 ml. of solution. The λ max. at 247 nm is typical of substituted aceto-

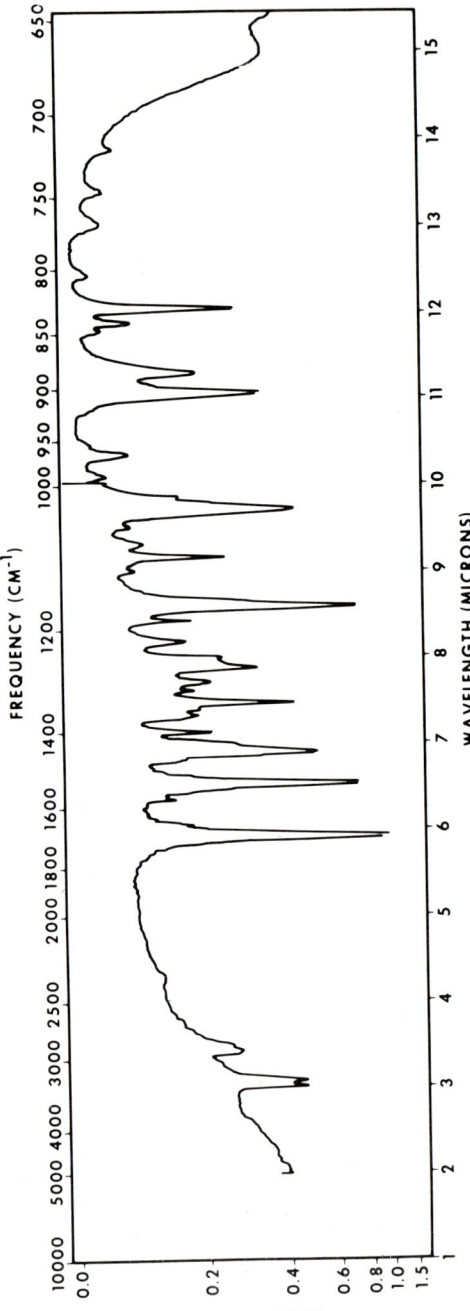

Fig. 1. Infrared spectrum

ACETOHEXAMIDE

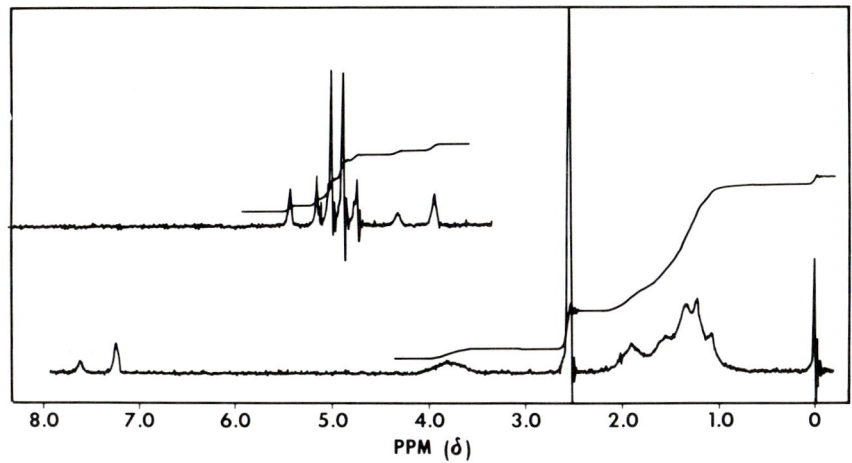

Fig. 2. NMR spectrum (sweep offset 000,200.0 Hz)

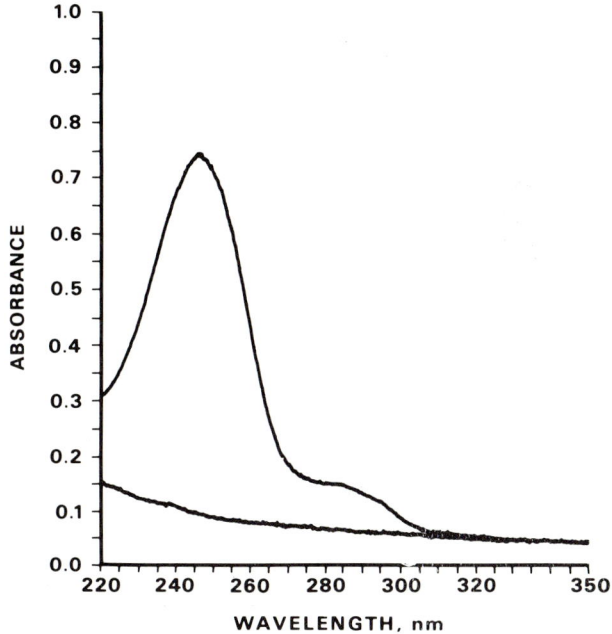

Fig. 3. Ultraviolet spectrum

phenone, and the λ max. at 284 nm is typical of a conjugated aromatic ring.

2.4 Melting Range[4]
Between 184° and 189° (Class Ia NF).

2.5 Differential Thermal Analysis
The curve in Figure 4 is a DTA spectrum of acetohexamide (Lilly Reference Standard, Lot No. 2KT47) as produced by a DuPont 900 D.T. analyzer. The spectrum shows a sharp phase transition at 192°C.

2.6 Thermogravimetric Analysis
The TGA spectrum of acetohexamide in Figure 5 (Lilly Reference Standard, Lot No. 2KT47) was produced with a DuPont 950 T.G. Analyzer. It indicates less than 1% weight loss to 191°C., approximately 5% weight loss at 198°C., and approximately 20% weight loss at 210°C.

2.7 Solubility[4]
Practically insoluble in water.
Practically insoluble in ether.
Slightly soluble in alcohol.
Slightly soluble in chloroform.
Soluble in pyridine.
Soluble in dilute solutions of
  alkali hydroxides.

3. Synthesis

Two examples for the preparation of acetohexamide are listed.

3.1 First Example[5]

$$CH_3CO-\underset{}{\bigcirc}-NH_2 \xrightarrow{NaNO_2 / HCl, AcOH}$$

ACETOHEXAMIDE

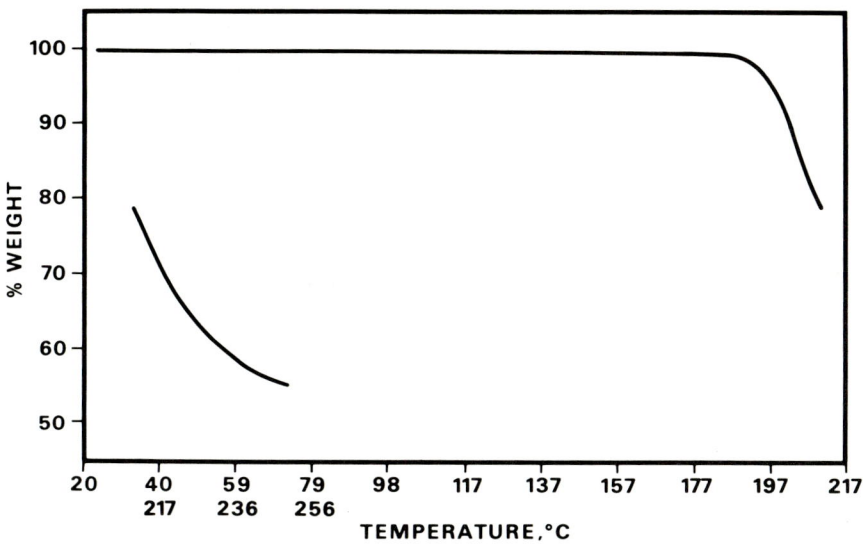

Fig. 4. DTA spectrum

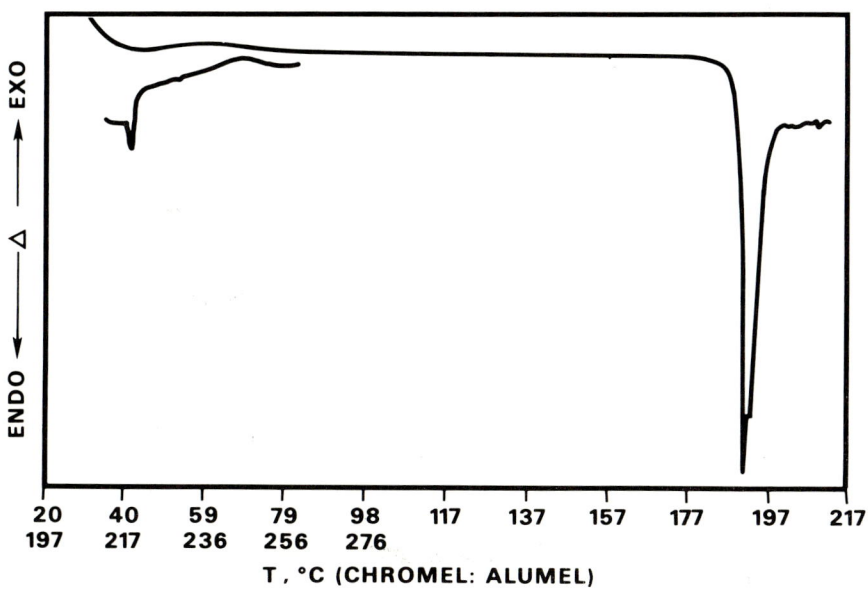

Fig. 5. TGA spectrum

7

CH$_3$CO—C$_6$H$_4$—N:HCl $\xrightarrow[\text{CuCl}_2\cdot 2\text{H}_2\text{O}]{\text{SO}_2, \text{AcOH}}$

CH$_3$CO—C$_6$H$_4$—SO$_2$Cl $\xrightarrow{\text{NH}_4\text{OH}}$

CH$_3$CO—C$_6$H$_4$—SO$_2$NH$_2$ $\xrightarrow[\text{C}_6\text{H}_{11}\text{—NCO}]{\text{K}_2\text{CO}_3}$

CH$_3$CO—C$_6$H$_4$—SO$_2$NHCONH—C$_6$H$_{11}$

3.2 Second Example[6]

CH$_3$CO—C$_6$H$_4$—Cl $\xrightarrow{\text{Na}_2\text{SO}_3}$

CH$_3$CO—C$_6$H$_4$—SO$_3$Na $\xrightarrow{\text{POCl}_3}$

CH$_3$CO—C$_6$H$_4$—SO$_2$Cl $\xrightarrow{\text{NH}_4\text{OH}}$

CH$_3$CO—C$_6$H$_4$—SO$_2$NH$_2$ $\xrightarrow[\text{K}_2\text{CO}_3]{\text{ClCO}_2\text{C}_2\text{H}_5}$

CH$_3$CO—C$_6$H$_4$—SO$_2$NHCO$_2$C$_2$H$_5$ $\xrightarrow{\text{C}_6\text{H}_{11}\text{—NH}_2}$

CH$_3$CO—C$_6$H$_4$—SO$_2$NHCONH—C$_6$H$_{11}$

4. Stability

Acetohexamide is stable under all normal storage conditions. Temperatures above 80°C. were required to produce measurable degradation[7] as indicated by solubility analysis, or by spectrophotometric assay at 249 nm[8]. Irradiation under a Hanovia lamp for one, and for three hours, showed about 25%, and 50%, decomposition by the spectrophotometer assay[7]. The compound p-acetylbenzenesulfonamide was detected as a degradation product using thin layer chromatography with several combinations of plates, solvents and visualization techniques[7].

4.1 Infrared Analysis[7]

Weigh 250 mg. of acetohexamide. Transfer to a 100-ml. volumetric flask. Add 10 ml. of 0.5 N sodium hydroxide solution and about 20 ml. of water. Shake for 30 minutes on an automated shaker, dilute to volume with water and mix well.

Transfer a 20-ml. aliquot of the above solution to a 125-ml. separatory funnel and heat on a steam bath for 10 minutes. Allow to cool, and make strongly acid with several drops of concentrated hydrochloric acid.

Extract the sample with three 25-ml. portions of chloroform. Drain the extract through anhydrous sodium sulfate and collect it in a 150-ml. beaker. Thoroughly wash the sodium sulfate with chloroform, and collect the wash with the extract.

Evaporate the sample, using a mild temperature and a stream of air, to a suitable volume for quantitative transfer to a 25-ml. volumetric flask. Use chloroform to transfer the sample, dilute to volume with chloroform, and mix well.

Using a suitable spectrophotometer, determine the absorbance of the final solution at the maximum at about 5.90 μ in 1.0 mm. sodium chloride cells using chloroform as the blank.

In the same manner determine the absorbance of 250 mg. of NF Acetohexamide Reference Standard.

(Sa. Abs./Std. Abs.) x (mg. Std./mg. Sa.) x 100% = percent acetohexamide.

4.2   Solubility Analysis[8]

Thermally degraded acetohexamide was evaluated using solubility analysis. Because of the high cost due to the large amount of time involved, and because the method is applicable only to the drug substance free of excipients, the use of solubility analysis as a routine procedure is limited.

5.   Drug Metabolic Products

The principal (urine) metabolite in man was found to be 1-(p-hydroxyethylbenzene-sulfonyl)-3-cyclohexylurea (hydroxyhexamide) by Welles, Root and Anderson[2]. Later, McMahon, Marshall and Culp[9] discovered other (urine) metabolites in man in which the cyclohexane ring was hydroxylated to form hydroxyaceto-hexamide and hydroxyhydroxyhexamide. (See the following structures with names.)

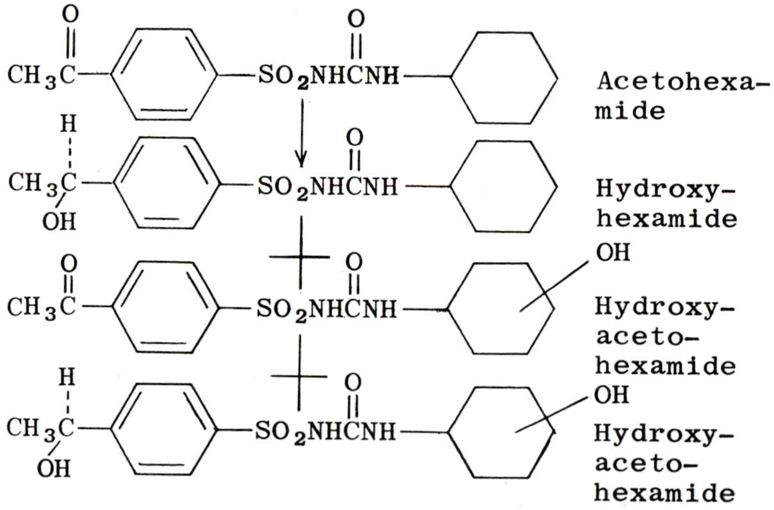

The metabolites of radioactive acetohexamide found in the urine of man by Galloway, McMahon, Culp, Marshall, and Young[10] were hydroxyhexamide, smaller amounts of hydroxyacetohexamide and hydroxyhydroxyhexamide, and small quantities of other hydroxylated isomers. Smith, Vecchio and Forist[11] determined the average biological half-lives of acetohexamide and its major metabolite hydroxyhexamide as 1.3 hours and 4.6 hours, respectively. The hydroxy metabolites also have hypoglycemic activity.

6. <u>Methods of Analysis</u>

    6.1 <u>Titrimetric</u>[4]
        This method is described in detail.

    6.2 <u>Ultraviolet Spectrophotometric</u>[7]
        <u>(alkali)</u>
        Transfer 50.0 mg. of acetohexamide to a 100-ml. volumetric flask. Add 15 ml. of purified water and 5.0 ml. of 0.5 N sodium hydroxide. Shake mechanically for at least 30 minutes. Dilute to volume with purified water and mix well. Transfer 2.0 ml. of the solution to a 100-ml. volumetric flask. Dilute to volume with purified water and mix well.
        Weigh 50.0 mg. of NF Acetohexamide Reference Standard, transfer to a 100-ml. volumetric flask. Dissolve and dilute just like the sample.
        Concomitantly determine the absorbance of the sample solution and of the standard solution in 1-cm. silica cells, at the maximum at about 249 nm, with a suitable spectrophotometer, using purified water as the blank.
        (Sa. abs./Std. abs.) x 100% = percent acetohexamide.

    6.3 <u>Ultraviolet Spectrophotometric</u>
        <u>(alcohol)</u>
        This method[12] uses absolute alcohol to dissolve 25.0 mg. of the acetohexamide and 25.0 mg. of the acetohexamide reference standard.

Each solution is diluted to 50.0 ml.; then 2.0 ml. of each are diluted to 100.0 ml. with absolute alcohol. The absorbances of the solutions are determined at the maximum at about 248 nm, using absolute alcohol as the blank. Calculation: (Sample abs./Standard abs.) x 100% = percent acetohexamide.

7. Pharmacokinetics

No pharmacokinetic studies have been reported in the literature. However, seventy to eighty percent of a single oral dose of 1 gram of acetohexamide was recovered as a metabolite within 24 hours in the urine of four human volunteers. It is suggested that acetohexamide is converted to hydroxyhexamide in the liver. Acetohexamide and hydroxyhexamide are probably converted to hydroxyacetohexamide and hydroxyhydroxyhexamide in both the liver and kidneys[9].

A multiple compartment system would probably be necessary to obtain a rate constant for metabolism, distribution rate constant, rate constant for absorption, etc. Sulfonylurea drugs may lower blood sugar by stimulating the beta cells of the pancreatic islets to release endogenous insulin. Also, it has been reported that the sulfonylureas block the degradation of insulin by the enzyme insulinase[13].

8. Identification

A. The X-ray diffraction pattern of acetohexamide conforms to the pattern stated in the NF[4]. No polymorphs have been observed and documented.[7] Infrared and a chemical method are listed for identification purposes[4].

B. A 1 in 100,000 solution of acetohexamide in 0.01 N sodium hydroxide exhibits an ultraviolet absorbance maximum at about 249 nm. Two other sulfonylurea compounds, chlorpropamide and tolbutamide, exhibit a maximum absorptivity at about 230 and about 228 nm, respectively, in the same medium[3].

C. Strickland[14] described a method for the separation and detection of four of the more important oral hypoglycemic agents using thin-layer chromatography. It was used for the identification of acetohexamide, chlorpropamide, tolbutamide, and phenformin hydrochloride.

References

1. USAN, J.A.M.A., **180**, 232 (1962).
2. Welles, J. S., Root, M. A., Anderson, R. C., Proc. Soc. Exp. Biol. and Med., **107**, 583-5 (1961).
3. Salim, E. F. and Hilty, W. W., J. Pharm. Sci., **56**, 385-6 (1967).
4. Amer. Pharm. Ass., National Formulary **XIII**, 19-21 (1970).
5. U.S. Patent 3,320,312 (Patented May 16, 1967).
6. Mfg. Chem., **34**, 454-6, 467 (1963).
7. Comer, J. P., The Lilly Research Laboratories, unpublished data.
8. Comer, J. P. and Howell, L. D., J. Pharm. Sci., **53**, 335-7 (1964).
9. McMahon, R. E., Marshall, F. J., and Culp, H.W., Pharmacol. Exptl. Therap., **149**, 272-9 (1965).
10. Galloway, J.A., McMahon, R. E., Culp, H. W., Marshall, F. J., and Young, E. C., Diabetes, **16**, 118 (1967).
11. Smith, D. L., Vecchio, T. J., and Forist, A. A., Metab., Clin. Exptl., **14**, 229-40 (1965).
12. Baltazar, J. and Ferreira Braga, M. M., Revista Portuguesa de Farmacia, **16**, 169-74 (1966).
13. Mirsky, I. A., Perisutti, G., and Diengott, D., Metab. Clin. Exptl., **5**, 156-61 (1956).
14. Strickland, R. D., J. Chromatog. **24**, 455-8 (1966).

The author expresses appreciation to Mr. C. D. Underbrink and Dr. A. D. Kossoy of the Analytical Development Department at Eli Lilly and Company for assistance in preparing and interpreting data in this profile.

CHLORDIAZEPOXIDE

A. MacDonald, A. F. Michaelis, and B. Z. Senkowski

Chemistry Reviewed by L. H. Sternbach

CONTENTS

Analytical Profile - Chlordiazepoxide

1. Description
   1.1 Name, Formula, Molecular Weight
   1.2 Appearance, Color, Odor

2. Physical Properties
   2.1 Infrared Spectrum
   2.2 Nuclear Magnetic Resonance Spectrum
   2.3 Ultraviolet Spectrum
   2.4 Mass Spectrum
   2.5 Optical Rotation
   2.6 Melting Range
   2.7 Differential Scanning Calorimetry
   2.9 Solubility
   2.10 Crystal Properties
   2.11 Dissociation Constant
   2.12 Distribution Coefficient

3. Synthesis

4. Stability Degradation

5. Drug Metabolic Products and Pharmacokinetics

6. Methods of Analysis
   6.1 Elemental Analysis
   6.2 Phase Solubility Analysis
   6.3 Chromatographic Analysis
       6.31 Thin-Layer Chromatographic Analysis
       6.32 Column Chromatographic Analysis
       6.33 Vapor Phase Chromatography
   6.4 Direct Spectrophotometric Analysis
   6.5 Colorimetric Analysis
   6.6 Polarographic Analysis
   6.7 Non-Aqueous Titration
   6.8 Gravimetric Method of Analysis

7. Acknowledgments

8. References

CHLORDIAZEPOXIDE

1. Description

   1.1 Name, Formula, Molecular Weight
   Chlordiazepoxide is 7-Chloro-2-(methylamino)-5-phenyl-3H-1,4-benzodiazepine 4-oxide.

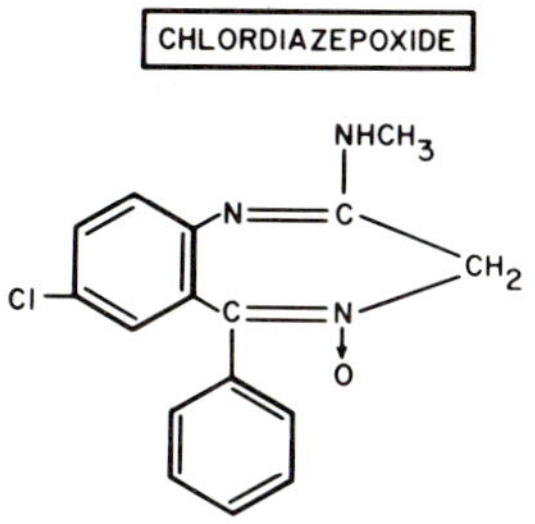

$C_{16}H_{14}ClN_3O$ Mol. Weight: 299.76

   1.2 Appearance, Color, Odor
   Slightly yellow, practically odorless, crystalline powder.

2. Physical Properties

   2.1 Infrared Spectrum
   The infrared spectrum of reference standard chlordiazepoxide is presented in Figure 1[1]. The spectrum was measured in a KBr pellet which contained 0.9 mg/300 mg KBr.

   The following bands (cm^{-1}) have been assigned for Figure 1[2].
   a. Characteristic for NH: 3270
   b. Characteristic for $N=C-NC_2H_5$: 1630
   c. Characteristic for aromatic groups: 1590, 1470

   2.2 Nuclear Magnetic Resonance Spectrum
   The NMR spectrum shown in Figure 2 was obtained by dissolving 50.3 mg of reference standard chlordiazepoxide in

FIGURE 1

Infrared Spectrum of Chlordiazepoxide

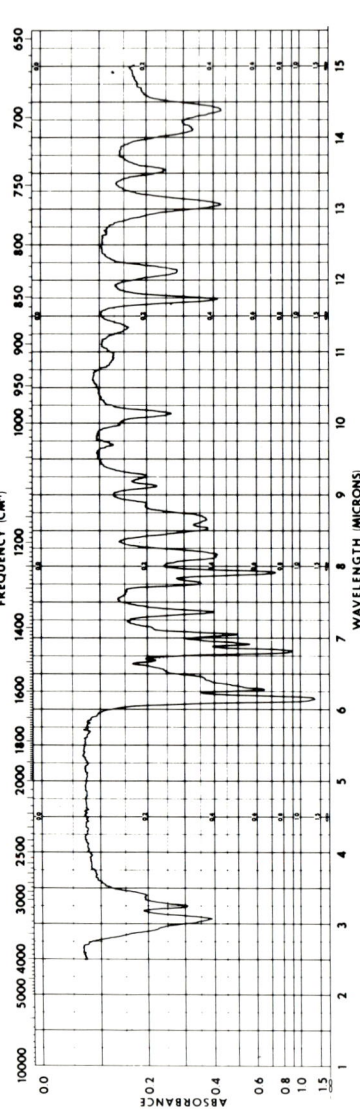

CHLORDIAZEPOXIDE

FIGURE 2

NMR Spectrum of Chlordiazepoxide

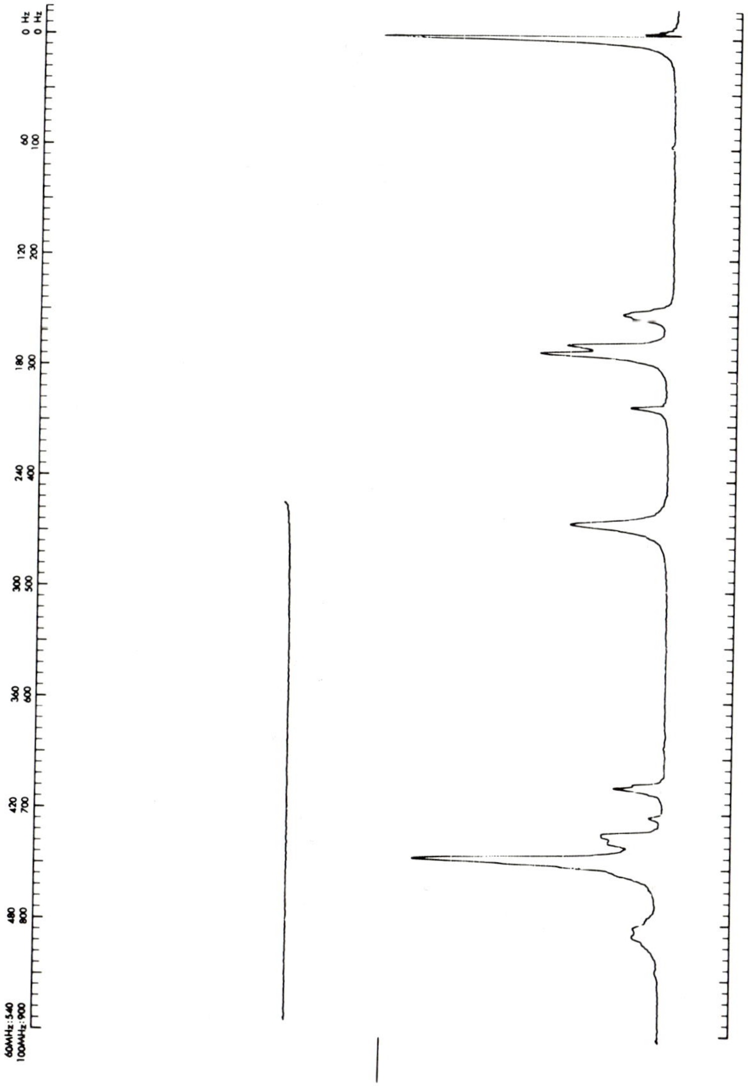

0.4 ml of DMSO-d_6 containing tetramethylsilane as internal reference. The spectral assignments are shown in Table I[3]. NMR studies by Nuhn and Bley at 100 Mhz and various temperatures indicate that the methylene protons at position three are not equivalent [4].

TABLE I

Chlordiazepoxide

Protons At	Chemical Shift τ (ppm)	Type (J in Hz)
C_3	5.60	5
C_5(Ph)	2.57	5
C_6, C_8	2.70	m(u)
C_9	3.24	5(c)
C_2(NH$C\underline{H}_3$)	7.21	d(5)
C_2(N$\underline{H}$)	1.94	d(b) (5+)

b = singlet; b(c) = complex singlet; d = doublet; d(b) = broad doublet; m(u) = unsymmetrical multiple

2.3 Ultraviolet Spectrum

Chlordiazepoxide when scanned between 420 and 210 nm (5.07 mg/L in acidified 3A alcohol) exhibits two maxima. These were located at 245-246 nm (a = 110.9) and 311-312 nm (a = 34.2). Minima were observed at 294-295 nm and 218-219 nm[5]. The spectrum shown in Figure 3 was obtained using a solution of 0.5 mg chlordiazepoxide/100 ml acidified alcohol (approximately 0.1NH_2SO_4 in 3A alcohol).

2.4 Mass Spectrum

The mass spectrum of chlordiazepoxide was obtained using a CEC 21-110 mass spectrometer with an ionizing energy of 70eV and a temperature of 190°C. Three low resolution spectra of chlordiazepoxide are shown in Figure 4. A regular change in intensities (e.g. at m/e 299, 282, 36) can be noted by comparing spectrum (a) to (c). This had been related to the time that the sample was exposed to the instrumental temperature, and is probably related to thermal effects. Hence, scan (a) should be considered most representative of the chlordiazepoxide mass spectral characteristics.

FIGURE 3

Ultraviolet Spectrum

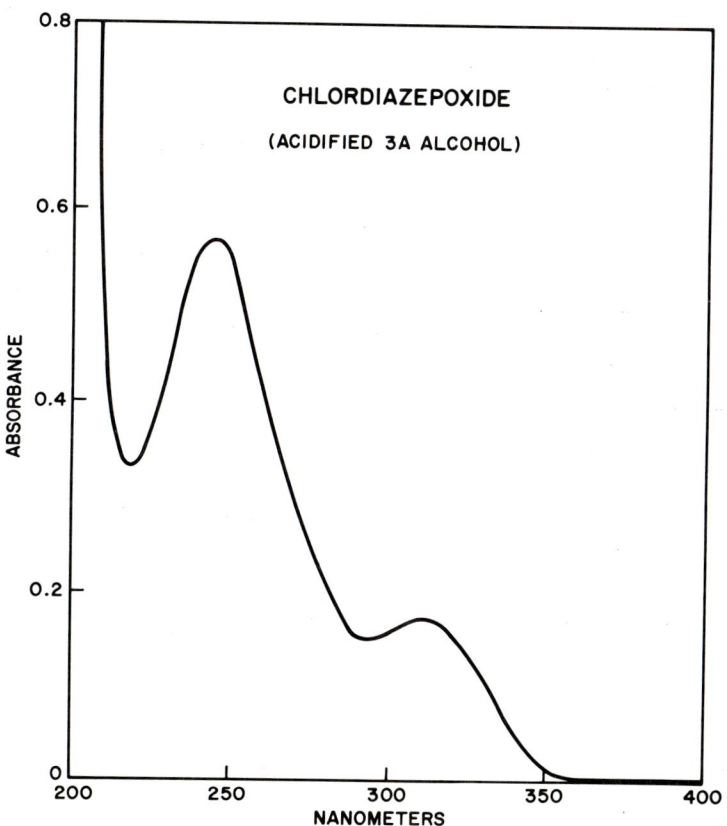

FIGURE 4

Mass Spectra of Chlordiazepoxide

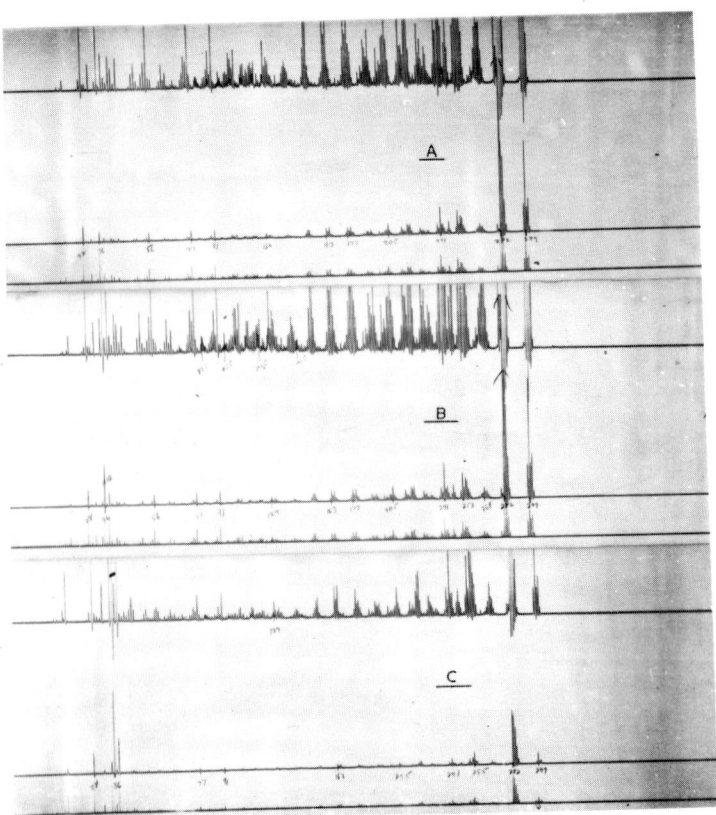

Table II lists the elemental compositions for the most diagnostic ions as determined by high resolution mass spectrometry[6]. The molecular ion for chlordiazepoxide was observed at m/e 299. The ions at m/e 283 and m/e 282 correspond to loss of O or OH as expected for an N-oxide. Loss of CH_4N leading to m/e 269 is best explained as loss of the methyl amino group. Other ions in Table II can be ascribed to losses of Cl, part of the seven-membered ring, and combinations of all these primary losses.

TABLE II

High Resolution Mass Spectrum of Chlordiazepoxide[a]

Found Mass	Calcd. Mass	C	H	N	O	Cl_{35}
299.0804	299.0868	16	14	3	1	1
298.0721	298.0790	16	13	3	1	1
283.0818	283.0874	16	14	3	0	1
282.0769	282.0796	16	13	3	0	1
269.0461	269.0525	15	10	2	1	1
268.0582	268.0640	15	11	3	0	1
253.0501	253.0531	15	10	2	0	1
247.1087	247.1107	16	13	3	0	0
241.0503	241.0531	14	10	2	0	1
218.0801	218.0842	15	10	2	0	0
205.0758	205.0764	14	9	2	0	0
190.0630	190.0655	14	8	1	0	0

a. Only peaks discussed are included in this table.

2.5 Optical Rotation

Chlordiazepoxide exhibits no optical activity.

2.6 Melting Range

A sharp melting point is not observed with chlordiazepoxide. The melting range is wide and depends on the rate of heating. The melting range reported in NF XIII is 240-244°C.

2.7 Differential Scanning Calorimetry

The DSC spectrum of chlordiazepoxide is shown in Figure 5. The endotherm observed at 253°C corresponds to the decomposition of the drug. The ΔH_f was found to be 7.3 kcal/mole[8].

FIGURE 5

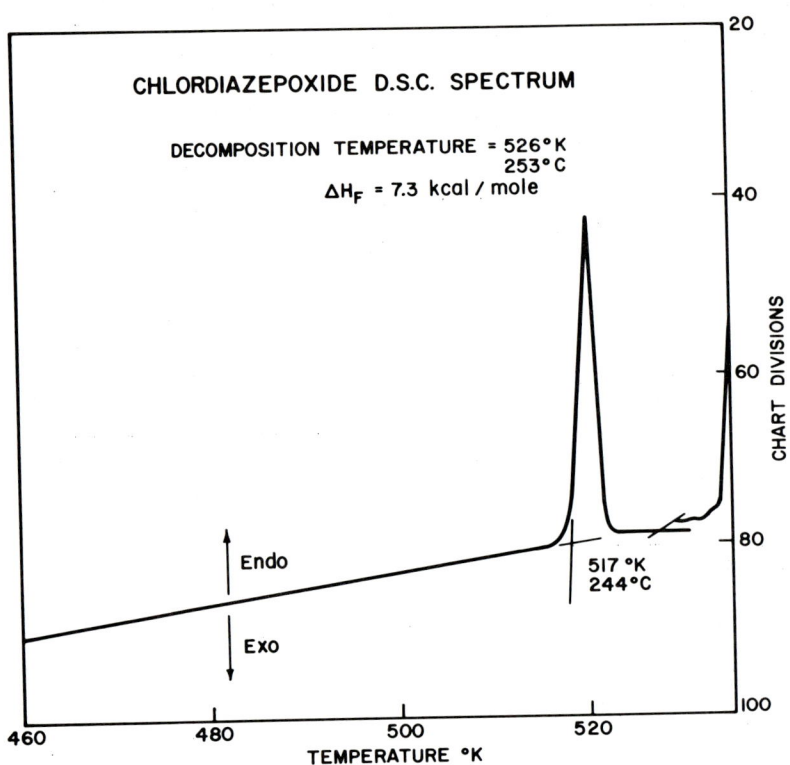

2.8 Thermogravimetric Analysis

A thermal gravimetric analysis performed on chlordiazepoxide exhibited no loss of weight when heated to 105°C[8].

2.9 Solubility

Approximate solubility data obtained at room temperature are given in the following table.

Solvent	Solubility mg/ml
petroleum ether (30°-60°)	insoluble
ether	1
water	2
isopropanol	11
3A alcohol	17
chloroform	17
95% ethanol	23
benzene	25
methanol	26

2.10 Crystal Properties

The x-ray powder diffraction pattern of chlordiazepoxide is presented in Table III[9].

Instrument Conditions

General Electric	Model XRD-6 Spectrogoniometer
Generator	50 KV 12-1/2 MA
Tube Target	Copper
Optics	0.2° Detector slit
	M.R. Sollor slit
	3° Beam slit
	0.0007" Ni filter
	4° take off angle
Goniometer	Scan at 0.4° 2Θ per minute
Detector	Amplifier - 16 coarse 817 fine (gain)
	Sealed proportional counter tube and DC voltage at plateau
	Pulse height selection E_L-5 volts; Eu - out
	Rate meter T.C.4
	2000 C/S full scale
Recorder	Chart speed 1 inch per 5 minutes

Samples prepared by grinding at room temperature.

TABLE III

Chlordiazepoxide

2θ	d^* Å	I/I_o^{**}
5.72°	15.45	43
7.00	12.63	4
8.08	10.94	6
9.00	9.83	6
10.96	8.07	15
11.48	7.71	100
11.88	7.45	17
12.28	7.21	10
13.92	6.36	31
14.64	6.05	10
14.96	5.92	10
15.76	5.62	23
16.12	5.50	17
17.28	5.13	59
17.64	5.03	28
18.76	4.73	38
19.20	4.62	4
19.48	4.56	11
19.96	4.45	4
20.36	4.36	20

*d = (interplanar distance) $\frac{n\lambda}{2 \sin \theta}$
**I/I_o = relative intensity (based on highest intensity of 1.00)

2.11 Dissociation Constant

The dissociation constant, pKa, for chlordiazepoxide has been determined spectrophotometrically to be 4.76 ± 0.05[10] and by titration using NaOH to be 4.9 (non-logarithmic plot)[11].

2.12 Distribution Coefficient

Reymond and Toome[12] found the distribution coefficient of chlordiazepoxide between n-octanol and pH 7.2 buffer to have a value of 171 at room temperature where $D = C_{octanol}/C_{buffer}$[12].

3. Synthesis

Chlordiazepoxide is prepared by the reaction scheme shown in Figure 6. 6-chloro-2-chloromethyl-4-phenyl-quinazoline 3-oxide is reacted with methylamine which presumably attacks the quinazoline at the 2 position. The nucleophilic attack is followed by enlargement of the ring to yield the 7-chloro-2-(methylamino)-5-phenyl-3H-1,4-benzodiazepine 4-oxide[13]. A complete review of the chemistry of benzodiazepines presents alternative pathways[14].

4. Stability Degradation

The known degradation products of chlordiazepoxide in aqueous solution are shown in Figure 7. In mild acid the lactam is formed while under strong acid treatment, the product of hydrolysis is 2-amino-5-chlorobenzophenone[15].

5. Drug Metabolic Products and Pharmacokinetics

The major metabolites in humans, the N-desmethyl metabolite and the lactam are shown in Figure 7. The analytical procedures for the metabolites have been described by deSilva[15].

A pharmacokinetic model for the disposition of chlordiazepoxide HCl in the dog has been presented in terms of a six-compartment open system[16]. The excellent agreement between the simulated and experimental data reflects the reliability of the assumption of first-order kinetics for all processes. The model provides a basis for the elucidation and quantitation of chlordiazepoxide and its pharmacologically active biotransformation products, the N-desmethyl metabolite and the lactam. The pathways in man[17-19] have been shown to be similar to those in the dog to the extent to which they are described in the model.

The main features of the physiological disposition of chlordiazepoxide HCl in the dog were: (a) its biotransformation to N-desmethyl as the exclusive route of drug elimination; (b) the elimination of N-desmethyl almost entirely by biotransformation, with up to 50% going to the lactam and the remainder going to an unidentified biotransformation product; and (c) the elimination of lactam by urinary excretion and further biotransformation.

FIGURE 6

SYNTHESIS OF CHLORDIAZEPOXIDE

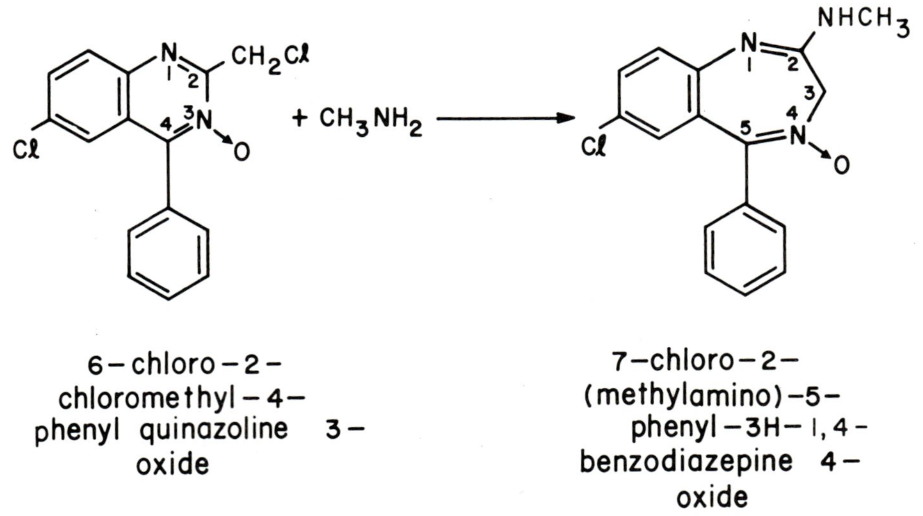

6-chloro-2-chloromethyl-4-phenyl quinazoline 3-oxide

+ CH_3NH_2 →

7-chloro-2-(methylamino)-5-phenyl-3H-1,4-benzodiazepine 4-oxide

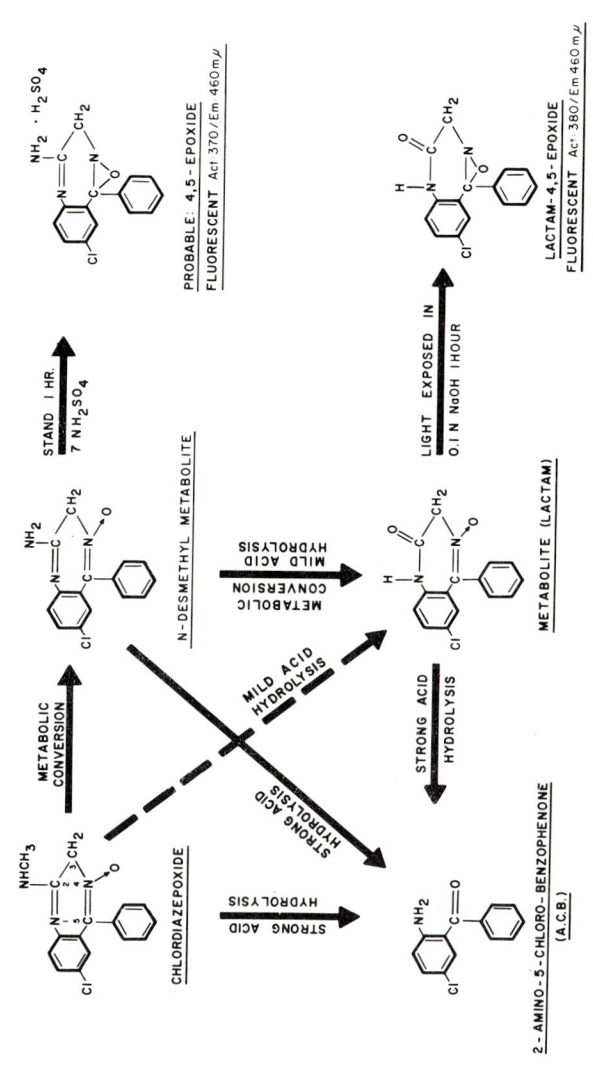

FIGURE 7

CHEMICAL REACTIONS OF Chlordiazepoxide

6. **Methods of Analysis**

   6.1 **Elemental Analysis**

Element	% Theory	Reported
C	64.11	64.38
H	4.71	4.66

   6.2 **Phase Solubility**
   Phase solubility analysis is carried out using isopropanol as the solvent. A typical example is shown in Figure 8 which also lists the conditions under which the analysis was carried out[20].

   6.3 **Chromatographic Analysis**
   Chromatographic analysis may be used to assess the stability and purity of chlordiazepoxide.

   6.31 **Thin-Layer Chromatographic Analysis**
   The following TLC procedure is useful for separating chlordiazepoxide from possible hydrolytic products[21]. Using silica gel G plates and ethyl acetate solvent systems, the sample containing 1 mg of chlordiazepoxide substance in chloroform, is spotted and subjected to ascending chromatography. After development of at least 10 cm, the plate is air dried and sprayed with 10% H_2SO_4. This is followed by heating at 105°C for 15 minutes and the following sequence of sprays:

   1. 0.1% sodium nitrite in water
   2. 0.5% ammonium sulfate in water
   3. 0.1% N-(1-Napthyl)-ethylenediamine dihydrochloride in water

   The lactam and benzophenone (ACB) products (Figure 7) will yield purple spots if present. Chlordiazepoxide does not react under this procedure and is subsequently visualized by platinum iodide spray. R_f's are as follows:

FIGURE 8

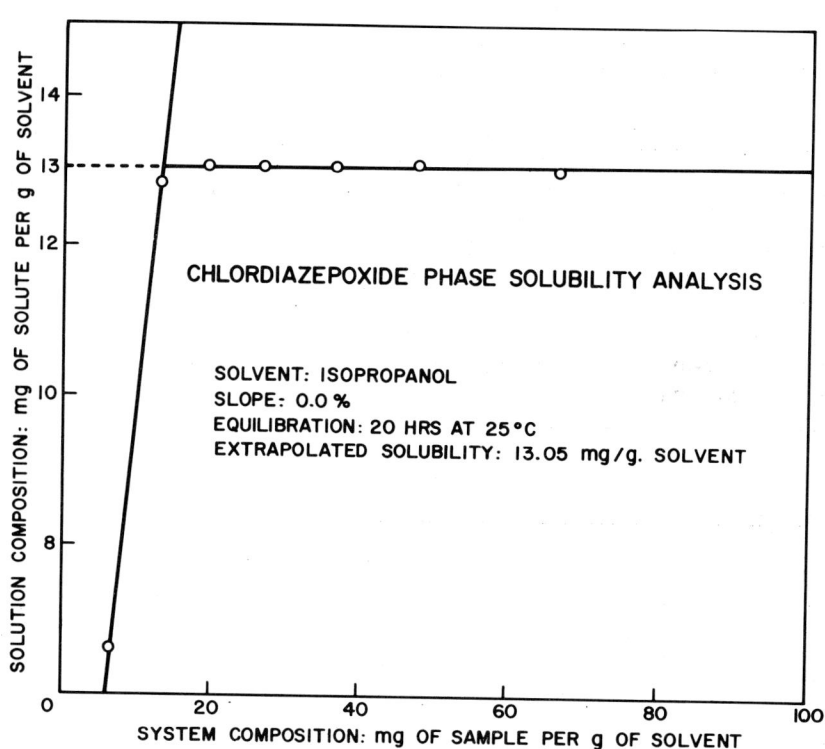

chlordiazepoxide	0.2
lactam (7-chloro-1,3-dihydro-5-phenyl-2H-1,4,benzodiazepin-2-one-4-oxide)	0.4 (sensitivity 0.1%)
benzophenone (5-chloro-2-aminobenzophenone)	0.9 (sensitivity 0.01%)

Additional TLC systems appearing in the literature are listed in Table IV.

6.32 Column Chromatographic Analysis

The analytical scale liquid-solid chromatographic separation of chlordiazepoxide has been reported by Scott and Bommer in their study of the separation of several benzodiazepines from each other and from biological media[27]. The important advantages indicated for this method of analysis are the mild operating conditions which enable collection of the compounds and also the simple sample preparation which does not require derivatization or hydrolysis of the sample.

The liquid solid chromatography was carried out using a Durapak "OPN" column (36-75 μ particle diameter). The mobile phase, which ranged from 100% hexane to 70:30 hexane-isopropanol (v/v), was pushed through the 1 mm i.d. stainless steel column using an air-driven pump at sufficient pressure to deliver a 1.0 ml/min. flow rate. The detection was an ultraviolet monitor with an 8 μl cell volume operated at 254 nm. Sensitivity in the microgram range was reported.

6.33 Vapor Phase Chromatography

J.A.F. deSilva has reported the determination of chlordiazepoxide in blood with a sensitivity limit of 0.5 to 0.8 μg/ml of blood[28]. The method involves the selective extraction of the compound into ether, hydrolysis to the 2-amino-5-chlorobenzophenone and subsequent determination by gas-liquid chromatography.

The GLC of unhydrolyzed chlordiazepoxide has been reported by Martin and Street[29] using acid-washed chromosorb W coated with dimethyldichlorosilane. The reported column temperature for chlordiazepoxide was 245°C.

TABLE IV

Thin-Layer Chromatography Systems for Chlordiazepoxide

System	Support	$R_f \times 100$	Detection	Reference
Methanol:acetone (12:88)	B	48	UV_{254}	22
96% ethanol	B	48	UV_{254}	22
Isopropanol:propyl ether	B	29	UV_{254}	22
Methanol:methyl-acetate:cyclohexane (17.8:48.6:33.6)	B	57	UV_{254}	22
Methanol:methyl-acetate:(17.1:82.3)	B	46	UV_{254}	22
Chloroform:methanol (10:1)	B	–	UV_{254}	23
Ethylacetate:ethanol (9:1)	B	–	UV_{254}	23
Benzene:dioxane: 28% ammonium hydroxide	A	38	Dragendorff Reagent	24
Acetone:cyclohexane: ethanol (4:4:2)	A	27	"	24
Benzene:acetone (4:1)	A	41	"	24
Methanol:acetone: ammonium hydroxide (50:50:1)	A	78	Marquis Reagent	25
Benzene:ethanol:25% ammonium hydroxide (50:10:5)	A	64	Dragendorff Reagent $KMnO_4$ H_2SO_4 - 10% EtOH 10% H_2O_2 HNO_3 - EtOH (10% sol.)	26

Adsorbant: A. Silica Gel G

  B. Silica Gel Gf

and under these conditions two peaks were obtained with retention times of 3.6 and 7.3 minutes, which indicates possible decomposition on the column. The sensitivity was reported to be 0.04 μg.

6.4 Direct Spectrophotometric Analysis

Direct spectrophotometric analysis of chlordiazepoxide has not been found to be applicable if significant quantities of the known hydrolytic contaminants are present. For material not containing the interfering species the reported maxima at 245-6 nm and 311-12 nm in acidified 3A alcohol may be used for quantitative measurements. The a values at these maxima are 110.9 and 34.2 respectively. The Technicon Autoanalyzer system for dosage form assays of chlordiazepoxide is based on the direct spectrophotometric assay.

6.5 Colorimetric Analysis

The colorimetric analysis of chlordiazepoxide is accomplished by acid hydrolysis of the compound followed by diazotization and coupling with N-(1-Napthyl)ethylenediamine dihydrochloride. The resultant reaction product exhibits a maxima at 540 nm. Similar procedures have been reported in the literature which employed other coupling reagents[30-31].

6.6 Polarographic Analysis

Polarographic behavior of chlordiazepoxide in both alcohol-water and non-aqueous media has been investigated[32,33,34]. General agreement is noted between authors. The following table summarizes some of the reported half-wave potentials.

The first wave has been ascribed to the reduction of the N→O, the second to the reduction of the 4,5 N=C$\stackrel{<}{}$ and the third to the reduction of the 1,2 N=C$\stackrel{<}{}$[4,5][32,33]. Proportionality between concentration and diffusion current was reported for the range 2×10^{-4} to 7×10^{-4}M. Dosage form analysis showed a precision of $\pm$ 2%[32].

CHLORDIAZEPOXIDE

TABLE V

Solvent	Wave 1	Wave 2	Wave 3	Reference
0.1N HCl containing 20% MeOH v/v	-0.250	-0.612	-1.127	32
Buffer at pH 4.6	-0.56	-0.84	-1.29	33
0.05 M NH_4Cl in 10% MeOH in absolute ethanol v/v	-0.826	-1.097	-1.605	32

6.7 Non-Aqueous Titration

Chlordiazepoxide may be titrated in $CHCl_3$ using $HClO_4$ in dioxane with a methyl red indicator. Each ml of 0.1N $HClO_4$ is equivalent to 29.98 mg of chlordiazepoxide[20,35].

6.8 Gravimetric Method of Analysis

The sample is dissolved in HCl and treated with 1% Reinecke salt. The precipitate is washed, dried at 105°C and weighed to constant weight or the dried precipitate may be dissolved in acetone for colorimetric determination at 530 mµ[36].

7. Acknowledgments

The authors wish to acknowledge the assistance of Dr. W. Benz, Dr. V. Venturella and Mr. T. Daniels in the preparation of this analytical profile.

8. References

1. Hawrylyshyn, M., Hoffmann-La Roche Inc., Personal Communication.
2. Traiman, S., Hoffmann-La Roche Inc., Personal Communication.
3. Johnson, J. and Venturella, V., Hoffmann-La Roche Inc., Personal Communication.
4. Nuhn, P. and Bley, W., *Pharmazie* 22a, 523 (1967).
5. Mahn, F., Hoffmann-La Roche Inc., Personal Communication.
6. Greeley, D. and Benz, W., Hoffmann-La Roche Inc., Personal Communications.
7. Venturella, V., Hoffmann-La Roche Inc., Personal Communication.
8. Donahue, J., Hoffmann-La Roche Inc., Personal Communication.
9. Sheridan, J.C., Hoffmann-La Roche Inc., Personal Communication.
10. Toome, V., Hoffmann-La Roche Inc., Personal Communication.
11. Yao, C., and Lau, E., Hoffmann-La Roche Inc., Personal Communication.
12. Reymond, G. and Toome, V., Hoffmann-La Roche Inc., Personal Communication.
13. Sternbach, L.H. and Reeder, E., *J. Org. Chem.*, 26, 111 (1961).
14. Archer, G. and Sternbach, L.H., *Chem. Reviews*, 68, 751 (1969).
15. deSilva, J.A.F. and D'Arconte, L.D., *J. Forensic Sci.*, 14, 184 (1969).
16. Kaplan, S.A., Lewis, M., Schwartz, M.A., Postma, E., Cotler, S., Abrusso, C.W., Lee, T.L., and Weinfeld, R.E., *J. Pharm. Sci.*, 59, 1569 (1970).
17. Koechlin, B.A., and D'Arconte, L., *Anal. Biochem.*, 5, 195 (1963).
18. Koechlin, B.A., Schwartz, M.A., Krol, G., and Oberhaens, W., *J. Pharmacol. Exp. Ther.*, 148, 339 (1965).
19. Schwartz, M.A., and Postma, E., *J. Pharm. Sci.*, 55, 1358 (1966).
20. MacMullan, E.A., Hoffmann-La Roche Inc., Personal Communication.

21. *National Formulary XIII*, 148-149 (1970).
22. Roeter, E. et al., *Zeit, f. Anal. Chem.*, 244, 45 (1969).
23. Beckstead, H.D., and Smith, S.J., *Arneim. Forsch.*, 18, 529 (1968).
24. Tomoda, M. *Kyoritsu Yakka Daigaku Kenkyv Neppo*, 10, 18 (1965).
25. Thomas, J.J. and Pyron, L.J., *J. Pharm. Belg.*, 19, 481 (1964).
26. Paulus, W., et al., *Arneimittel-Forsch*, 13, 609 (1963).
27. Scott, C.G., and Bommer, P., *J. Chrom. Sci.*, 8, 446 (1970).
28. deSilva, J.A.F., in *Theory and Application of GLC in Industry and Medicine*, Kromen and Bender, Eds., Grune and Stratton, Inc., p. 252, (1958).
29. Martin, C. and Street, H.V., *J. Chromatog.*, 22, 274 (1966).
30. Balogh, E. and Ajtay, K., *Rev. Med. (Targu-Mures)*, 14, 159 (1968).
31. Randall, L.O., *Dis. Nervous Syst. Suppl.*, 22, 1 (1961).
32. Senkowski, B.Z., et al., *Anal. Chem.*, 36, 1991 (1964).
33. Oelschlaeger, H., et al., *Arch. Pharm.*, 300, 250 (1967).
34. Oelschlaeger, H., et al., *Arch. Pharm.*, 296, 396 (1963).
35. Beral, H., et al., *Rev. Chim. (Bucharest)*, 16, 169 (1965).
36. Grev. I., and Barku, S., *Farmacia (Bucharest)*, 16, 199 (1968).

CHLORDIAZEPOXIDE HYDROCHLORIDE

A. MacDonald, A. F. Michaelis, and B. Z. Senkowski

CONTENTS

Analytical Profile - Chlordiazepoxide Hydrochloride

1. Description
   1.1 Name, Formula, Molecular Weight
   1.2 Appearance, Color, Odor

2. Physical Properties
   2.1 Infrared Spectrum
   2.2 Nuclear Magnetic Resonance Spectrum
   2.3 Ultraviolet Spectrum
   2.4 Mass Spectrum*
   2.5 Optical Rotation
   2.6 Melting Range
   2.7 Differential Scanning Calorimetry
   2.8 Thermogravimetric Analysis
   2.9 Solubility
   2.10 Crystal Properties
   2.11 Dissociation Constant*
   2.12 Distribution Coefficient*

3. Synthesis*

4. Stability Degradation*

5. Drug Metabolic Products and Pharmacokinetics*

6. Methods of Analysis
   6.1 Elemental Analysis
   6.2 Phase Solubility Analysis
   6.3 Chromatographic Analysis
      6.31 Thin-Layer Chromatography*
      6.32 Column Chromatographic Analysis*
      6.33 Vapor Phase Chromatography*
   6.4 Direct Spectrophotometric Analysis
   6.5 Colorimetric Analysis*
   6.6 Polarographic Analysis*
   6.7 Non-Aqueous Titration
   6.8 Gravimetric Method of Analysis*

7. Acknowledgments

8. References
   * Refer to Analytical Profile on Chlordiazepoxide.

1. Description

1.1 Name, Formula, Molecular Weight
Chlordiazepoxide hydrochloride is 7-chloro-2-(methylamino)-5-phenyl-3H-1,4-benzodiazepine 4-oxide hydrochloride.

$C_{16}H_{14}ClN_3O \cdot HCl$ 	Mol. Wt. 336.22

1.2 Appearance, Color, Odor
White or practically white, odorless, crystalline powder.

2. Physical Properties

2.1 Infrared Spectrum
The infrared spectrum of chlordiazepoxide hydrochloride is presented in Figure 1[1]. The spectrum was measured in a KBr pellet, and contained 1.0 mg/300 mg KBr. The following bands (cm^{-1}) have been assigned for Figure 1[2].

a. Characteristic for NH_2^+: 3100-2550
b. Characteristic for N=C-N$^+$H$_2$: 1680
c. Characteristic for aromatic and C=N: 1610, 1550

FIGURE 1

Infrared Spectrum of Chlordiazepoxide Hydrochloride

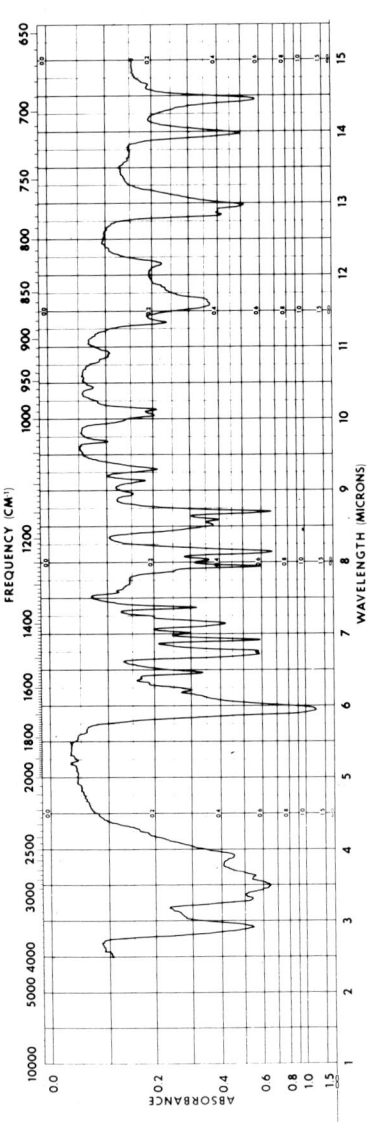

2.2 Nuclear Magnetic Resonance Spectrum

The NMR spectrum shown in Figure 2 was obtained by dissolving 49.9 mg chlordiazepoxide hydrochloride in 0.5 ml of DMSO-d_6 containing tetramethylsilane as internal reference. The spectral assignments are shown in Table I[3]. NMR studies by Nuhn and Bley at 100 MHz and various temperatures indicate that the methylene protons at position three are not equivalent[4].

TABLE I

Chlordiazepoxide Hydrochloride

Protons at	Chemical Shift τ (ppm)	Type (J in Hz)
C_3	5.06	S(b)
C_5 (Ph)	2.52	S(c)
C_6	2.12	S
C_8	2.27	d(3)
C_9	3.02	d(3)
C_2 (HN$\underline{CH}_3$)	6.74	S(b)
C_2 (NH)	-0.72	S(b)
$\underline{H}$ (HCl)	1.98	S

S = singlet; S(b) = broad singlet; S(c) = complex singlet; d = doublet

2.3 Ultraviolet Spectrum

Chlordiazepoxide hydrochloride measured between 360 and 210 nm in acidified 3A alcohol (0.1N H_2SO_4) exhibits maxima at 245-6 nm (a = 96.5) and 311-12 nm (a = 30.5). Minima were observed at 218 and 295-6 nm[5]. The spectrum is shown in the profile for chlordiazepoxide[6].

2.4 Mass Spectrum

The mass spectrum of chlordiazepoxide hydrochloride is identical to that reported for chlordiazepoxide[6].

2.5 Optical Rotation

Chlordiazepoxide hydrochloride exhibits no optical activity.

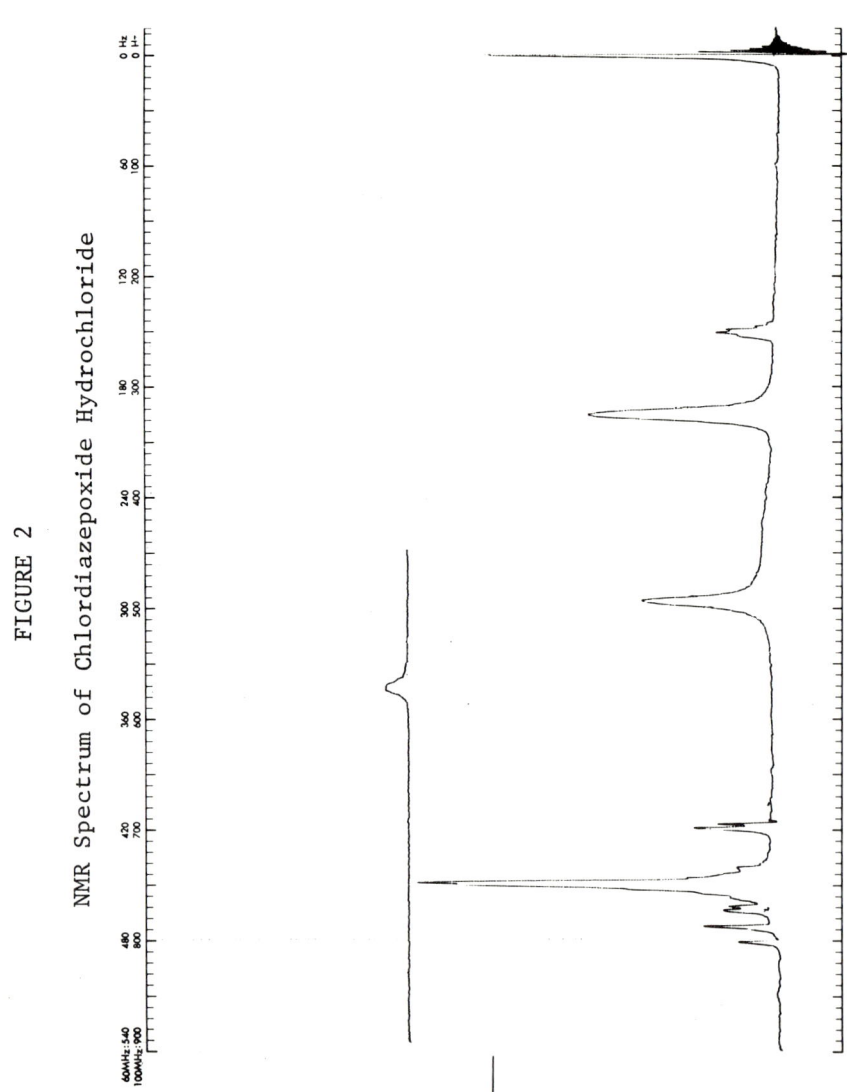

FIGURE 2

NMR Spectrum of Chlordiazepoxide Hydrochloride

2.6 Melting Range
Chlordiazepoxide melts with decomposition. The range described in USP XVIII is 212-218°C using Class I method[7].

2.7 Differential Scanning Calorimetry
The DSC spectrum of chlordiazepoxide hydrochloride is shown in Figure 3. The exotherm observed at 233°C corresponds to the melting of the drug followed by decomposition at approximately 241°C. The ΔH_f was found to be approximately 25.5 kcal/mole[8].

2.8 Thermogravimetric Analysis
A TGA scan performed on chlordiazepoxide hydrochloride exhibited no loss of weight when heated to 105°C[8].

2.9 Solubility
Approximate solubility data obtained at room temperature are given in the following table (decomposition may occur on standing in solution):

TABLE II

Solvent	Solubility mg/ml
Petroleum ether (30-60°)	insoluble
Benzene	insoluble
Ether	insoluble
Isopropanol	1
Chloroform	2
95% ethanol	57
Methanol	116
3A alcohol	138
Water	165

2.10 Crystal Properties
The x-ray powder diffraction pattern of chlordiazepoxide hydrochloride is presented in Table III[9].

FIGURE 3

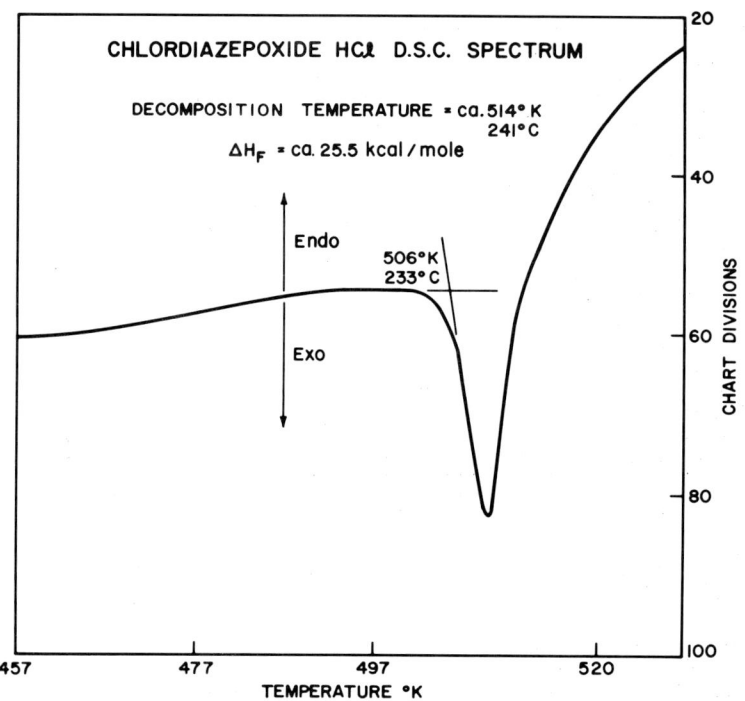

TABLE III

Chlordiazepoxide Hydrochloride

2θ	d(Å)*	I/I_o**
6.96°	12.70	24%
12.24	7.23	4
12.60	7.03	21
13.08	6.77	49
14.08	6.29	13
14.8	5.99	5
16.00	5.54	9
17.56	5.05	3
18.84	4.71	17
19.12	4.64	5
20.00	4.44	2
20.52	4.33	21
21.04	4.22	53
21.2	4.19	43
21.80	4.08	19
22.16	4.01	35
22.80	3.90	73
23.68	3.76	21
24.32	3.66	28
25.08	3.55	7
25.88	3.44	5
26.16	3.41	5
26.60	3.35	6
27.16	3.28	4
28.04	3.18	17
28.52	3.13	100
29.20	3.06	20
29.80	3.00	5

*d $= \dfrac{n \lambda}{2 \Sigma \sin \theta}$ (interplanar distance)

**I/I_o = relative intensity (based on highest intensity of 100)

Instrument Conditions

General Electric Model XRD-6 Spectrogoniometer

Generator	50 KV 12-1/2 MA
Tube target	Copper
Optics	0.2° Detector Slit
	MR sollor slit
	3° Beam slit
	0.0007" Ni Filter
	4° take off angle
Goniometer	Scan at 0.4° 2θ per minute
Detector	Amplifier - 16 coarse, 8.7 fine (gain)
	Sealed proportional counter tube and DC voltage at plateau.
	Pulse height selector E_L - 5 volts; Eu - out
	Rate Meter T.C. 4
	2000 C/S fullscale
Recorder	Chart Speet - 1 inch per 5 minutes

Samples prepared by grinding at room temperature.

6. Methods of Analysis

6.1 Elemental Analysis

Element	% Theory	% Reported
C	57.15	57.20
H	4.50	4.37

6.2 Phase Solubility Analysis

Phase solubility analysis is carried out using isopropanol as the solvent. A typical example is shown in Figure 4 which also lists the conditions under which the equilibration took place[10]. However, the inherent instability of acidic solutions of chlordiazepoxide such as the hydrochlorides may result in some decomposition during phase analysis.

FIGURE 4

Chlordiazepoxide HCl Phase Solubility Analysis

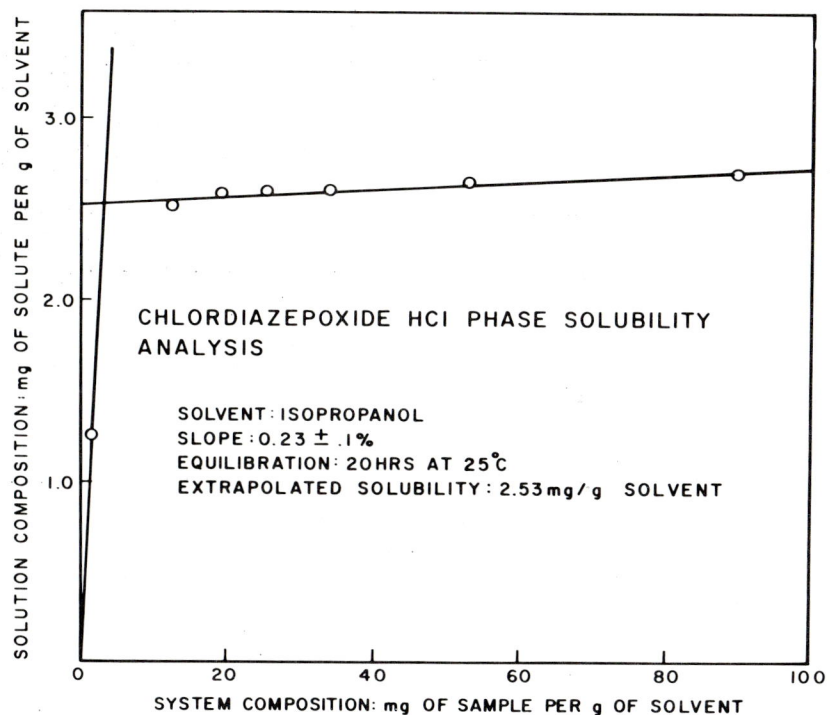

6.4 Direct Spectrophotometric Analysis

Direct spectrophotometric analysis of chlordiazepoxide hydrochloride has not been found to be applicable if any of the known hydrolytic contaminants are present.

For material not containing the interfering species, the reported maxima at 245-6 nm and 311-12 nm in acidified 3A alcohol may be used for quantitative measurements. The a values at these maxima are 96.5 and 30.5 respectively.

6.7 Non-Aqueous Titration

Chlordiazepoxide hydrochloride may be titrated directly in non-aqueous solvents.

A suitable solvent for dissolution of the compound is glacial acetic acid. The quantitative titration is performed after addition of approximately 10 ml of 15% mercuric acetate in glacial acetic acid to complex the hydrochloride anion. The end point is determined potentiometrically or colorimetrically, utilizing crystal violet indicator. One equivalent of the compound is titrated under the conditions described.

7. Acknowledgments

The authors wish to acknowledge the assistance of Dr. W. Benz, Dr. V. Venturella, and Mr. T. Daniels in the preparation of this analytical profile.

8. References

   1. Hawrylyshyn, M., Hoffmann-La Roche Inc., Personal Communication.

   2. Traiman, S., Hoffmann-La Roche Inc., Personal Communication.

   3. Johnson, J. and Venturella, V., Hoffmann-La Roche Inc., Personal Communication.

   4. Nuhn, P. and Bley, W., *Pharmazie*, <u>22a</u>, 523 (1967).

   5. Mahn, F., Hoffmann-La Roche Inc., Personal Communication.

   6. Chlordiazepoxide Analytical Profile.

   7. *United States Pharmacopia XVIII*, p. 113 (1970).

   8. Donahue, J., Hoffmann-La Roche Inc., Personal Communication.

   9. Sheridan, J.C., Hoffmann-La Roche Inc., Personal Communication.

   10. MacMullan, E.A., Hoffmann-La Roche Inc., Personal Communication.

CYCLOSERINE

J. W. Lamb

CONTENTS

1. Description
    1.1 Name, Formula, Molecular Weight
    1.2 Appearance, Color, and Odor
2. Physical Properties
    2.1 Infrared Spectrum
    2.2 Nuclear Magnetic Resonance Spectrum
    2.3 Ultraviolet Spectrum
    2.4 Optical Rotation
    2.5 Melting Range
    2.6 Differential Thermal Analysis
    2.7 Thermogravimetric Analysis
    2.8 Solubility
    2.9 Crystal Properties
3. Synthesis
    3.1 Chemical Synthesis
    3.2 Biosynthesis
4. Stability - Degradation
5. Drug Metabolic Products
6. Methods of Analysis
    6.1 Elemental Analysis
    6.2 Spectrophotometric Analysis
    6.3 Colorimetric Analysis
    6.4 Chromatographic Analysis
        6.41 Paper Chromatographic Analysis
        6.42 Thin Layer Chromatographic Analysis
        6.43 Bioautographic Analysis
    6.5 Microbiological Analysis
        6.51 High Level Plate System
        6.52 Low Level Plate System
        6.53 Photometric System
7. References

CYCLOSERINE

1. Description

    1.1 Name, Formula, Molecular Weight
    Cycloserine is D-4-amino-3-isoxazolidone.

$$H_2N-CH-C=O$$
$$||$$
$$CH_2NH$$
$$\diagdown O \diagup$$

$C_3H_6N_2O_2$ Mol. Wt. 102.09

    1.2 Appearance, Color, and Odor
    White to slightly yellow, practically odorless crystalline powder.

2. Physical Properties

    2.1 Infrared Spectrum
    The infrared spectrum of cycloserine presented in Fig. 1 was taken in a KBr pellet. A spectrum of the same sample taken in a Nujol Mull is essentially identical to the one presented. Hidy[1], Kuehl[2], and Stammer[3] showed that the solid state spectrum of cycloserine has two ionizable groups with pKa_1, equal to 4.4 - 4.5 and pKa_2 equal to 7.4. [1]Spectral bands typical of an amino acid zwitterion (2200 cm^{-1} assigned to the $-NH_3^+$) and a resonance stabilized hydroxamate anion (1600 to 1500 cm^{-1}) are in agreement with the peaks represented in Fig. 1.

    2.2 Nuclear Magnetic Resonance Spectrum
    The NMR Spectrum as presented in Fig. 2 was obtained by preparing a 10% solution of cycloserine in D_2O. The assignments in Fig. 2 are in agreement with those of Stammer[3]. Cycloserine shows a complex absorption in the 4.3 - 5.0 ppm and 3.8 - 4.2 ppm regions in acidic and basic solutions respec-

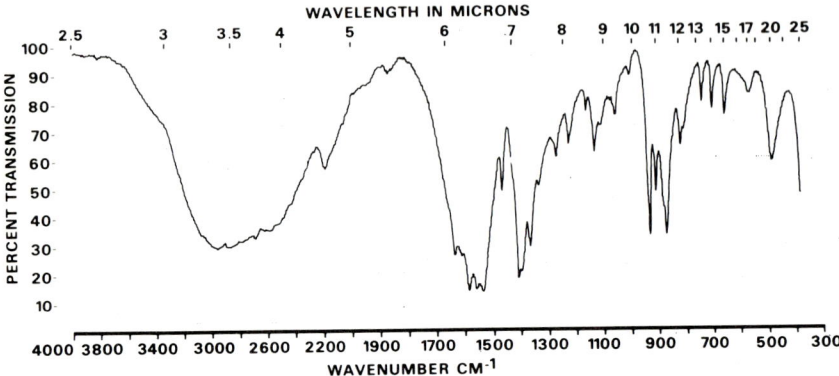

Fig. 1. Infrared absorption spectrum of cycloserine

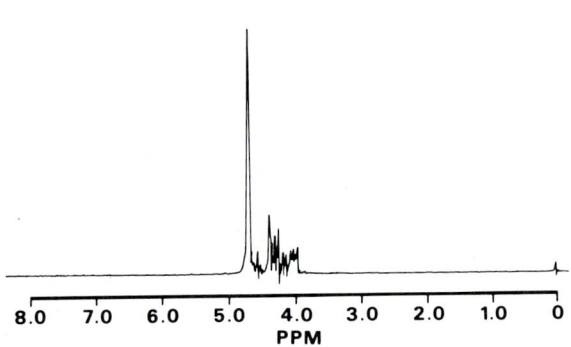

Fig. 2. Nuclear magnetic resonance cycloserine

tively. The complex group of six peaks centered around 4.68 ppm are the three magnetically nonequivalent isoxazolidone ring protons.

2.3 <u>Ultraviolet Spectrum</u>
Cycloserine is reported to exhibit an absorption band peaking at 226 nm in water[2], and a single band peaking at 219 nm when scanned in 0.1N HCl[4].

2.4 <u>Optical Rotation</u>
The following rotations have been reported:

$[\alpha]_D^{25}$ + 112 (C, not specified)[1]

$[\alpha]_D^{25}$ + 116 (C, 1.17 in water)[2]

$[\alpha]_D^{22}$ + 115 (C, 1.0 in water)[3]

$[\alpha]_D^{25'}$ + 109 - 113 (C, not specified)[4]

2.5 <u>Melting Range</u>
The following melting (decomposition) point temperatures have been reported:
156 C[1]
154 - 155 C[2]
156 C[5]

2.6 <u>Differential Thermal Analysis</u>
A differential thermal analysis was performed on cycloserine. A melting endotherm, followed by a rapid exotherm was observed.
At a heating rate of 20 C/min. the endotherm peaked at 152 C and the exotherm at 160 C.

2.7 <u>Thermogravimetric Analysis</u>
A TGA performed on cycloserine indicated a 1.0% weight loss at 147 C. Weight loss occurred rapidly as the temperature approached the melting (decomposition) point. The measurement was performed under nitrogen

sweep at a heating rate of 5 C/min.

2.8 Solubility

The following solubility data were obtained from Weiss[6]. Cycloserine is essentially insoluble in common organic solvents but readily soluble in water.

- 100 mg/ml in water
- 1.95 mg/ml in methanol
- 0.85 mg/ml in acetone
- 0.90 mg/ml in pyridine
- 1.60 mg/ml in formamide
- 1.50 mg/ml in ethylene glycol monomethyl ether
- 1.00 mg/ml in benzyl alcohol

2.9 Crystal Properties

The crystallographic properties of cycloserine were determined by Pepinsky[7]. Bond distances are normal and the five-membered ring is nearly planar.

3. Synthesis

3.1 Chemical Synthesis

Cycloserine has been synthesized by several workers including Stammer[3] and Evans[8]. The method of Evans will be briefly described. Evans[8] reported cycloserine can be synthesized by converting DL-Serine to its methyl ester hydrochloride by Fischer esterification.

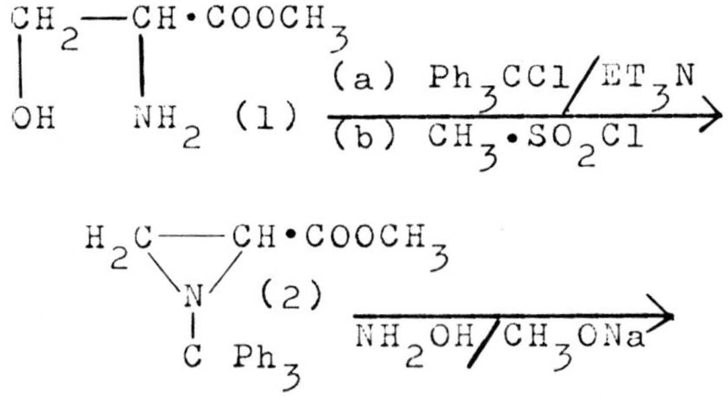

$$\underset{(1)}{\overset{CH_2-CH\cdot COOCH_3}{\underset{OHNH_2}{||}}} \xrightarrow[\text{(b) } CH_3\cdot SO_2Cl]{\text{(a) } Ph_3CCl/ET_3N}$$

$$\underset{(2)}{\overset{H_2C-CH\cdot COOCH_3}{\underset{\underset{CPh_3}{|}}{\diagdown N \diagup}}} \xrightarrow{NH_2OH/CH_3ONa}$$

$$\underset{(3)}{\underset{\underset{CPh_3}{|}}{\overset{H_2C-CH\cdot CO\cdot NH\cdot OH}{\diagdown N \diagup}}} \xrightarrow{HCl} \underset{(4)}{\overset{CH_2-CH\cdot NH_2}{\underset{Cl\quad CO\cdot NH\cdot OH}{|\quad\quad |}}} \xrightarrow{\text{basic resin}}$$

$$\underset{(5)}{\overset{H_2C-CH-NH_2}{\underset{\underset{H}{\overset{|}{N}}}{\overset{|\quad\quad|}{O\quad C=O}}}}$$

D-Serine methyl ester is converted into the N-triphenylmethyl derivative which, when heated in the presence of methane sulphonyl chloride, yielded the substituted ethyleneamine (2). Reaction of (2) with hydroxylamine and sodium methoxide gives the corresponding hydroxamic acid (3). This product is converted, by the action of hydrochloric acid, into D-α amino-β-chloro-N-hydroxypropionamide (4), which undergoes cyclization to D-cycloserine (5) when treated with a strongly basic ion exchange resin.

3.2 Biosynthesis

Cycloserine is produced by <u>Streptomyces orchidaceus</u>. According to Harned[5] the isolation from the culture filtrate is accomplished by: (1) adsorption on a strong base anion exchange resin, (2) elution with H_2SO_4, and (3) formation of a water insoluble, crystalline, silver salt. The free acid is prepared by decomposition of the silver salt with HCl and crystallization from the filtrate with acetone or alcohol.

4. Stability - Degradation

The dry crystalline solid (0.2% or less moisture) is stable at 100 C for 24 hours and can be stored for long periods in a desiccator at room temperature without measurable loss of potency.[9] Craig[9] reported that dilute solutions (0.1-1.0 mg/ml) can be stored under refrigeration without loss of potency. Cummings[10] found that at concentrations of 10-40 mcg/ml there is a progressive drop in drug potency over a 15 day period as determined by a chemical method in a synthetic assay medium. A day old 1 mg/ml aqueous solution of cycloserine maintained at 5 C showed a 7.5% loss of potency when assayed turbidimetrically with Klebsiella pneumonia. After one week the potency loss was about 45%. Using the same test solutions, both photometric and agar diffusion assays with Staphylococcus aureus and the colorimetric assay of Jones[11] did not show significant potency loss.

Concentrated aqueous solutions, crystalline solids containing significant amounts of moisture, and crystalline solids exposed to humid atmosphere are not stable.[9] Inactivation is due to the formation of the dimer $(C_3H_6O_2N_2)_2$ and in concentrated aqueous solutions can be prevented by the addition of an equivalent concentration of a strong alkali such as sodium hydroxide.[9] Craig[9] reported that concentrated aqueous solutions of 100 mg/ml containing one equivalent weight of sodium hydroxide are stable for at least 30 days at 25 C. Cycloserine is unstable in acid. Treatment with 6N HCl at 60 C gives β-aminoxy-D-alanine hydrochloride where as cycloserine in methanol and HCl gives β-aminoxy-D-alanine methyl ester dehydrochloride.[12]

5. Drug Metabolic Products
Robson[12] determined that cycloserine is well absorbed when administered orally. About 65% was excreted unchanged in the urine while 35% was metabolized to unknown substances.

6. Methods of Analysis

  6.1 Elemental Analysis

CYCLOSERINE

Element	% Theory	Reported			
		Ref. 14	Ref. 2	Ref. 1	Ref. 5
C	35.29	35.27	35.75	35.4	35.5
H	5.96	6.04	5.56	5.98	6.0
N	27.44	27.01	27.19	26.9	26.6

6.2 <u>Spectrophotometric Analysis</u>

The ultraviolet absorption band at 219 nm of 4-amino-3-isoxazolidone is a function of the carbonyl group (see Sec. 2.3). This analysis is useful as a measure of purity and is used as a quantitative test for cycloserine in formulations. The dimer (2,5-bis-(aminoxymethyl-1-3, 6-diketopiperazine) absorbs at 288 nm due to the loss of hydroxylamine when treated with alkali reagent.

6.3 <u>Colorimetric Analysis</u>

The routine chemical assay for cycloserine is the colorimetric method of Jones. (Refer to CFR 148d.1(b)). Cycloserine reacts with sodium nitropentacyanoferrate in a slightly acidic aqueous solution to give an intense blue-colored complex suitable for quantitative measurement at 625 nm[4]. This assay is specific for the ring structure of cycloserine.

According to Craig[9], this method has been applied successfully to the determination of cycloserine in biological fluids, such as blood, urine, cerebrospinal fluid and to the determination of crystalline cycloserine. No naturally occurring amino acids have been found to interfere with assay results. The minimum assayable level is about 100 ppm in tissue or solid samples such as animal feeds, and about 25 ppm in liquid samples.

6.4 <u>Chromatographic Analysis</u>

Qualitative chromatographic methods can be used for identification of cycloserine and for separation of cycloserine and dimer.

6.41 **Paper Chromatographic Analysis**

Solvent systems and corresponding Rf values are reported in the following table:

	Solvent System	Rf
A.	Propanol/water 7:3[13]	.50
B.	Butanol/water/acetic acid 3:1:1[13]	.15
C.	Acetone/water 2:1[13]	.70
D.	80% Ethanol/water[14]	.40
E.	Butanol/acetic acid/water 4:1:5[3]	.68
F.	Methyl ethyl ketone/pyridine/water 4:1:6[3]	.76
G.	Tert. butanol/n. butanol saturated with 0.8N NH_4OH 1:1[15]	.82

Detection system: Brownish-yellow spot when treated with Ninhydrin reagent.

6.42 **Thin Layer Chromatographic Analysis**

TLC systems and corresponding Rf values for cycloserine found in our laboratory are as follows:

	Solvent System	Rf
A.	Methanol/water 4:1[13]	.57
B.	Methanol/ethyl acetate/water 6:3:2[13]	.40

Detection system: Ninhydrin reagent

6.43 **Bioautographic Analysis**

Bioautography is a qualitative measure of biologically active cycloserine by its bioreactivity and by its mobility, but is not suitable for the detection of dimer.

Solvent system: propanol/water 1:9
Test organism: *S. aureus* ATCC 6538P
Rf value: 0.5

6.5 Microbiological Analysis

Craig[9] describes two microbiological plate systems suitable for the assay of cycloserine. The first is a high level assay for solutions or solids containing greater than 500 ppm. The test organism can be S. aureus ATCC 6538P or E. coli NRRL 4348, and the reference standard is 50 mcg/ml. The low level assay is performed using a more sensitive assay organism, Bacillus megatherium ATCC 25833. Relative activities of dimer to cycloserine for the three bacterial systems are: S. aureus 1:100; E. coli 1:20; and B. megatherium 1:600. The minimum no-effect ratios of dimer to cycloserine for the three organisms are 15, 2, and 10, for S. aureus, E. coli, and B. megatherium respectively. The low level plate assay is capable of determining levels as low as 0.1 ppm in liquid and 1-3 ppm in tissue. Sensitivity is achieved through use of a sensitive culture and a medium relatively free of alanine. Assay solutions prepared with 12.5, 25, and 50 fold greater concentrations of DL-alanine than cycloserine resulted in values that were 90, 82, and 72% respectively of those values obtained without alanine.

A suitable photometric assay is available for the assay of cycloserine materials that have a potency of 0.02 mg/gm or more. This system measures only the cycloserine isomers whereas S. aureus measures both dimer and cycloserine (See Sec. 4).

6.51 High Level Plate System
Refer to Code of Federal Regulations 148d.1(e).

6.52 Low Level Plate System
Refer to Craig[9].

6.53 Photometric System
Refer to CFR 141.111 substituting K. pneumoniae ATCC 10031 for S. aureus.

7. References

1. P.H. Hidy, E.B. Hodge, J. Am. Chem. Soc. **77**, 2345 (1955).
2. F.A. Kuehl, Jr., J. Am. Chem. Soc. **77**, 2344 (1955).
3. C. Stammer, J. McKinney, J. Org. Chem. **30**, 3436 (1965).
4. Analytical Laboratories, Eli Lilly and Company.
5. R.L. Harned, P.H. Hidy, and E.K. Baro, Antibiotics and Chemotherapy **5**, 204 (1955).
6. P.J. Weiss, M.L. Andrew, and W.W. Wright, Antibiotics Chemotherapy **7**, 374-377 (1957).
7. R. Pepinsky, Record Chem. Progr. **17**, 145 (1956).
8. R.M. Evans, The Chemistry of the Antibiotics Used in Medicine, p. 12-13. Oxford, New York: Pergamon Press 1965.
9. G.H. Craig, and R.L. Harned, In Press.
10. M. Cummings, R.A. Patnade, and P.C. Hudgins, Antibiotics and Chemotherapy, **5**, 198 (1955).
11. L. Jones, Anal. Chem. **28**, 39 (1956).
12. J. Robson, F. Sullivan, Pharmacol. Rev. **15**, 195 (1963).
13. R. Hussey, Personal Communication.
14. D.H. Harris, M. Rugar, M. Riagan, F.J. Wolf, R. Peck, H. Wallick, and H.W. Woodruff, Antibiotics and Chemotherapy, **5**, 183 (1955).
15. S.M. Conzelman, Jr., Antibiotics and Chemotherapy, **5**, 444 (1955).

CYCLOTHIAZIDE

C. D. Wentling

CONTENTS

1. Description
   1.1 Name, Formula, Molecular Weight
   1.2 Appearance, Color, Odor
2. Physical Properties
   2.1 Infrared Spectrum
   2.2 Nuclear Magnetic Resonance Spectrum
   2.3 Ultraviolet Spectrum
   2.4 Mass Spectrum
   2.5 Melting Range
   2.6 Differential Thermal Analysis
   2.7 Thermogravimetric Analysis
   2.8 pKa
3. Synthesis
4. Stability - Degradation
5. Drug Metabolic Products
6. Methods of Analysis
   6.1 Elemental Analysis
   6.2 Titrimetric Analysis
   6.3 Direct Spectrophotometric Analysis
   6.4 Thin Layer Chromatographic Analysis
7. References

CYCLOTHIAZIDE

1. ## Description

    1.1 ### Name, Formula, Molecular Weight
    Cyclothiazide is 6-chloro-3,4-dihydro-3-(5-norbornen-2-yl)-2H-1,2,4-benzothiadiazine-7-sulfonamide 1,1-dioxide. It is also known as 6-chloro-3,4-dihydro-3-(5-norbornen-2-yl)-7-sulfamoyl-1,2,4-benzothiadiazine-1,1-dioxide; 3-(bicyclo-[2,2,1]-hept-2'-ene-6'-yl)-6-chloro-7-sulfamyl-3,4-dihydro-1,2,4-benzothiadiazine-1,1-dioxide; 6-chloro-3-(5-bicyclo[2.2.1]hept-2-enyl)-7-sulfamoyl-3,4-dihydro-1,2,4-benzothiadiazine-1,1-dioxide and by many slight variations of the particular nomenclature.

$C_{14}H_{16}ClN_3O_4S_2$ Mol. Wt.: 389.88

    1.2 ### Appearance, Color, Odor
    It is a white to off-white, essentially odorless powder.

2. ## Physical Properties

    2.1 ### Infrared Spectrum
    The infrared spectrum of cyclothiazide (Lilly Working Standard, Lot No. 9JZ42, crystallized from alcohol-water) is presented in Figure 1. The spectrum is of a sample in a KBr pellet, taken on a Beckman IR-12 spectrophotometer. A spectrum of a sample in a Nujol Mull is essentially the same[1]. Figure 1 is also in agreement with other published spectra of cyclothiazide such as that of a sample recrystallized from alcohol-heptane[2] or that published by the Drug

Standard Laboratory[3] both of which are in KBr pellets.

C. Underbrink[4] assigns the following bands (cm^{-1}) to cyclothiazide:
a. characteristic for NH or NH_2: 3390, 3260
b. characteristic for SO_2-N: 1350, 1310, 1180, 1160
c. probably characteristic for NH_2 of SO_2-NH_2: 1560

Whitehead et al.[5] assign the intense absorption band at approximately 6.2 μ (1600 cm^{-1}) as characteristic for 3,4-dihydro-3-substituted 7-sulfamoyl-1,2,4-benzothiadiazine 1,1-dioxides.

2.2 Nuclear Magnetic Resonance Spectrum

A nuclear magnetic resonance spectrum of cyclothiazide in DMSO-d_6 is presented in Figure 2. Whitney et al.[6], through the interpretation of an NMR spectrum, estimated the material they were using to be about 80% endo and 20% exo. H. Boaz[7], who interpreted the above spectrum, supplies in Table I the specific assignments for Figure 2.

2.3 Ultraviolet Spectrum

Salim and Hilty[3] reported maxima at 271 and 315 nm in methanol.

A scan of the Lilly Working Standard in methanol (0.01 mg. per ml.) from 350 to 210 nm. produced maxima at 227, 271 and 315 nm[8]. A similar scan in ethanol yielded maxima at 227, 272 and 315 nm. and in alkaline media produced maxima at 274 and 324 nm[1].

2.4 Mass Spectrum

The mass spectrum of cyclothiazide (Lilly Working Standard) was determined using a Perkin-Elmer Hitachi RMU-6D mass spectrometer[9]. The intensities were measured from a low resolution mass spectrum and are summarized as a bar graph in Figure 3. The molecular ion (M^+389) for Cl^{35} is visible along with that of the molecular ion for Cl^{37} (M^+391). The base peak (normalized intensity = 100) is at m/e 66.

CYCLOTHIAZIDE

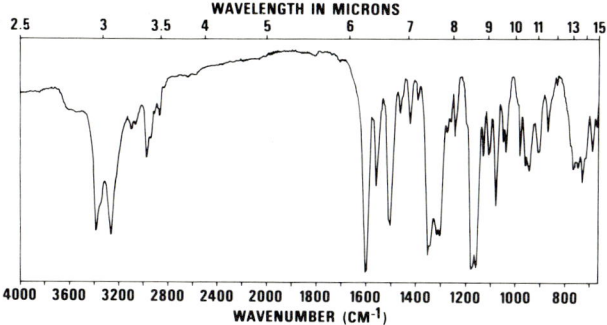

Fig. 1. Infrared spectrum of cyclothiazide taken in a KBr pellet on a Beckman IR-12 spectrophotometer

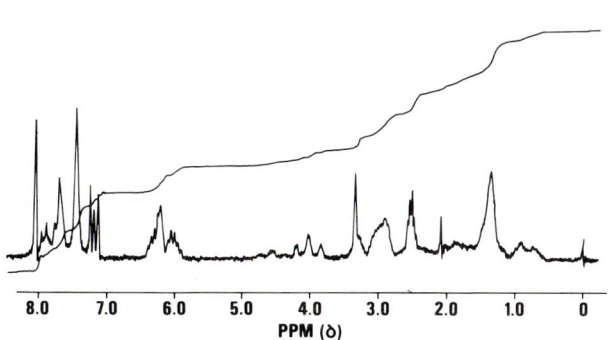

Fig. 2. Nuclear magnetic resonance spectrum of cyclothiazide taken on a Varian Associates A-560 spectrophotometer in dimethylsulfoxide-d_6

TABLE I
NMR SPECTRAL ASSIGNMENTS FOR CYCLOTHIAZIDE

PROTONS AT	CHEMICAL SHIFT (δ)	ISOMER FORM	COUPLING CHARACTERISTICS
8	8.05	two exo	s,u
	8.03	two endo	
4	7.75–7.90		d; $J = 11.5$
2	7.67		s, broad
7 NH$_2$	7.44		s, broad
5	7.23, 7.12	endo	s, sharp
	7.18	two exo u	
5'	6.30, 6.27	endo	q,u; $J_{5',6'} = 5.5$; $J_{5',4'} = 2.5$
	6.20	exo	
6'	6.01, 5.95	endo	q,u; $J_{5',6'} = 5.5$; $J_{6',1'} = 2.5$
	6.20	exo	
3	4.54	exo	t; $J_{3,2'} \simeq J_{3,4} \simeq 11.5$
	4.01	endo	

s ≡ singlet; d ≡ doublet; t ≡ triplit, q ≡ quartet
u ≡ unresolved; J ≡ coupling constant in Hz

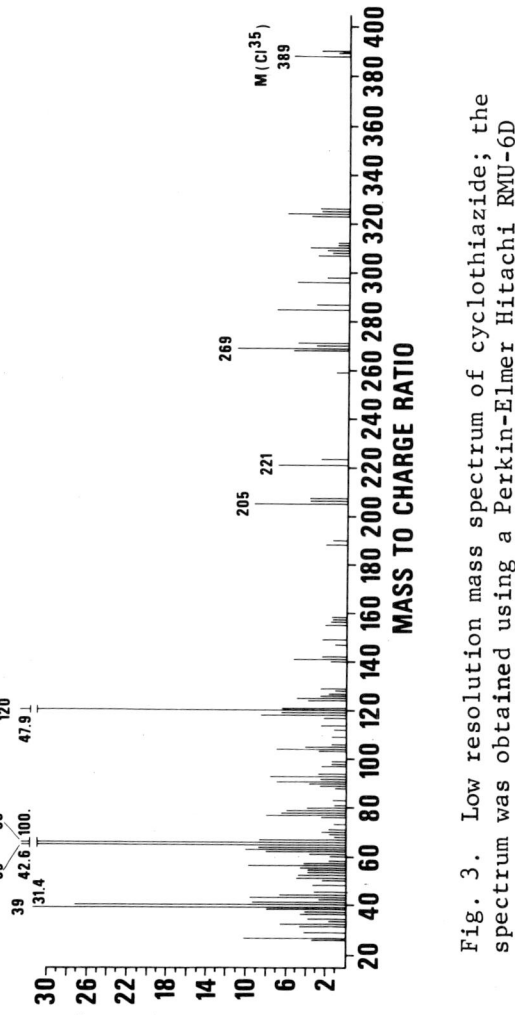

Fig. 3. Low resolution mass spectrum of cyclothiazide; the spectrum was obtained using a Perkin-Elmer Hitachi RMU-6D mass spectrometer

2.5 Melting Range
The melting point or range for cyclothiazide has been reported over a rather wide span and the temperatures (°C.) are presented below:
235 (approximately)[10]
234[5]
229-230 and 226-230[11]
222.5-223.5[8]
220 (approximately with decomposition)[3] USP Class I

2.6 Differential Thermal Analysis
A differential thermal analysis of the Lilly Working Standard was performed using a DuPont 900 Differential Thermal Analyzer at a heating rate of 20°C. per min. with a nitrogen atmosphere[8]. The thermogram shows an endotherm at approximately 241°C. indicating decomposition.

2.7 Thermogravimetric Analysis
A thermal gravimetric analysis of the Lilly Working Standard was performed using a DuPont 950 Thermogravimetric Analyzer at a heating rate of 5°C. per minute and a nitrogen atmosphere. The sample maintained a constant weight through 217°C. after which weight was rapidly lost.

The disparity between the results of the differential thermal analysis and the thermogravimetric analysis was noted and apparently is real.

2.8 pKa
Whitehead et al.[5] report that in aqueous 66% N,N-dimethylformamide, 3,4-dihydro-3-substituted-7-sulfamoyl-1,2,4-benzothiadiazine 1,1-dioxides are characterized by two pKa's of 11.0-11.4 and 13.0-13.3. Novello and Sprague[12] reported pKa's of 9.1 and 10.5 for cyclothiazide. In the latter case the pKa represents the pH at half neutralization in 30% aqueous ethanol determined potentiometrically.

3. Synthesis

Cyclothiazide can be prepared by the addition of an excess of ammonia to 5-norbornylenyl-carboxaldehyde. This reaction mixture is then added to a solution of 4-chloro-6-fluorobenzene-1,3-disulfonamide, and the product is precipitated by addition to dilute acid[10]. Alternate processes similar to the above involve use of the aldehyde-ammonia complex or the aldimine produced in the first step above by variations of the reaction media.

Other syntheses reported are those of Whitehead et al.[5] and Müller et al.[11]. In both of these processes the starting materials are 4-amino-6-chlorobenzene-1,3-disulfonamide and 5-norbornylenylcarboxaldehyde. The conditions under which the reaction is performed vary somewhat.

The syntheses are presented in Figure 4.

4. Stability - Degradation

Cyclothiazide appears to be very stable in the solid state and under ordinary ambient conditions. Cyclothiazide is rapidly decomposed when heated in boiling acidic or basic alcohol solutions and is more rapid in the acidic solution[13]. By the thin layer chromatography method of Koch[13], one of the decomposition products has the same R_f value as 4-amino-6-chlorobenzene-1,3-disulfonamide.

5. Drug Metabolic Products

No report of metabolic products related to cyclothiazide is recorded.

6. Methods of Analysis

6.1 Elemental Analysis

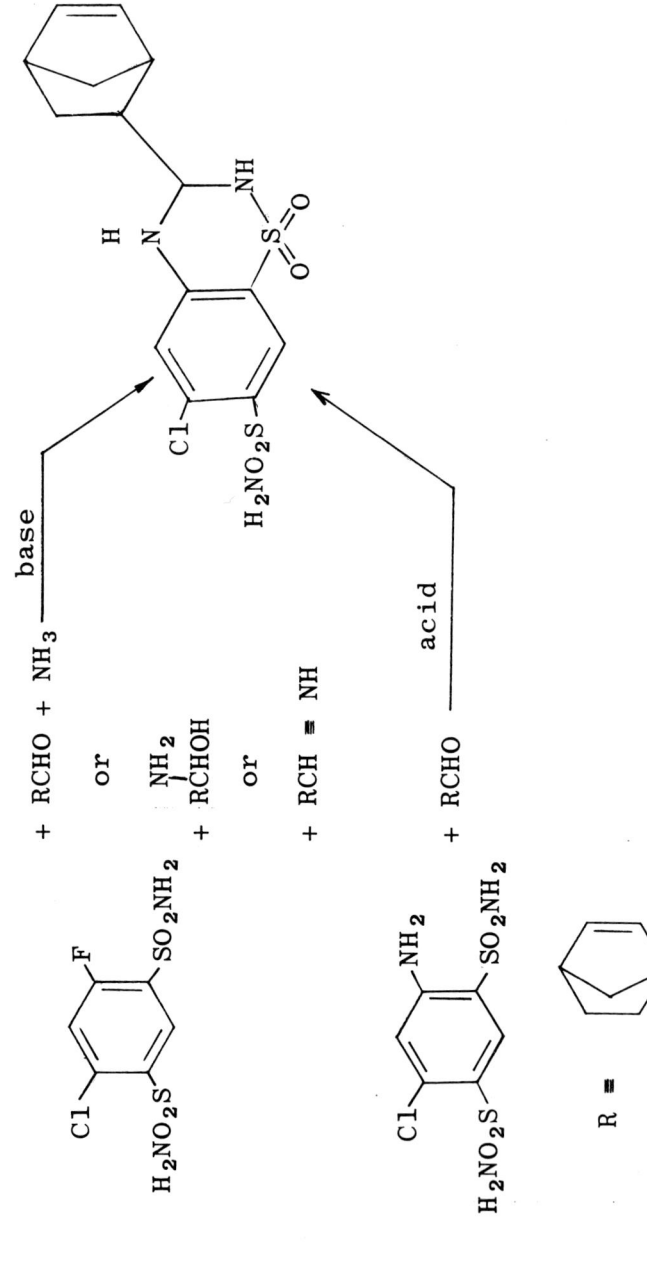

FIGURE 4 ROUTES OF SYNTHESIS OF CYCLOTHIAZIDE

Element	% Theory	Reported[5]
C	43.13	43.13
H	4.14	3.89
N	10.78	10.69

6.2 Titrimetric Analysis

Cyclothiazide has a sulfonamide group which facilitates nonaqueous titrations. Salin and Hilty[3] in their discussion of the titration using sodium methoxide indicate the compound exhibits two titratable groups when dissolved in ethylenediamine with o-nitroaniline indicator. In dimethylformamide with thymol blue indicator cyclothiazide exhibits only one acidic group. The former conditions are those of the NF XIII cyclothiazide assay.[14]

6.3 Direct Spectrophotometric Analysis

Cyclothiazide exhibits several absorption maxima in the ultraviolet range under various conditions (see Section 2.3). While this absorbance will not differentiate cyclothiazide from some starting materials, it does lend itself to the assay of formulated batches[3,14], and facilitates a means of detection for thin layer chromatographic work[13,15].

6.4 Thin Layer Chromatographic Analysis

The most notable work in this area is probably that of Duchêne and Lapière[15]. By means of a two dimensional chromatogram on Alumina GF 254 Merck, they separated 15 components of a mixture of 18 therapeutically active diuretics. The first solvent is ethylacetate with 1.5% water, developed at 22°C. and, after drying, a solvent system of butanol/chloroform, 30:100 is used with development at 5-8°C. Ultraviolet light is employed for visualization of the separated components. The cyclothiazide used in this work shows three components when chromatographed in the first direction with R_f's given as 0.59/0.62/0.67. In the second direction the R_f is given as 0.54. The chromatogram, however, shows two spots. Variations of the system for the first direction are also reported.

Koch[13], in routine analytical work, uses Brinkman precoated plates, Silica Gel F254, with a solvent system of acetonitrile/chloroform, 2:1. Cyclothiazide is visualized by means of ultraviolet light and has an R_f of 0.64. This system has been used to observe induced decomposition as stated in Section 4.

Whitney et al.[6] chromatographed the compound on Kieselgel G (Merck) with a methylene chloride/methanol, 65:25, solvent system. Visualization is by the chloride-iodide-starch method, positive for N-H bonds. By this system an R_f of 0.86 was reported.

References

1. D. Woolf, personal communication, Eli Lilly and Co., Indianapolis, Indiana, 46206.
2. O. R. Sammul, W. L. Brannon, A. L. Hayden, J. Ass. Offic. Agr. Chem. $\underline{47}$, 918-91 (1964).
3. E. F. Salim and W. W. Hilty, J. Pharm. Sci. $\underline{56}$, 518-19 (1967).
4. C. Underbrink, personal communication, Eli Lilly and Co., Indianapolis, Indiana, 46206.
5. C. W. Whitehead, J. J. Traverso, H. R. Sullivan, and F. J. Marshall, J. Org. Chem. $\underline{26}$, 2814-18 (1961).
6. P. L. Whitney, G. Fölsch, P. O. Nyman, and B. G. Malmström, J. Biol. Chem. $\underline{242}$, 4206-11 (1967).
7. H. Boaz, personal communication, Eli Lilly and Co., Indianapolis, Indiana, 46206.
8. F. E. Gainer, personal communication, Eli Lilly and Co., Indianapolis, Indiana, 46206.
9. A. Kossoy, personal communication, Eli Lilly and Co., Indianapolis, Indiana 46206.
10. C. W. Whitehead and J. J. Traverso, U. S. Patent 3,419,552 (1968).
11. E. Müller and K. Hasspacher, U. S. Patent 3,275,625 (1966).
12. F. C. Novello and J. M. Sprague, Ind. Chim. Belge. 32 (spec. no.), 222-5 (1967).
13. W. Koch, personal communication, Eli Lilly and Company, Indianapolis, Indiana 46206.
14. "The National Formulary" 13th edition, Mack Publishing Co., Easton, Pa. 18042 (1970) p. 191-2.
15. M. Duchêne and C. L. Lapière, J. Pharm. Belg. $\underline{20}$, 275-84 (1965).

DIAZEPAM

A. MacDonald, A. F. Michaelis, and B. Z. Senkowski

CONTENTS

Analytical Profile - Diazepam

1. Description
    1.1   Name, Formula, Molecular Weight
    1.2   Appearance, Color, Odor

2. Physical Properties
    2.1   Infrared Spectrum
    2.2   Nuclear Magnetic Resonance Spectrum
    2.3   Ultraviolet Spectrum
    2.4   Mass Spectrum
    2.5   Optical Rotation
    2.6   Melting Range
    2.7   Differential Scanning Calorimetry
    2.9   Solubility
    2.10 Crystal Properties
    2.11 Dissociation Constant
    2.12 Distribution Coefficient

3. Synthesis

4. Stability Degradation

5. Drug Metabolic Products and Pharmacokinetics

6. Methods of Analysis
    6.1   Elemental Analysis
    6.2   Phase Solubility Analysis
    6.3   Chromatographic Analysis
        6.31   Thin Layer Chromatographic Analysis
        6.32   Column Chromatographic Analysis
        6.33   Vapor Phase Chromatography
    6.4   Direct Spectrophotometric Analysis
    6.5   Polarographic Analysis
    6.6   Non-Aqueous Titration

7. References

DIAZEPAM

1. Description

   1.1 Name, Formula, Molecular Weight
   Diazepam is 7-chloro-1,3-dihydro-1-methyl-5-phenyl-2H-1,4-benzodiazepin-2-one.

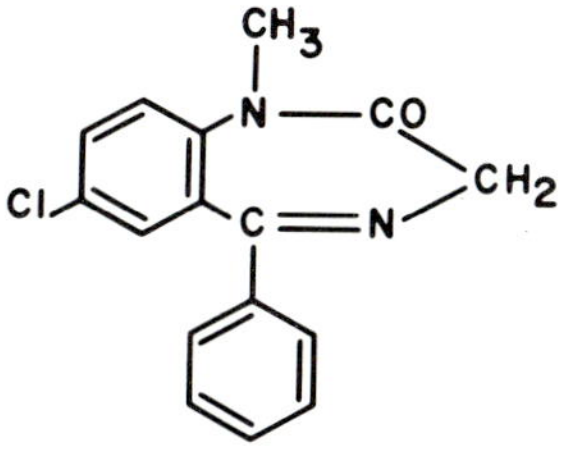

DIAZEPAM

$C_{16}H_{13}ClN_2O$ Mol. Wt. 284.75

   1.2 Appearance, Color, Odor
   Off-white to yellow, practically odorless, crystalline powder.

2. Physical Properties

   2.1 Infrared Spectrum
   The infrared spectrum of reference standard diazepam is presented in Figure 1[1]. The spectrum was measured in a KBr pellet which contained 1 mg/400 mg KBr.

   The following bands (cm^{-1}) have been assigned for Figure 1[2].
   a. Characteristic for NH: 3390
   b. Characteristic for -C=O: 1680
   c. Characteristic for aromatic groups: 1560, 1480

   2.2 Nuclear Magnetic Resonance Spectrum
   The NMR spectrum shown in Figure 2 was obtained by dissolving 47 mg of reference standard diazepam in 0.5 ml of CDCl$_3$ containing tetramethylsilane as internal

Figure 1

Infrared Spectrum of Diazepam

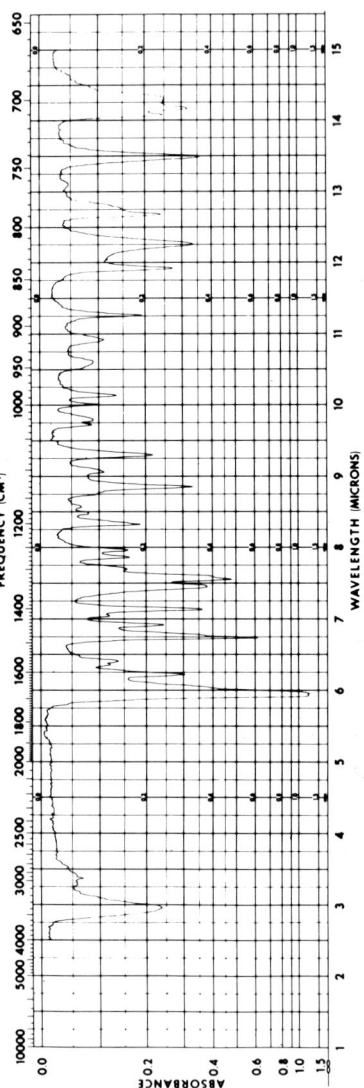

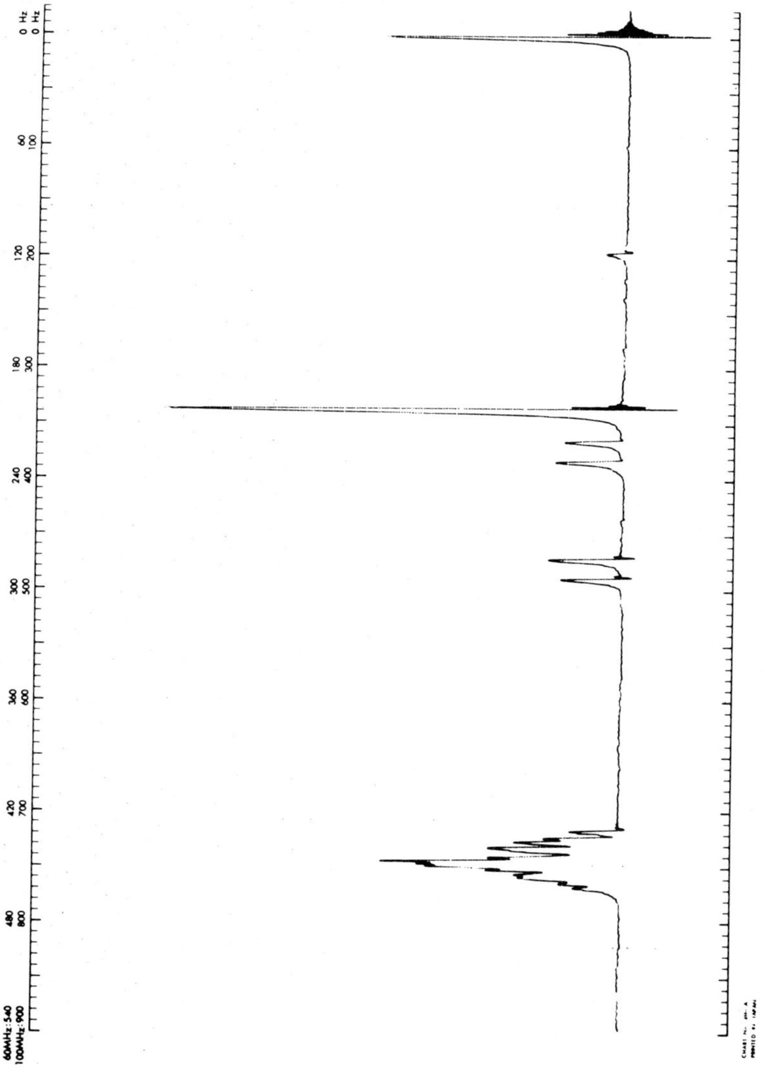

DIAZEPAM

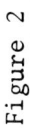

Figure 2

NMR Spectrum of Diazepam

reference. The spectral assignments are shown in Table I[3]. Nuhn and Bley[4] reported that at room temperature in C_6D_6 both methylene protons exhibit an AB spectrum. With increasing temperature the doublet was converted to a single line indicating rapid inversion of the ring. This observation was confirmed by Linscheid and Lehn[5].

TABLE I

Diazepam

Protons at	Chemical Shift τ (ppm)	Type (J in Hz)
C_1 methyl	6.62	s
C_3(a)	5.19	d(11)
C_3(b)	6.25	d(11)
C_6, C_8, C_9, Ph	2.55	m

s = singlet; d = doublet; m = multiplet

2.3 Ultraviolet Spectrum

Diazepam when scanned between 420 and 210 nm in acidified 3A alcohol exhibits 3 maxima as shown in Figure 3. These were located at 242 $\pm$ 2 nm (a = 100), 285 $\pm$ 2 nm (a = 43.7) and 368 $\pm$ 2 nm (a = 14.5). Minima were observed at 221 $\pm$ 2 nm, 266 $\pm$ 2 nm and 334 $\pm$ 2 nm[6].

2.4 Mass Spectra

The mass spectrum shown in Figure 4 was obtained using a CEC 21-110 mass spectrometer with an ionizing energy of 70 eV and a temperature of 190°C. Table II lists the elemental compositions for the most diagnostic ions as determined by high resolution mass spectrometry[7]. The molecular ion for diazepam was observed at m/e 284. The ions at m/e 256 and m/e 255 correspond to a loss of CH_2N and HCO respectively with the loss of chlorine shown by the ion m/e 249. Other ions in Table II can be ascribed to losses of Cl, and parts of the seven membered ring.

DIAZEPAM

Figure 3

Ultraviolet Spectrum of Diazepam

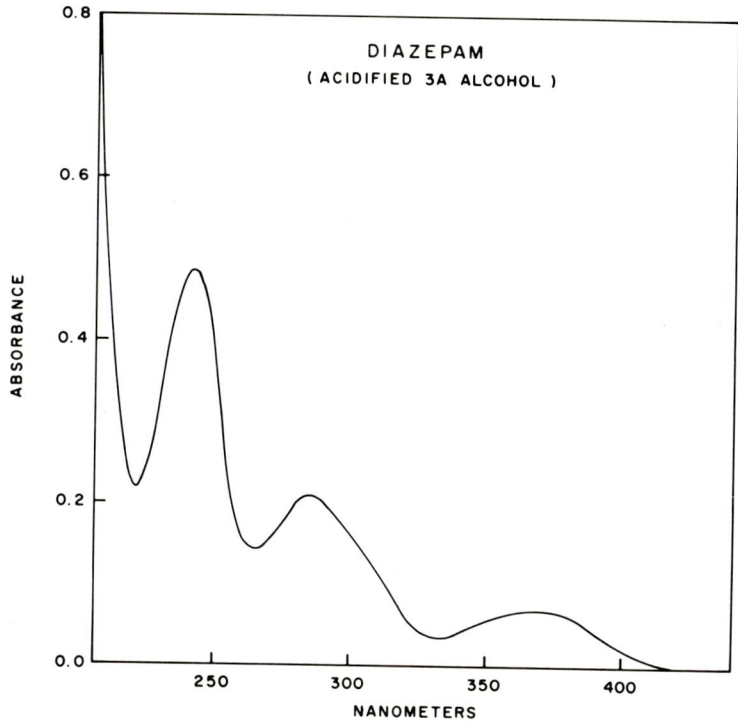

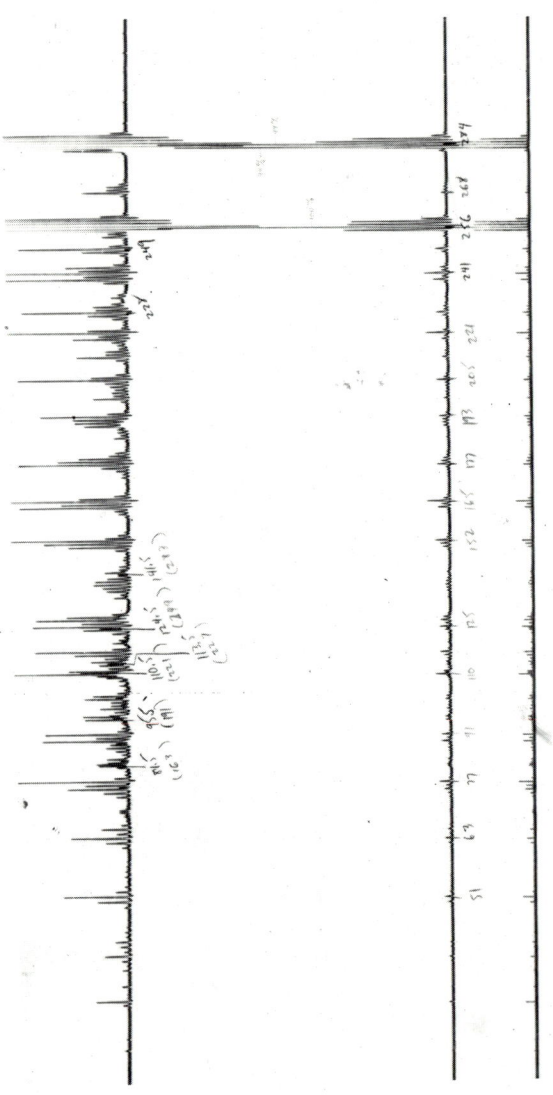

Figure 4

Mass Spectrum of Diazepam

TABLE II

High Resolution Mass Spectrum of Diazepam[a]

Found Mass	Calcd. Mass	C	H	N	O	Cl$_{35}$
284.0681	284.0759	16	13	2	1	1
283.0630	283.0681	16	12	2	1	1
256.0521	256.0572	15	11	1	1	1
255.0680	255.0687	15	12	2	0	1
249.1009	249.1071	16	13	2	1	0
241.0520	241.0531	14	10	2	0	1
239.0374	239.0375	14	8	2	0	1
228.0582	228.0578	14	11	1	0	1
221.0817	221.0884	15	11	1	1	0
219.0917	219.0920	15	11	2	0	0
213.0353	213.0344	13	8	1	0	1
205.0758	205.0764	14	9	2	0	0
199.0300	199.0313	13	8	0	0	1
186.0236	186.0235	12	7	0	0	1

[a] Only peaks discussed are included in this table.

2.5 **Optical Rotation**
Diazepam exhibits no optical activity.

2.6 **Melting Range**
The melting range reported in NF XIII is 131-135°C.

2.7 **Differential Scanning Calorimetry**
The DSC spectrum of diazepam is shown in Figure 5. The edotherm observed at 129°C corresponds to the melt with a ΔH_f of 5.9 kcal/mole[8]. The decomposition temperature is 180°C.

2.8 **Thermogravimetric Analysis**
A thermal gravimetric analysis performed on diazepam exhibited no loss of weight when heated to 105°C[8].

2.9 **Solubility**
Approximate solubility data obtained at room temperature are given in the following table.

Figure 5

Diazepam D.S.C. Spectrum

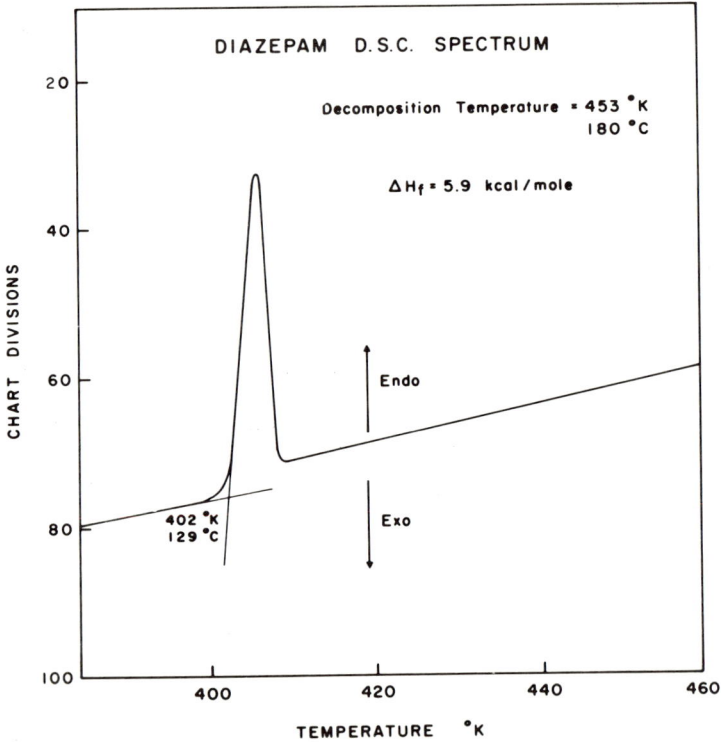

DIAZEPAM

Solvent	Solubility mg/ml
Water	.05
Petroleum Ether (30°-60°)	.9
Propylene Glycol	17
Ether	18
Isopropanol	20
3A Alcohol	32
95% Ethanol	41
Methanol	49
Acetone	125
Benzene	220
Dimethylacetamide	296
Chloroform	>500

2.10 Crystal Properties

The x-ray powder diffraction pattern of diazepam is presented in Table III[9].

Instrument Conditions

General Electric	Model XRD-6 Spectrogoniometer
Generator	50 KV, 12-1/2 MA
Tube target	Copper
Optics	0.2° Detector Slit
	MR soller slit
	3° Beam slit
	0.0007" Ni Filter
	4° take off angle
Goniometer	Scan at 0.4° 2θ per minute
Detector	Amplifier - 16 coarse, 8.7 fine (gain)
	Sealed proportional counter tube and DC voltage at plateau
	Pulse height selector E_L - 5 volts; Eu - out
	Rate meter T.C. 4
	2000 C/S full scale
Recorder	Chart speed - 1 inch per 5 minutes

Samples prepared by grinding at room temperature.

TABLE III

Diazepam

2Θ	d(Å)*	I/I₀**
9.44°	9.37	15%
11.00	8.04	7
13.20	6.71	3
13.60	6.51	18
14.60	6.07	3
17.20	5.16	5
17.52	5.06	10
18.88	4.70	100
20.96	4.24	3
21.56	4.12	6
22.04	4.03	15
22.80	3.90	65
23.80	3.74	27
24.44	3.64	25
26.12	3.41	4
26.68	3.34	13
26.88	3.32	10
27.48	3.25	11
28.32	3.15	2
28.88	3.09	2
29.44	3.03	14
29.68	3.01	16

*d $= \dfrac{n \lambda}{2 \sin \Theta}$ (interplanar distance)

**I/I₀ = relative intensity (based on highest intensity of 1.00)

2.11 <u>Dissociation Constant</u>
The pKa for diazepam has been determined spectophotometrically to be 3.4[10].

2.12 <u>Distribution Coefficient</u>
The distribution coefficient of diazepam between 1-octanol and pH 7.2 phosphate buffer is 382 at room temperature where D = $C_{octanol}/C_{buffer}$[10].

3. Synthesis
Diazepam may be prepared by the reaction scheme shown in Figure 6 with the reaction of 2-methylamino-5-chlorobenzophenone and ethyl glycinate to form diazepam[11]. A complete review of the chemistry of benzodiazepines describes other synthetic routes[12].

4. Stability-Degradation
The acid hydrolysis products for both diazepam and its major metabolite are shown in Figure 7. The acid hydrolysis of chlordiazepoxide is also included since it is the same as for the major diazepam metabolite[13].

5. Drug Metabolic Products and Pharmacokinetics
The major metabolites of diazepam in humans are shown in Figure 8[14]. The major metabolic pathways were shown to consist of demethylation at the nitrogen in position 1, addition of a hydroxyl group at carbon 3, and conjugation of the respective derivatives. The primary metabolite in blood is the N-desmethyl diazepam and in urine the oxazepam-glucuronide. Analytical procedures for the metabolites have been published using ultraviolet spectrometry[14], thin-layer chromatography[14] and gas chromatography[14,15].

6. Methods of Analysis

   6.1   Elemental Analysis

Element	% Theory	Reported
C	67.49	67.33
H	4.60	4.63

   6.2   Phase Solubility
Phase solubility analysis is carried out using isopropanol:hexane (1:1) as the solvent. A typical example is shown in Figure 9 which also lists the conditions under which the analysis was carried out[16].

   6.3   Chromatographic Analysis
      6.31   Thin-Layer Chromatographic Analysis
The following TLC system is useful as an identity test and evaluation of diazepam substance. Using silica gel G plates and a mixed solvent system of

Figure 6

Synthesis of Diazepam

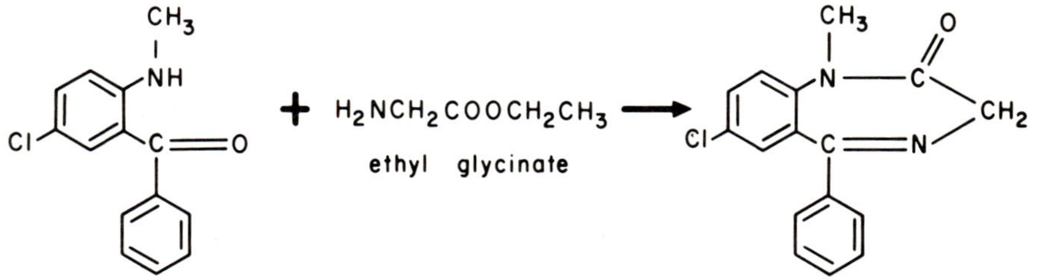

2-methylamino-5-chloro-benzophenene + H₂NCH₂COOCH₂CH₃ (ethyl glycinate) → (7-chloro-1,3-dihydro-1-methyl-5-phenyl-2H-1,4-benzodiazepine-2-one)

Figure 7

Hydrolysis of Diazepam

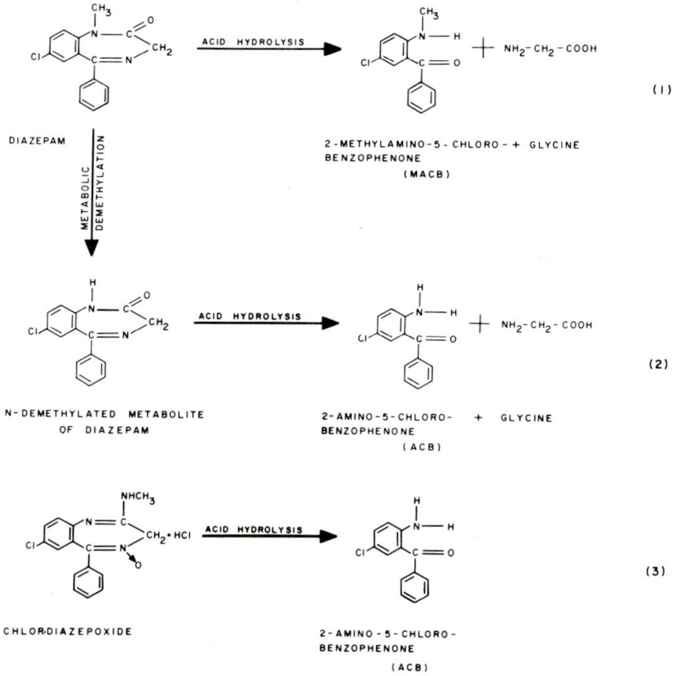

Figure 8

Metabolism of Diazepam

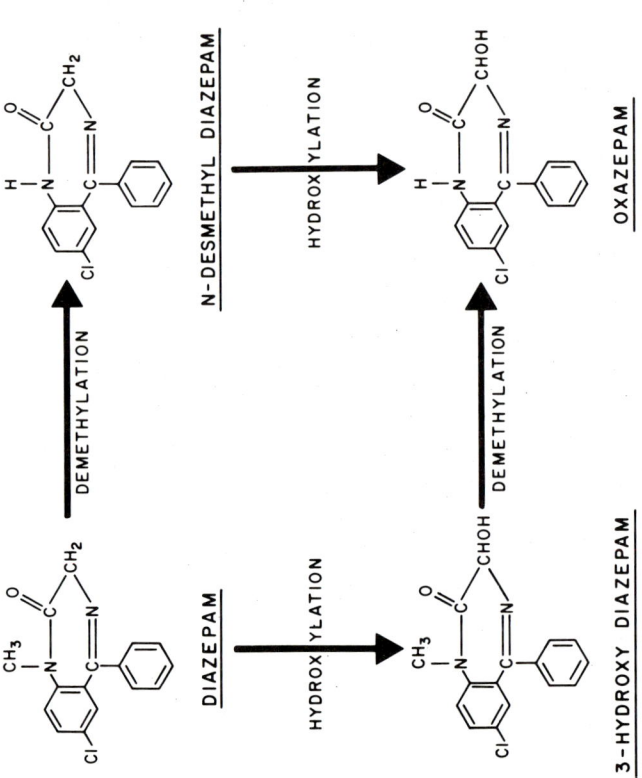

Figure 9

Phase Solubility Analysis of Diazepam

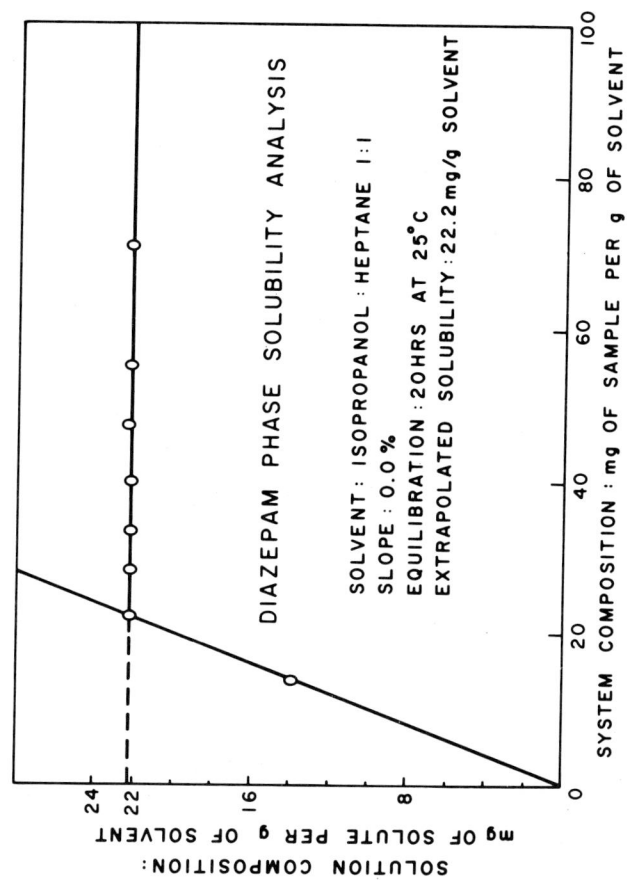

n-heptane:ethyl acetate (1:1 v/v), the sample containing 0.5 mg of diazepam substance in acetone is spotted and subjected to ascending chromatography. After development for at least 10 cm, the plate is air-dried and sprayed with potassium iodoplatinate solution. Diazepam appears as a purple spot with an approximate R_f of 0.3-0.4[17]. The second system is useful for identification of diazepam and its metabolites in extracts of biological materials[14,15]. The method uses two dimensional development of Brinkmann [F254] precoated silica gel plates using chloroform: heptane:ethanol (10:10:1 v/v) for the first dimension and chloroform:acetone (90:10 v/v) for the second dimension. The R_f ranges for each compound with the respective systems are tabulated below. The spotting solvent is acetone: hexane (20:80 v/v), the final extract.

TABLE IV

Compound	R_f System I Chloroform:Heptane:Ethanol 10 10 1	R_f System II Chloroform: Acetone 90 10
Diazepam	0.38-0.43	0.39-0.41
N-Desmethyl-diazepam	0.20-0.23	0.16-0.21
3-Hydroxy-diazepam	0.29-0.30	0.33-0.33
Oxazepam	0.09-0.11	0.08

6.32 Column Chromatographic Analysis

The analytical scale liquid-solid chromatographic separation of diazepam and its metabolites has been reported by Scott and Bommer in their study of the separation of benzodiazepines from each other and from biological media[18]. The liquid-solid chromatography was carried out using Durapak "OPN" (36-75μ particle diameter) 100 cm column with a 1 mm inside diameter and hexane:isopropanol (95:5 v/v) as solvent. The flow rate was 1.0 ml/min using an air driven pump. The detector was an ultraviolet monitor set at 254 nm. The cell volume was 8 µl. Complete separation of diazepam, n-desmethyl diazepam, 3-hydroxy diazepam and oxazepam is reported at 2 µg sensitivity per compound.

6.33 Vapor Phase Chromatography

The acid hydrolysis of blood extracts containing diazepam and its metabolite (Figure 7) has been used by deSilva[13] to prepare the respective benzophenones as volatile derivatives for gas chromatography. The column used was a 2 foot, 1/4", 2% Carbowax 20 M on silanized Gas Chrom P at 190° using a tritium electron capture detector. The disadvantages of this approach i.e. the benzophenones obtained in acid hydrolysis are not specific for a single benzodiazepine have been eliminated in a recent paper by deSilva[15] where the benzodiazepines have been chromatographed directly.

The direct method uses a 4 foot, 4 mm, column at 230°C packed with 3% OV-17 on 60/80 mesh Gas Chrom Q. The detector was a Ni^{63} electron capture detector operated at 310°. The use of stable high temperature phases and the high temperature Ni^{63} detector has enabled the quantitation of intact diazepam and its metabolites in blood and urine extracts at the following sensitivities.

Compound	Sensitivity, nanograms
Diazepam	1.0
N-desmethyl diazepam	2.0
Oxazepam	1.0

6.4 Direct Spectrophotometric Analysis

Direct spectrophotometric analysis of diazepam is applicable provided significant quantities of the hydrolytic contaminants are not present. For material not containing the interfering species the reported maxima at 368 $\pm$ 2 nm in 0.1N alcoholic sulfuric acid with an absorptivity value of 14.5 may be used for quantitative measurement. The Technicon Autoanalyzer system for dosage form assays of diazepam is based on the direct spectrophotometric assay.

6.6 Polarographic Assay

A single reduction wave for diazepam has been observed by several investigators in aqueous system[19,20,21]. The single wave has been attributed to the reduction of the -C=N- moiety and the diffusion current is proportional to concentration in the range of 2×10^{-4} to 7×10^{-4}M.

6.7 Non-Aqueous Titration

Diazepam may be titrated in acetic anhydride using $HClO_4$ in glacial acetic acid and nile blue hydrochloride indicator. Each ml of 0.1N $HClO_4$ is equivalent to 28.48 mg of diazepam[17].

8. References

1. Hawrylyshyn, M., Hoffmann-La Roche Inc., Personal Communication.
2. Traiman, S., Hoffmann-La Roche Inc., Personal Communication.
3. Johnson, J. and Venturella, V., Hoffmann-La Roche Inc., Personal Communication.
4. Nuhn, P. and Bley, W., *Pharmzie*, 22a, 523 (1967).
5. Linscheid, P. and Lehn, J. J., *Bull. Soc. Chim. Fr.*, 1967 (3), 992.
6. Colarusso, R., Hoffmann-La Roche Inc., Personal Communication.
7. Greeley, D. and Benz, W., Hoffmann-La Roche Inc., Personal Commjnication.
8. Donahue, J., Hoffmann-La Roche Inc., Personal Communication.
9. Sheridan, J. C., Hoffmann-La Roche Inc., Personal Communication.
10. Toome, V., Hoffmann-La Roche Inc., Personal Communication.
11. Sternbach, L. H., Fryer, R. I., Metlesics, W., Reeder, E., Sach, G., Saucy, G. and Stempel, A., *J. Org. Chem.*, 27, 3788 (1962).
12. Archer, G. and Sternbach, L. H., *Chem. Rev.*, 68, 751 (1969).
13. deSilva, J. A. F., Schwartz, M. A., Stefanovic, V., Kaplan, J. and D'Arconte, L., *Anal. Chem.*, 11, 2099 (1964).
14. deSilva, J. A. F., Koechlin, B. A. and Bader, G., *J. Pharm. Sci.*, 55, 692 (1966).
15. deSilva, J. A. F., and Puglisi, C. V., *Anal. Chem.*, 42, 1725 (1970).
16. MacMullan, E. A., Hoffmann-La Roche Inc., Personal Communication.
17. National Formulary XIII, 221 (1970).
18. Scott, C. G. and Bommer, P., *J. Chrom. Sci.*, 8, 446 (1970).
19. Senkowski, B. Z., et al., *Anal. Chem.*, 36, 1991, (1964).
20. Oelschlaeger, H., et al., *Arch. Pharm.*, 297, 431 (1964).
21. Oelschlaeger, H., et al., *Collection Czech. Chem. Commun.*, 31, 1264 (1966).

ERYTHROMYCIN ESTOLATE

J. M. Mann

CONTENTS

1. Description
    1.1 Name, Formula, Structure, and Molecular Weight
    1.2 Appearance, Color, and Odor
2. Physical Properties
    2.1 Solubilities
    2.2 Infrared Spectrum
    2.3 Ultraviolet Spectrum
    2.4 Mass Spectrum
    2.5 X-ray Powder Diffraction
    2.6 Melting Point
    2.7 Nuclear Magnetic Resonance
    2.8 pKa
    2.9 Crystallinity
    2.10 Differential Thermal Analysis
3. Synthesis
4. Stability-Degradation
5. Drug Metabolic Products
6. Methods of Analysis
    6.1 Infrared Analysis
    6.2 Ultraviolet Analysis
    6.3 Microbiological Analysis
        6.31 Bioautographic Analysis
    6.4 Thin Layer Chromatography
7. References

ERYTHROMYCIN ESTOLATE

1. ## Description

   1.1 ### Name, Formula, Structure, and Molecular Weight

   Synonyms for this compound include erythromycin propionate lauryl sulfate; erythromycin propionate dodecyl sulfate; lauryl sulfate salt of the propionic ester of erythromycin; monopropionylerythromycin lauryl sulfate; and propionyl erythromycin lauryl sulfate.

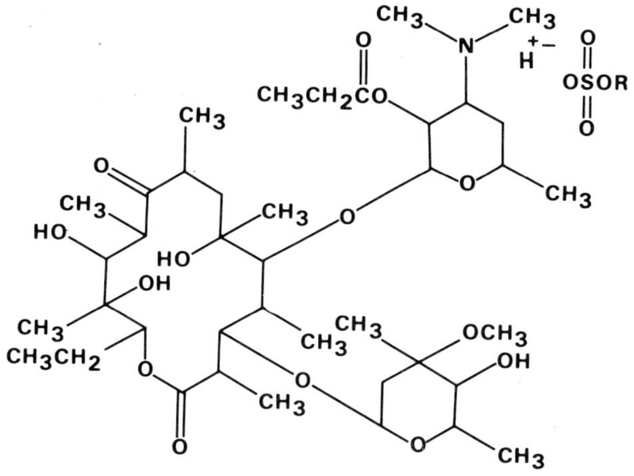

   Esthromycin estolate

   $C_{52}H_{97}NO_{18}S$ Mol. wt.: 1056.43

   R in the structural formula above represents "lauryl", which is predominately a C_{12} aliphatic hydrocarbon. The compound has a theoretical erythromycin base activity of 694.9 mcg/mg.

1.2 Appearance, Color, and Odor

The compound is a white crystalline powder which is essentially odorless and tasteless.

2. Physical Properties

2.1 Solubilities

Marsh and Weiss[1] have reported the solubilities of erythromycin estolate shown in Table I.

2.2 Infrared Spectrum

The infrared spectrum of erythromycin estolate is the most commonly accepted method for compound identification. The spectrum of a 10 mg/ml chloroform solution of erythromycin estolate from 850-4000 cm^{-1} is shown in Figure 1. The most characteristic difference between the infrared spectra of erythromycin base and estolate and that of anhydroerythromycin is that the latter is lacking the keto-carbonyl band at 1685 cm^{-1} (5.93 µ). If a sample of erythromycin base or estolate contains at least 5 percent anhydroerythromycin, a decrease in the intensity at 1685 cm^{-1} should be observed. This decrease can most readily be detected by measuring the ratio of the absorbance at 1685 cm^{-1} (5.93 µ) to that of the ester absorbance at 1735 cm^{-1} (5.76 µ). The amount of water in the sample can also be evaluated by the band at 1610 cm^{-1} (6.2 µ).[2]

Stephens[3] has reported infrared absorption bands for monopropionylerythromycin in chloroform.

2.3 Ultraviolet Spectrum

The maximum ultraviolet absorption of aqueous solutions of monopropionyl erythromycin is at 285 nm. Monopropionyl erythromycin was used because propionyl erythromycin lauryl sulfate is practically insoluble. Murphy[4] has reported that the ultraviolet spectrum of the esters of erythromycin are not significantly

TABLE I

Solubilities of Erythromycin Estolate

Solvent	mg/ml
Water	0.160
Methanol	>20
Ethanol	>20
Isopropanol	>20
Isoamyl alcohol	>20
Cyclohexane	0.080
Petroleum ether	0.058
Benzene	0.922
Iso octane	0.050
Carbon tetrachloride	0.058
Ethyl acetate	>20
Isoamyl acetate	1.250
Acetone	>20
Methyl ethyl ketone	>20
Diethyl ether	0.228
Ethylene chloride	>20
Chloroform	>20
Carbon disulfide	0.088
Pyridine	>20
Formamide	>20
Ethylene glycol	>20
Propylene glycol	>20
Dimethyl sulfoxide	>20
1,4 Dioxane	>20
0.1N NaOH	12.330
0.1N HCl	0.168

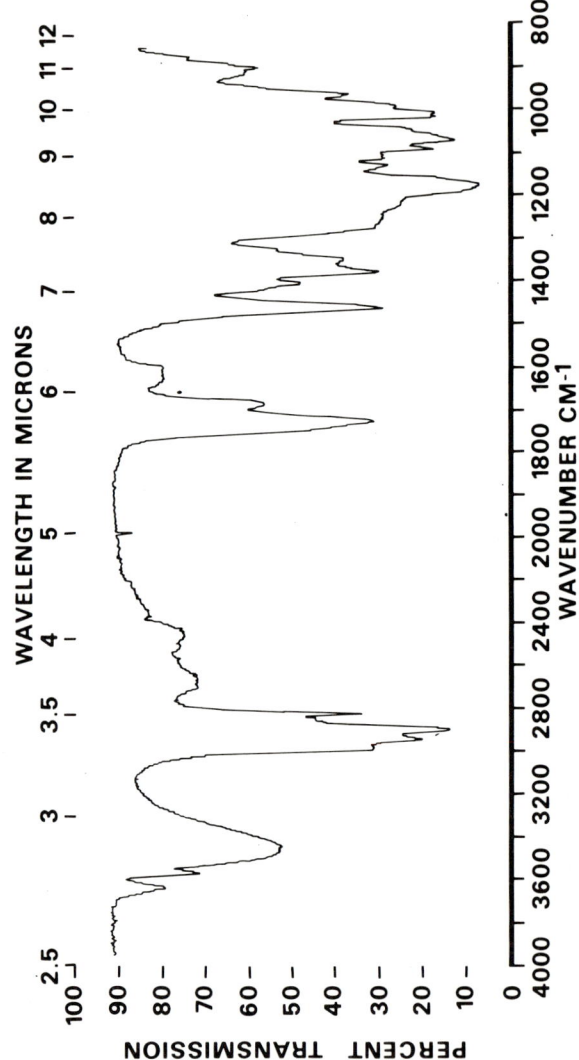

Fig. 1. Infrared absorption spectrum of erythromycin estolate.

different from that of erythromycin except in compounds containing the benzene nucleus in the acid moiety.

2.4 Mass Spectrum
Spectral data are not reported here since the lauryl sulfate radical precludes obtaining useful information.

2.5 X-ray Powder Diffraction
The x-ray powder diffraction pattern of erythromycin estolate ($\lambda = 1.5405$ Å) is shown in Table II.

2.6 Melting Point
The melting range of erythromycin estolate is 135-140 C and is accompanied by decomposition.[5]

2.7 Nuclear Magnetic Resonance
DeMarco[6] has made spectral assignments for erythromycin estolate. The spectrum shown in Figure 2 was obtained from a pyridine d_5/D_2O preparation at 100 MHZ. Assignments are shown in Table III.

2.8 pKa
The pKa for erythromycin estolate in 66% dimethylformamide/34% water is 6.9.

2.9 Crystallinity
When mounted in mineral oil and examined by means of a polarizing microscope, propionyl erythromycin lauryl sulfate exhibits birefringence and extinction positions when the microscope stage is revolved.[7]

2.10 Differential Thermal Analysis
A differential thermal analysis of erythromycin estolate at a heating rate of 20 C/min. exhibits a melting endotherm at 147 C.

3. Synthesis
The propionyl ester of erythromycin is pre-

TABLE II

X-ray Powder Diffraction Data

Erythromycin Estolate

d	I/I_1
22.1	0.16
19.2	0.16
16.4	0.20
14.5	0.16
13.6	0.16
11.0	1.00
9.9	0.30
8.9	0.30
7.2	0.20
7.0	0.30
6.5	0.16
5.5	0.80
5.1	0.16
4.9	0.30
4.7	0.20
4.5	0.20
4.3	0.16
3.9	0.12
3.8	0.12
3.6	0.16
3.4	0.12
3.3	0.08
3.03	0.04
2.90	0.04
2.79	0.08
2.63	0.04
2.47	0.04
2.41	0.04
2.34	0.08
2.20	0.04

TABLE III

NMR Spectral Assignments Of Erythromycin Estolate

Proton	Resonance (ppm)	
CH_3 of lauryl		
CH_3 of propionyl	1.4	superimposed
$-CH_2-$ of lauryl	1.1-1.3	superimposed
All CH_3 of aglycone and sugar other than those mentioned below	1.1-1.6	superimposed
C-6	1.73	singlet
$N(CH_3)_2$	2.80	singlet
OCH_3	3.57	singlet
C-7	2.6	doublet of doublet
C-2, 8, 10 C-3, 5;	3.0-3.5	superimposed
C-amine sugar	4.4	multiplet
C-13	5.61	doublet of doublet

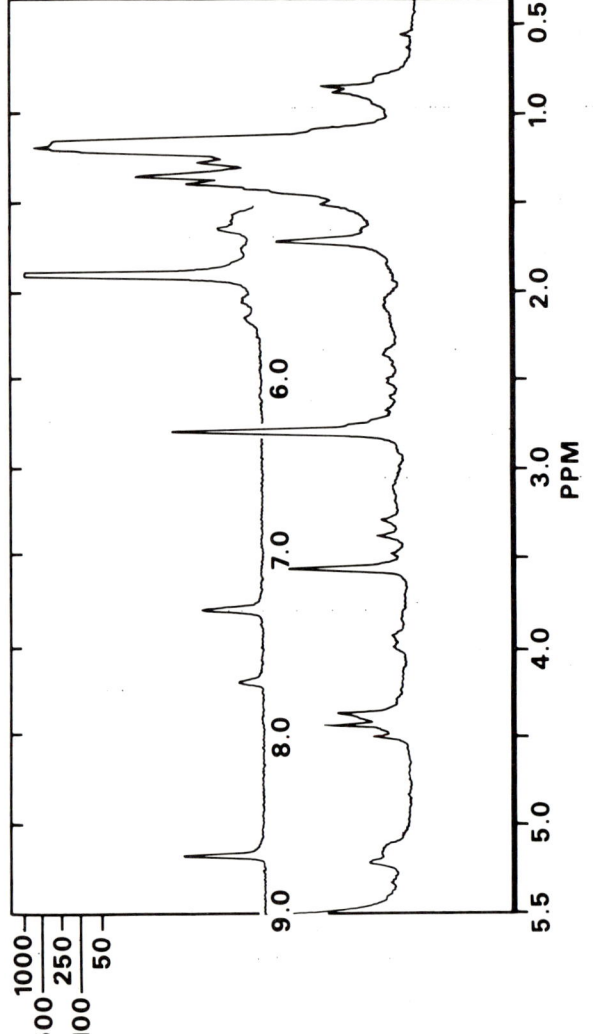

Fig. 2. NMR spectrum of erythromycin estolate

pared[3] by the reaction of either propionic anhydride or propionyl chloride and erythromycin in the presence of sodium bicarbonate or potassium carbonate as shown in step 1. Erythromycin estolate (step 2) is formed[8] by the addition of sodium lauryl sulfate to the ester dissolved in acidic acetone, and is isolated by diluting with water.

(1) $C_{37}H_{67}NO_{13}$ + $C_6H_{10}O_3$

    (Erythromycin)   (Propionic anhydride)

$\xrightarrow{\text{Base}}$ $C_{40}H_{71}NO_{14}$

    (Erythromycin propionate)

(2) $C_{40}H_{71}NO_{14}$ + $C_{12}H_{25}O_4SNa$

    (Erythromycin   (Sodium lauryl
    propionate)   sulfate)

$\xrightarrow[\text{H+}]{\text{Aqueous}}$ $C_{52}H_{97}NO_{18}S$

    (Propionyl erythromycin lauryl sulfate)

4. **Stability - Degradation**

Erythromycin estolate differs from other forms of erythromycin in that it is extremely stable to acid hydrolysis.[5] Erythromycin liberated from the ester by mild alkaline hydrolysis is subject to rapid decomposition in strongly acid solutions.[9] Kavanagh[10] has stated that deterioration of erythromycin increases with an increase in temperature and decreases with an increase in pH up to 8.0. Buffered aqueous solutions of erythromycin base are quite stable at this pH. Acetone solutions of the ester form are stable, while acetone solutions of the propionyl erythromycin lauryl sulfate preparation are not. Powders and dry formulations are stable for at least five years. Liquid preparations become unacceptable after 2 years due to undesirable taste.

5. Drug Metabolic Products

Both erythromycin and propionyl erythromycin are present *in vivo* after the therapeutic use of erythromycin estolate.[11] In studies using rabbit microsomes, Mao and Tardrew[12] have reported that erythromycin is demethylated to des-N-methyl erythromycin and formaldehyde. Propionyl erythromycin was also demethylated to propionyl des-N-methyl erythromycin; however, the rate of demethylation was less than that of erythromycin. Propionyl des-N-methyl erythromycin could subsequently be converted to des-N-methyl erythromycin. Stephens *et al.*[11] have shown that of the total antibacterial activity in whole blood, serum, plasma, and urine of individuals 2 hours after the fifth 250 mg dose of erythromycin estolate, 20-25% was present as erythromycin and 75-80% as propionyl erythromycin. This ratio remains relatively constant during the course of therapy. In studies involving nonfasting subjects, total erythromycin activity averaged 2-4 mcg/ml of whole blood, of which 0.7-1.0 mcg/ml was base activity.

6. Methods of Analysis

6.1 Infrared Analysis

Washburn[13] has reported that the potency of erythromycin may be readily determined by infrared absorbance at 10.46 µ. Investigation of bands located at 7.29, 9.02, 9.88, and 10.46 µ showed that only the latter consistently agreed with activities obtained by the microbiological methods.

Erythromycin estolate has also been quantitated by determining infrared absorbance at 9.9 µ, provided that anhydroerythromycin is absent. This compound, if present, increases the apparent potency of erythromycin.[14] In this method, 120.0 mg of sample is added to a 125 ml separatory funnel containing 30 ml 1N NaOH and 30 ml chloroform. After shaking for 1 minute, the chloroform layer is collected. The extraction is repeated with one 20 ml and two 10 ml portions and added to the first extract. The chloroform solution is evaporated under nitrogen, and the residue is dried in a desiccator in vacuo for 2 hours. The dried extract is dissolved in 20.0 ml chloroform, transferred to a 1.0 mm cell, and scanned versus a chloroform blank in a suitable spectrophotometer between 9.6 and 10.2 microns. Absorbance at 9.9 microns is determined. Dissolve 120.0 mg of erythromycin reference powder in 20.0 ml chloroform, and proceed as indicated above for sample preparation.

$$\frac{\text{Absorbance of sample at } 9.9\ \mu}{\text{Absorbance of standard at } 9.9\ \mu} \times \text{potency of standard (mcg/mg)} = \text{mcg of erythromycin base per mg.}$$

6.2   Ultraviolet Analysis

The ultraviolet analysis for erythromycin estolate is basically that described by Kuzel et al.[15] Alkali, buffer, and reference standard solutions must be prepared prior to assaying.

Alkali reagent is prepared by slurrying 42.0 g $Na_3PO_4 \cdot 12H_2O$ in 125 ml 0.5N NaOH and adding 100 ml purified water. After heating on a steam bath to facilitate dissolution, the solution is further diluted to 250 ml with purified water and filtered.

Phosphate buffer (pH 7.0) is prepared by dissolving 13.55 g KH_2PO_4 and 27.20 g K_2HPO_4 in 5 liters purified water.

The reference standard solution is prepared by dissolving 70.0 mg erythromycin in 200 ml methanol. This solution is diluted to 500.0 ml with phosphate buffer (pH 7.0). The stock is stable for 7 days when refrigerated.

A 10.0 ml aliquot of the standard is added to each of three 25 ml volumetric flasks followed by the addition of 1.0 ml 0.5N H_2SO_4 to one of these. After mixing, this flask is set aside at room temperature for at least 60 minutes. Two ml of the alkali reagent is added

to each of the remaining flasks, and they are
then heated in a 60 C water bath for 15 minutes.
These solutions are then cooled to room temperature in an ice bath. After diluting to volume
with purified water, the absorbance in a 1-cm
silica cell at 236 nm is determined using a
Beckman DU spectrophotometer or equivalent.
One ml of 0.5N NaOH is added to neutralize the
acid-treated standard, which is then processed
as indicated for the other two aliquots of
standard starting with the addition of alkali
reagent. The acid-treated standard represents
the standard blank. The H_2SO_4 is used to destroy the erythromycin base, making it possible
to measure any absorbance in the blank that may
be due to excipients. Solutions of erythromycin
base have negligible UV absorbance, but the
alkali reagent and heat convert erythromycin to
a UV-absorbing specie.
Samples are prepared by dissolving in
methanol an amount of erythromycin estolate that
would approximate in potency the standard solution. Sufficient methanol is added so that the
dilution will contain 40% methanol when brought
to volume with the phosphate buffer solution.
Hydrolysis of the ester to the base is accomplished by allowing the dilution to stand overnight at room temperature or heating for 2 hours
in a controlled 60 C water bath equipped with
a circulator. The solution is filtered and diluted to an appropriate assay concentration with
40% methanol/60% phosphate buffer. Methanol is
needed for dissolution of erythromycin estolate,
and the phosphate buffer provides a neutral
medium for hydrolysis. Aliquots containing
10.0 ml are transferred to three 25 ml volumetric flasks. One flask is treated with H_2SO_4
and two with alkali reagent as described above
for the processing of standard.

Calculations:

$$\frac{\text{Absorbance of sample} - \text{Absorbance of sample blank}}{\text{Absorbance of standard} - \text{Absorbance of standard}}$$

$$\frac{}{\text{blank}} \times \frac{70 \;[\text{potency of standard (mcg/mg)}]}{500 \times 1000}$$

$$\times \frac{10}{25} \times \text{dilution factor} = \text{mg erythromycin base in sample.}$$

6.3 Microbiological Analysis

Methods for the determination of base activity from erythromycin estolate are given in *Analytical Microbiology*, Vols. 1 & 2.[10,16] Photometric assays are conducted using *Staphylococcus aureus* ATCC 9144 with a test range from 0.05 to 2.0 mcg of erythromycin activity/ml in the assay tubes. *Sarcina lutea* ATCC 9341, is employed for the cylinder-plate assay of erythromycin liberated from erythromycin estolate. The assay is a two layer system of Grove & Randall[17] Agar No. 11 utilizing an assay range of 0.5-2.0 mcg/ml of sample. The test is satisfactory for the estimation of activity in body fluids.

Erythromycin estolate should be dissolved in a small quantity of methanol and brought to volume with pH 8.0 phosphate buffer[17]. Overnight hydrolysis at 25 C or at 60 C for 2 hours is necessary to liberate erythromycin base before determining microbiological activity.

6.31 Bioautographic Analysis

Stephens *et al.*[11] have reported a two step method for the separation of propionyl erythromycin from erythromycin in body fluids. In this procedure, the chromatogram is first developed in absolute methanol to separate the antibiotic from fluid protein. The chromatogram is then developed by the descending technique in a system containing NH_4Cl, NaCl, dioxane, and methyl ethyl ketone. Resultant zones of inhibition are visualized by applying the chromatogram to nutrient agar containing *S. lutea* as the indicator microorganism.

Bulk erythromycin estolate or finished products can be examined by using only the second system.

6.4   <u>Thin Layer Chromatography</u>
Erythromycin and erythromycin estolate can be separated and quantitated by thin layer chromatography.[18] Samples or standards should approximate 50 mcg erythromycin estolate and 5-10 mcg erythromycin base when spotted on silica gel G-254 plates. Approximately 120 ml of the developing solvent (methanol A.R.) is placed into the developing tank and allowed to equilibrate. The plate is developed until the solvent is approximately 15 cm from the origin The plate is removed and allowed to air dry. Antibiotic spots are visualized by spraying the plate with a fresh mixture of 95% ethanol/anisaldehyde/conc. sulfuric acid, 90:5:5(v/v) followed by heating of the plate at 110 C for 10 minutes. The R_f values of erythromycin estolate and erythromycin base are approximately 0.7 and 0.35 respectively.

7. References
   1. J.R. Marsh, and P.J. Weiss, J.A.O.A.C., 50, 457-462 (1967).
   2. C.D. Underbrink, personal communication, Lilly Research Laboratories.
   3. V.C. Stephens, U.S. Patent 2,993,833, July 15, 1961.
   4. H.W. Murphy, Antibiotics Ann., 500-513 (1953-1954).
   5. V.C. Stephens, J.W. Conine, and H.W. Murphy, J. Am. Pharm., 48, 620-622 (1959).
   6. P.V. DeMarco, personal communication, Lilly Research Laboratories.
   7. Code of Federal Regulations, 21, §141.504.
   8. M.D. Bray, and V.C. Stephens, U.S. Patent 3,000,870, Sept. 19, 1961.
   9. E. Korecká, Proc. Symposia Antibiotics, Praha, 355 (1960).
   10. F. Kavanagh, *Analytical Microbiology*, Vol. II, Academic Press (In Press).
   11. V.C. Stephens, C.T. Pugh, N.E. Davis, M.M. Hoehn, S. Ralston, M.C. Sparks, and L. Thompkins, J. Antibiot., 12, 551-557 (1969).
   12. J.C.H. Mao, and P.L. Tardrew, Biochem. Pharm., 14, 1049-1058 (1965).
   13. W.H. Washburn, J. Amer. Pharm. Assoc., Sc. Ed., 1, 48-49 (1954).
   14. G.W. Wallace, personal communication, Lilly Research Laboratories.
   15. N.R. Kuzel, J.M. Woodside, J.P. Comer, and E.E. Kennedy, Antib. and Chemo., 6, 1234-2141 (1954).
   16. F. Kavanagh, *Analytical Microbiology*, Vol. I, Academic Press (1963).
   17. D.C. Grove, and W.A. Randall, *Assay Methods of Antibiotics*, Medical Encyclopedia Inc., (1955).
   18. C. Bloom, and Analytical Staff, personal communication, Lilly Research Laboratories.

HALOTHANE

R. D. Daley

CONTENTS

1. Description
   1.1 Name, Formula, Molecular Weight
   1.2 Appearance, Color, Odor
2. Physical Properties
   2.1 Infrared Spectra
   2.2 Nuclear Magnetic Resonance Spectra
   2.3 Ultraviolet Spectra
   2.4 Mass Spectra
   2.5 Optical Rotation
   2.6 Vapor Pressure and Boiling Point
   2.7 Density
   2.8 Refractive Index
   2.9 Solubility
3. Synthesis
4. Stability-Degradation
5. Drug Metabolic Products
6. Methods of Analysis
   6.1 Analysis for Halothane in Mixtures
      6.11 Gas Chromatography
      6.12 Infrared Absorption
      6.13 Ultraviolet Absorption
      6.14 Other Methods
   6.2 Analysis for Impurities in Halothane
      6.21 Gas Chromatography
      6.22 Infrared Absorption
      6.23 Mass Spectrometry
   6.3 Analysis for Thymol
7. Determination in Body Fluids and Tissues
   7.1 Gas Chromatographic Methods
      7.11 Methods Using Prior Extraction
      7.12 Methods Using Prior Distillation or Gas Phase Partitioning
      7.13 Direct Injection Methods
   7.2 Absorptiometric Methods
      7.21 Turbidimetric Method
      7.22 Infrared Absorption
      7.23 Ultraviolet Absorption
   7.3 X-ray Spectrography
8. References

HALOTHANE

1. Description

1.1 Name, Formula, Molecular Weight

Halothane is 2-bromo-2-chloro-1,1,1-trifluoroethane. Commercial halothane contains 0.01 percent thymol as a stabilizer.

$$\begin{array}{c} F Cl \\ | | \\ F-C-C-H \\ | | \\ F Br \end{array}$$

C_2HF_3ClBr Mol. Wt.: 197.39

1.2 Appearance, Color, Odor

Colorless, mobile liquid, with an odor resembling that of chloroform.

2. Physical Properties

2.1 Infrared Spectra

Fig. 1 shows the infrared spectrum of halothane (Ayerst Laboratories Inc. Batch No. 1CKB). The spectrum is that of undiluted halothane in a 0.104 mm. potassium bromide cell vs. a potassium bromide plate. Also, because some of the absorption bands are quite intense, Fig. 1 shows the spectrum of a 4.0 volume percent solution of halothane in carbon disulfide, in a 0.104 mm. potassium bromide cell, vs. a 0.1 mm. cell filled with carbon disulfide. A Beckman Model IR-12 instrument was used. Considering the variety of sample handling techniques used, this spectrum and other published spectra (1-3) are the same.

Theimer and Nielsen (1) made a detailed study of the infrared and Raman spectra of halothane. They assigned bands of the gas phase infrared spectrum to fundamental vibrations as follows: CF_3 deformation-520, 552, 665 cm.$^{-1}$; CBr stretching-718 cm.$^{-1}$; CCl stretching-814 cm.$^{-1}$; CC stretching-863 cm.$^{-1}$; CH bending-1133, 1198 cm.$^{-1}$; CF stretching-1179, 1273, 1313 cm.$^{-1}$; CH stretching-3017 cm.$^{-1}$. Their paper contains the gas phase spectrum of halothane in the 1.5 to 35 micron region, using LiF, NaCl, and CsBr prisms.

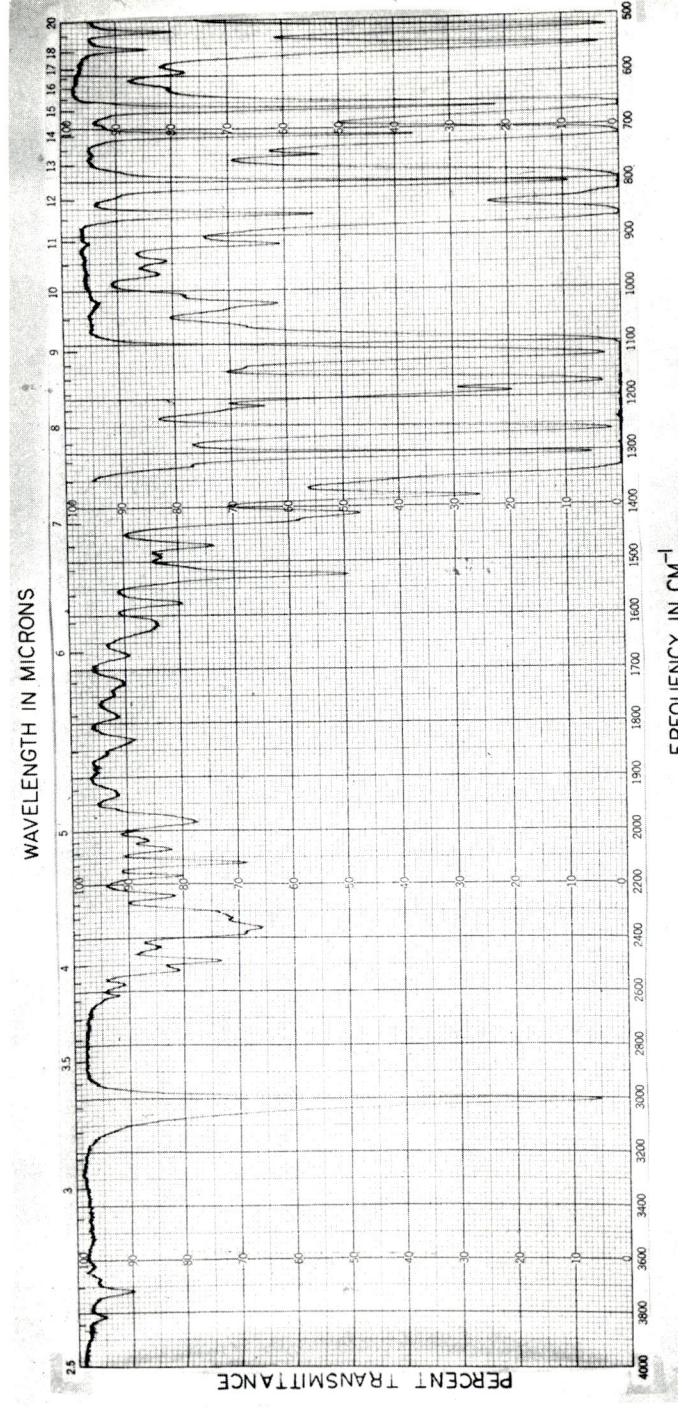

Fig. 1. Infrared spectrum of halothane, Ayerst Laboratories Inc. Batch 1CKB; 4000 to 500 cm.$^{-1}$, undiluted; 1350 to 500 cm.$^{-1}$, 4% (v/v) in CS$_2$ vs. CS$_2$; 0.1 mm. cells.

Kalow (2) published a gas phase spectrum of halothane in the 7 to 16 micron region. Kakac and Hudlicky (3) published the spectrum of a carbon tetrachloride solution of halothane in the 1600 to 450 cm.$^{-1}$ region.

2.2 Nuclear Magnetic Resonance Spectra

Fig. 2 shows the proton magnetic resonance spectrum of halothane. This spectrum was obtained on a carbon tetrachloride solution, using a Varian HA-100 instrument, with a tetramethylsilane reference. The quartet centered at 5.76 p.p.m. is assigned to CH adjacent to a CF_3 group (4).

2.3 Ultraviolet Spectra

Fig. 3 shows the ultraviolet spectrum of halothane (Ayerst Laboratories Inc. Batch No. 1CKB) in 2,2,4-trimethylpentane solution. The solution was run vs. the solvent on a Cary Model 14 instrument, using 1.0 mm. cells. Fig. 3 also shows a scan of solvent vs. solvent in the same cells for comparison. The discontinuity in the solution scan at 221 nanometers is a change in absorbance scale; the absorbance scale is 0 to 1 for the 350 to 221 nanometer region and 1 to 2 for the 221 to 200 nanometer region, for the solution. The halothane concentration is 0.40 volume percent, or 7.4 grams per liter. The wavelength of maximum absorption is 203 nanometers, with an absorptivity of 2.50 liter/g. cm., **or a molar absorptivity** of 490 liter/mole cm. This absorption is presumably due to the C-Br structure.

Kalow (2) has published the ultraviolet spectrum of halothane gas in air (195 to 280 nanometers). His data indicate the wavelength of maximum absorption to be 206 nanometers.

2.4 Mass Spectra

Fig. 4 shows a plot of the mass spectrum of halothane. The data were obtained with an AEI MS-9 mass spectrometer. The assignments and compositions are as follows:

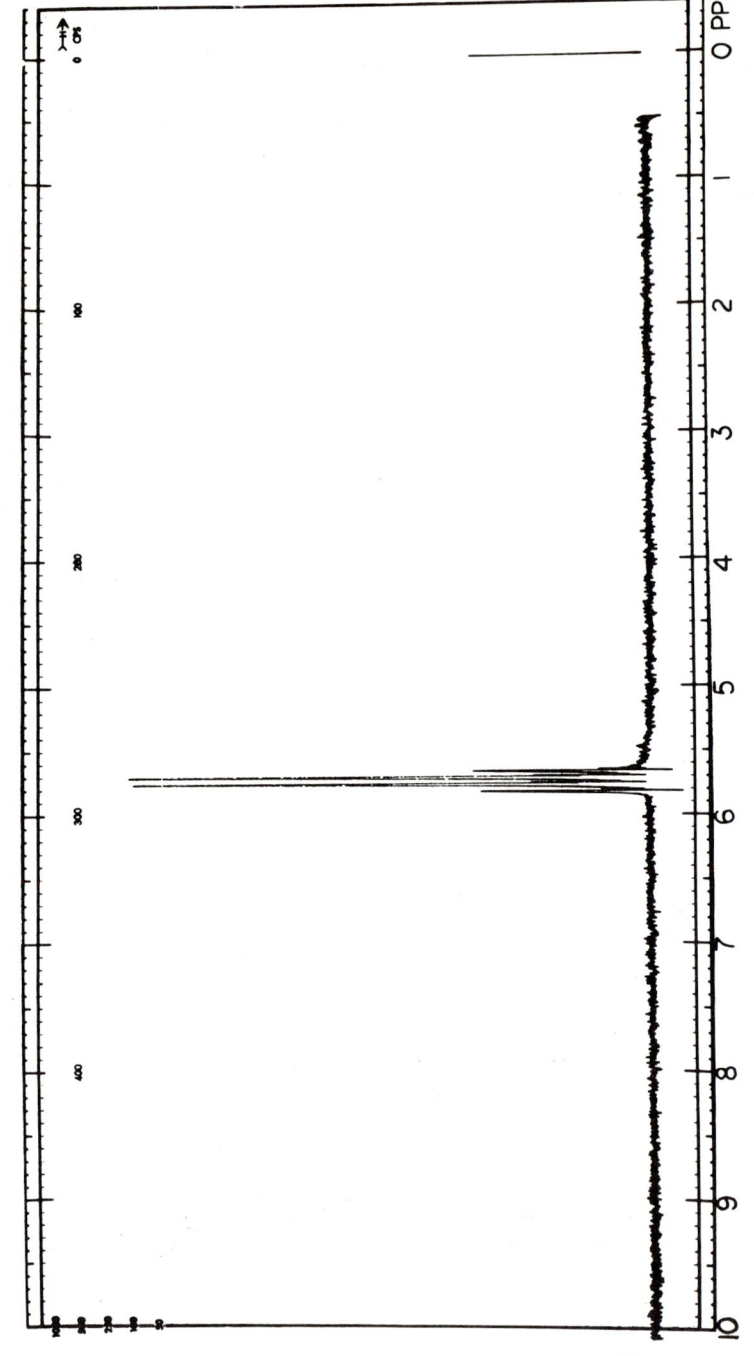

Fig. 2. Nuclear magnetic resonance spectrum of halothane in carbon tetrachloride, tetramethylsilane reference (courtesy of Dr. J. M. Pryce).

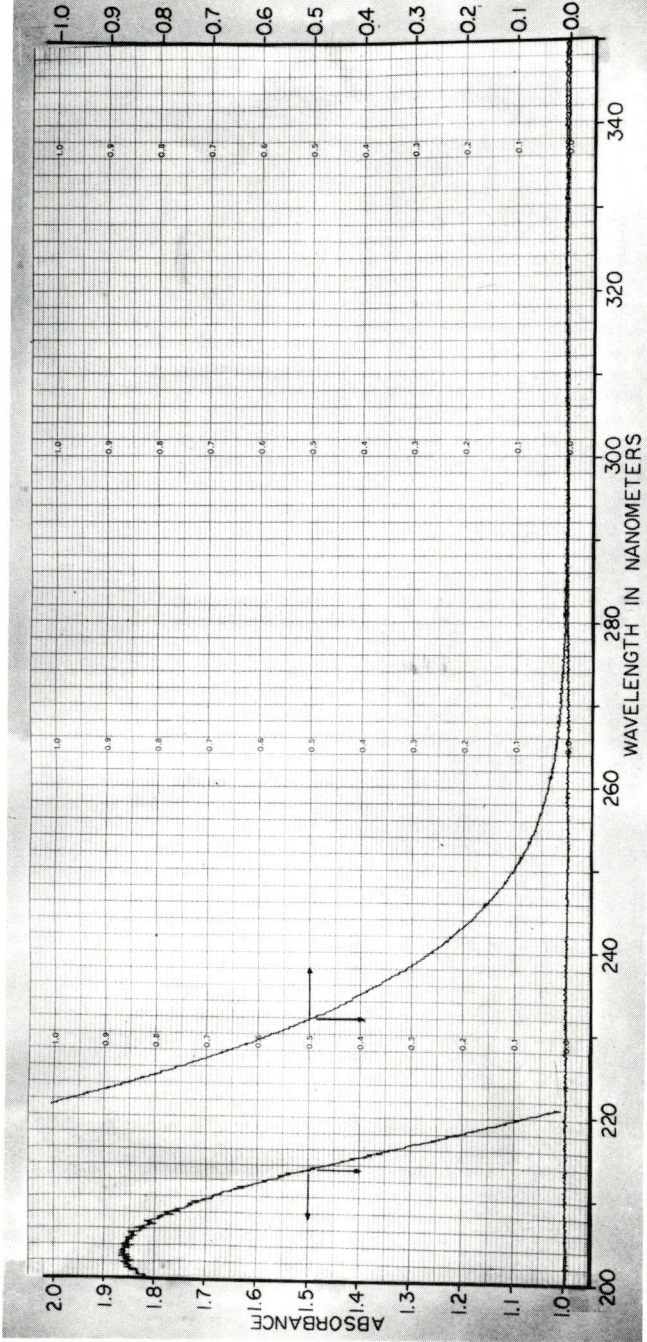

Fig. 3. Ultraviolet spectrum of halothane, Ayerst Laboratories Inc. Batch 1CKB, 0.40% (v/v) in 2,2,4-trimethylpentane, 1.0 mm. cells.

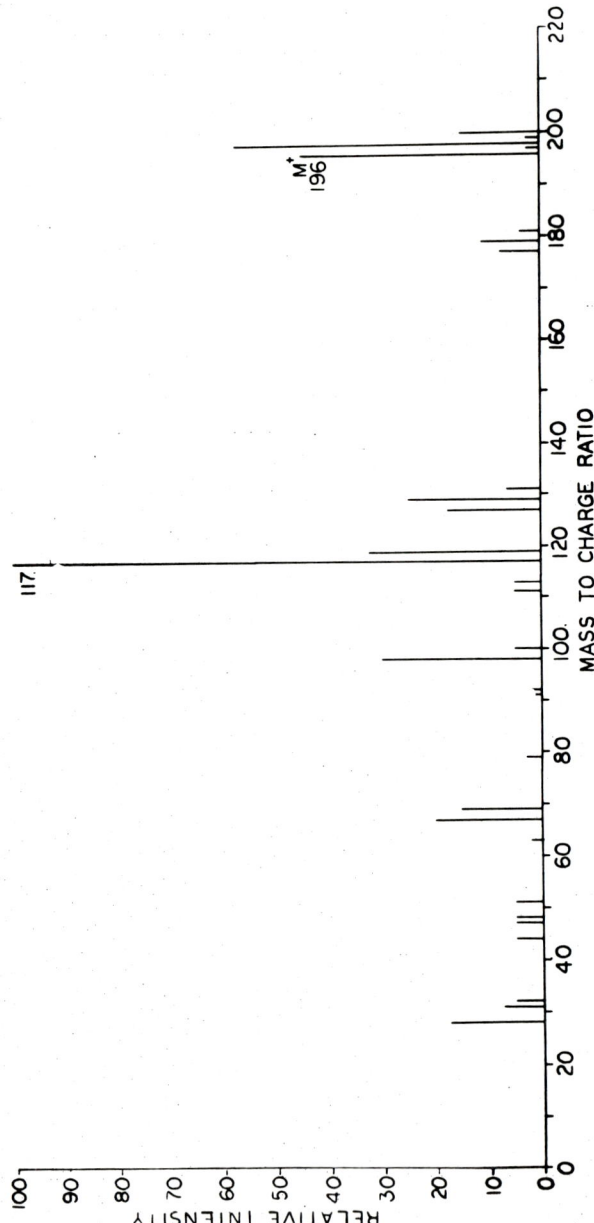

Fig. 4. Mass spectrum of halothane (data courtesy of Dr. J. M. Pryce).

HALOTHANE

m/e	Assignment	Composition
196	M$^+$	$CF_3CHClBr$
177	(M-F)$^+$	$CF_2CHClBr$
161*	(M-Cl)$^+$	CF_3CHBr
160*	(M-HCl)$^+$	CF_3CBr
142*	(161-F)$^+$	CF_2CHBr
141*	(160-F)$^+$	CF_2CBr
127	(M-CF$_3$)$^+$	$CHClBr$
117	(M-Br)$^+$	CF_3CHCl
111		$CFHBr$
98		C_2F_2HCl
92		$CHBr$
91		CBr
79		Br
69		CF_3
67		$CFHCl$
63		C_2HF_2

*Relative intensities of these mass spectral peaks are 1.0 or less.

The molecular ion and those at m/e 127 and m/e 69 support the proposed structure. The ions at m/e 111, 67, and 63 are rearrangement ions (4).

2.5 Optical Rotation

Although the halothane molecule has an asymmetric carbon atom, the commercial product is a racemic mixture; resolution of the mixture has not been reported (5).

2.6 Vapor Pressure and Boiling Point

The vapor pressure of halothane (6) can be calculated from the equation

$$\log_{10} p = \frac{6.8513 - 1082.495}{t + 222.44}$$

where p is the vapor pressure in mm. Hg and t is the temperature in degrees C. Some vapor pressure values are as follows:

Temperature, degrees C.	Vapor Pressure, mm. Hg (6)
20	243.4
30	365.8
50.19	760.

Some reported boiling points (degrees C) are:

50.0 - 50.5 at 760 mm. Hg (7)

50.2 at 760 mm. Hg (8, 9)

50.1 at 754 mm. Hg (10)

49.5 - 49.8 at 740 mm. Hg (11)

50.15 at 760 mm. Hg (12)

2.7 Density

The density of halothane has been reported as 1.871 g./ml. (9), 1.872 g./ml. (11), and 1.8692 g./ml. (6) at 20 degrees C.

2.8 Refractive Index

The refractive index n_D^{20} of halothane has been reported as 1.3697 (9) and as 1.3700 (6, 8, 11).

2.9 Solubility

Halothane is slightly soluble in water (0.345 parts in 100 parts water) (13) and is miscible with the following solvents: methanol, ethanol, chloroform, carbon disulfide, diethyl ether, hydrocarbons, and fixed oils.

3. Synthesis

Halothane can be prepared by: (a) brominating 2-chloro-1,1,1-trifluoroethane, or chlorinating 2-bromo-1,1,1-trifluoroethane (7); rearrangement of 1-bromo-2-chloro-1,1,2-trifluoroethane with (b) aluminum chloride (10) or (c) aluminum bromide (14); (d) treatment of 1,2-dibromo-1,1,2-trichloroethane with hydrogen fluoride and antimony tri- and pentachlorides (8); (e) treatment of 1,2-dibromo-2-chloro-1,1-difluoroethane with hydrogen fluoride and either antimony tri- and pentachloride or antimony pentachloride alone (15); treatment of 2,2-dibromo-2-chloro-1,1,1-trifluoroethane with (f) iron and hydrochloric acid or acetic acid (16) or (g) sodium sulfite and sodium hydroxide (12), or (h) 2-chloro-1,1,1-trifluoroethane (17); (i) rearrangement of 2-bromo-1-chloro-1,1,2-trifluoroethane with aluminum chloride (18). These reactions are shown in Fig. 5, where the letters correspond to the various methods mentioned above.

4. Stability-Degradation

Halothane is stable when stored in amber glass bottles or when 0.01 weight percent thymol is added. Exposure of unstabilized halothane to light causes slow decomposition with formation of volatile acids and bromine (5, 13). It has been reported that 2,3-dichlorohexafluoro-2-butene is formed when halothane is heated in the presence of copper and oxygen (19, 20) or evaporated at room temperature in the presence of air and copper (20); other investigators either dispute or do not confirm these observations (21-25).

5. Drug Metabolic Products

It has been reported that halothane is partially metabolized to bromide, chloride, trifluoroacetate, and, to a

FIGURE 5
PREPARATION OF HALOTHANE

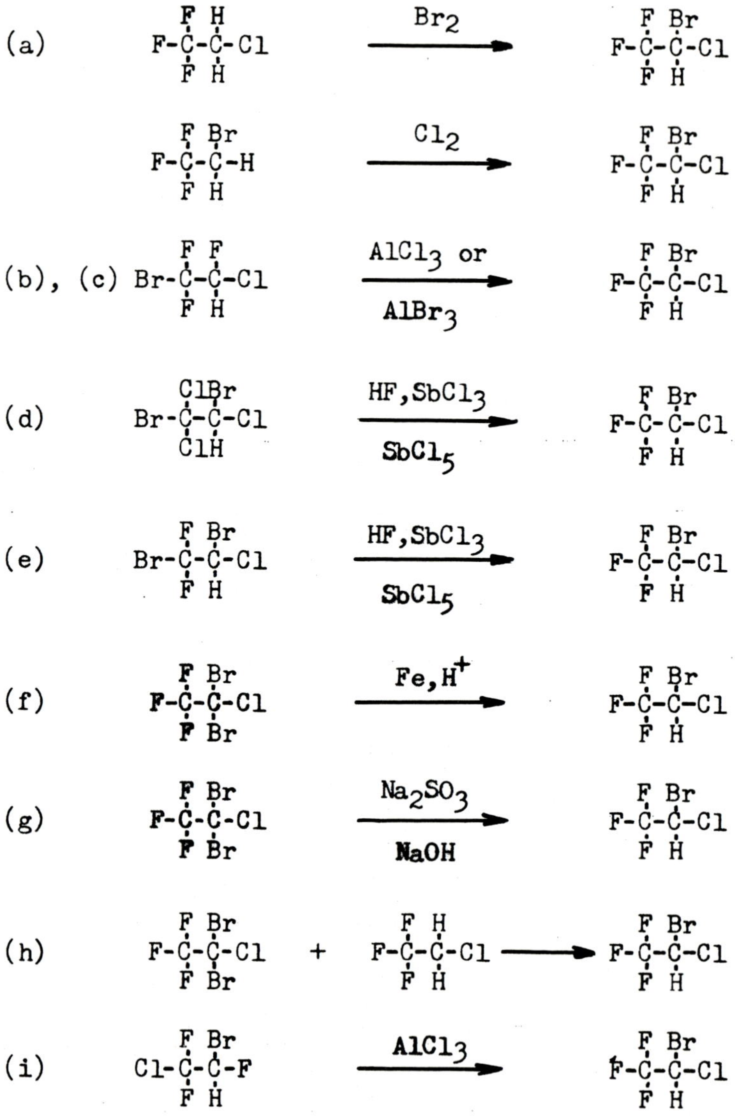

(a) $\mathrm{F\text{-}\underset{F}{\overset{F}{C}}\text{-}\underset{H}{\overset{H}{C}}\text{-}Cl} \xrightarrow{Br_2} \mathrm{F\text{-}\underset{F}{\overset{F}{C}}\text{-}\underset{H}{\overset{Br}{C}}\text{-}Cl}$

$\mathrm{F\text{-}\underset{F}{\overset{F}{C}}\text{-}\underset{H}{\overset{Br}{C}}\text{-}H} \xrightarrow{Cl_2} \mathrm{F\text{-}\underset{F}{\overset{F}{C}}\text{-}\underset{H}{\overset{Br}{C}}\text{-}Cl}$

(b), (c) $\mathrm{Br\text{-}\underset{F}{\overset{F}{C}}\text{-}\underset{H}{\overset{F}{C}}\text{-}Cl} \xrightarrow[\mathrm{AlBr_3}]{\mathrm{AlCl_3\ or}} \mathrm{F\text{-}\underset{F}{\overset{F}{C}}\text{-}\underset{H}{\overset{Br}{C}}\text{-}Cl}$

(d) $\mathrm{Br\text{-}\underset{Cl}{\overset{ClBr}{C}}\text{-}\underset{H}{\overset{}{C}}\text{-}Cl} \xrightarrow[\mathrm{SbCl_5}]{\mathrm{HF,SbCl_3}} \mathrm{F\text{-}\underset{F}{\overset{F}{C}}\text{-}\underset{H}{\overset{Br}{C}}\text{-}Cl}$

(e) $\mathrm{Br\text{-}\underset{F}{\overset{F}{C}}\text{-}\underset{H}{\overset{Br}{C}}\text{-}Cl} \xrightarrow[\mathrm{SbCl_5}]{\mathrm{HF,SbCl_3}} \mathrm{F\text{-}\underset{F}{\overset{F}{C}}\text{-}\underset{H}{\overset{Br}{C}}\text{-}Cl}$

(f) $\mathrm{F\text{-}\underset{F}{\overset{F}{C}}\text{-}\underset{Br}{\overset{Br}{C}}\text{-}Cl} \xrightarrow{\mathrm{Fe,H^+}} \mathrm{F\text{-}\underset{F}{\overset{F}{C}}\text{-}\underset{H}{\overset{Br}{C}}\text{-}Cl}$

(g) $\mathrm{F\text{-}\underset{F}{\overset{F}{C}}\text{-}\underset{Br}{\overset{Br}{C}}\text{-}Cl} \xrightarrow[\mathrm{NaOH}]{\mathrm{Na_2SO_3}} \mathrm{F\text{-}\underset{F}{\overset{F}{C}}\text{-}\underset{H}{\overset{Br}{C}}\text{-}Cl}$

(h) $\mathrm{F\text{-}\underset{F}{\overset{F}{C}}\text{-}\underset{Br}{\overset{Br}{C}}\text{-}Cl} + \mathrm{F\text{-}\underset{F}{\overset{F}{C}}\text{-}\underset{H}{\overset{H}{C}}\text{-}Cl} \longrightarrow \mathrm{F\text{-}\underset{F}{\overset{F}{C}}\text{-}\underset{H}{\overset{Br}{C}}\text{-}Cl}$

(i) $\mathrm{Cl\text{-}\underset{F}{\overset{F}{C}}\text{-}\underset{H}{\overset{Br}{C}}\text{-}F} \xrightarrow{\mathrm{AlCl_3}} \mathrm{F\text{-}\underset{F}{\overset{F}{C}}\text{-}\underset{H}{\overset{Br}{C}}\text{-}Cl}$

small extent, to carbon dioxide. These metabolic products have been reported in various species of animals as follows:

Man   Trifluoroacetate (26, 27); Bromide (27, 28)

Rabbit Trifluoroacetate (29)

Rat   Chloride (30), Bromide (31), Carbon dioxide (32)

In addition to the above, unidentified non-volatile metabolites have been reported in mice (33, 34).

6. Methods of Analysis

  6.1 Analysis for Halothane in Mixtures

    6.11 Gas Chromatography

    A number of investigators have described the analysis for halothane in mixtures by gas chromatography. Adlard and Hill (35) analyzed mixtures of halothane, ether, oxygen, nitrous oxide, carbon dioxide, and cyclopropane in respired anesthetic mixtures. Rutledge et al (36) analyzed for halothane in gas samples. Theye (37) analyzed respiratory gases for oxygen, carbon dioxide, and halothane to study gas exchange during halothane anesthesia. Lowe (38, 39) reported halothane analyses at 10 second intervals during clinical anesthesia, taking advantage of the fact that usually a single anesthetic was used, so that little or no chromatographic separation was needed. Good separation of halothane from ethanol, methanol, chloroform, dichloromethane, dichloroethane, diethyl ether, vinyl ether and other organic vapors has been described (40). Rehder et al (41) analyzed mixtures of halothane, oxygen, and carbon dioxide to determine halothane and oxygen intake in the anesthetized patient. Analysis for halothane in gas mixtures is also described by Mirolyubova (42) and by Wortley et al (43). Tiengo analyzed for halothane in alveolar air (44). Patzelova determined oxygen, carbon dioxide, nitrous oxide, and halothane in respiratory gases (45). Table I shows systems used by some of these investigators.

    6.12 Infrared Absorption

    Kalow (2) investigated the feasibility of

TABLE 1

Gas Chromatographic Systems Used for Halothane in Gas Mixtures

Reference Number	Column	Carrier Gas	Column Temp., °C	Detector
35	(a) 2 ft. long x ¼ in. 15% dinonyl phthalate on 52-60 mesh firebrick, in parallel with (b) 20 ft. long x ¼ in. 20% dimethyl sulfoxide on 52-60 mesh firebrick	H_2	(a) 75 (b) 20	Thermal Conductivity
36	6 ft. long x ¼ in. Tide detergent	He	90	Thermal Conductivity
37	(a) 6 ft. long x 3/16 in. molecular sieve in parallel with (b) 12 ft. long x 3/16 in. 10% Amine 220 and 5% Carbowax 400 on 30-60 mesh T-6 Teflon	He	64.5	Thermal Conductivity
38, 39	12 in. long x ¼ in. 60-80 mesh Chromosorb P, water saturated	N_2	Room temp.	Flame Ionization
41	(a) 60 cm. long x 3/16 in., 20% dioctyl sebacate on firebrick, in parallel with (b) 120 cm. long x 3/16 in. 10% dioctyl sebacate on silica gel, both columns in	H_2	50	Thermal Conductivity

TABLE 1
(cont'd)

Reference Number	Column	Carrier Gas	Column Temp., °C	Detector
	series with (c) 183 cm. long x 3/16 in. 5A molecular sieve			
43	6.5 ft. long x 0.062 in. I.D. silicone fluid MS 550 on Chromosorb P, 60-80 mesh, 25:85	air	88	Flame Ionization
45	(a) 50 cm. long x 4 mm. I.D. 15% Kel-F oil 10 on 0.16 to 0.2 mm. Chromaton N-AW in series with (b) 270 cm. 100-120 mesh Porapak Q	H_2	Room temp.	Thermal Conductivity

monitoring halothane vapor concentrations by ultraviolet and infrared absorption. He concluded that either technique could be used.

Kakac and Hudlicky (3, 46) used infrared absorption to analyze liquid mixtures of halothane and 1-bromo-2-chloro-1,1,2-trifluoroethane resulting from aluminum chloride rearrangement of the latter. Davies et al (47) and Sechzer et al (48) used infrared absorption to analyze for halothane in respiratory gases during and after anesthesia. Rehder et al (27) and Larson et al (49) used commercial infrared halothane analyzers for measuring halothane in gas mixtures.

6.13 Ultraviolet Absorption

As mentioned in 6.12, Kalow (2) investigated the use of ultraviolet absorption for monitoring halothane vapor concentrations (see 2.3, Fig. 3). He recommended a wavelength of 228 nanometers for measurement, since it was free from interference by nitrous oxide. However, the ready availability of mercury lamps makes measurements at the 254 nanometer mercury line attractive, and relatively simple photometric systems suffice to measure halothane vapor concentrations of 0 to 5 percent (50). Commercial ultraviolet halothane meters have been used to monitor halothane vapor concentrations and to check the calibration of vaporizers (51, 52).

6.14 Other Methods

Duncan (53) extracted halothane from gas mixtures with petroleum ether, treated the extract with lithium aluminum hydride to liberate halide, precipitated the halides with silver ion, and determined the amount of precipitate photometrically. This method is an adaptation of Goodall's method (54) for halothane in blood.

Interferometers have been used to measure halothane vapor concentrations and to check ultraviolet halothane meter calibrations (51, 55).

6.2 Analysis for Impurities in Halothane

Halothane as furnished for anesthesia is a material of high purity. However, in addition to the 0.01 weight percent thymol stabilizer, it may contain traces of volatile impurities. Also, U.S.P. XVIII (56) allows a non-volatile residue of 1 mg. per 50 ml. after 2 hours drying at 105°C; thymol is volatile at this temperature. In anesthetic use, halothane is evaporated, along with any volatile impurities. Thymol is not volatile under conditions existing in anesthetic vaporizers, and it will accumulate in the residue.

The discussion of impurity analysis in 6.21 to 6.23 below relates only to halothane as marketed for anesthesia. Halothane is an excellent solvent and, like any high purity material, can be easily contaminated unless handled carefully. Analyses of drain samples from vaporizers must be interpreted with caution; as an example of possible pitfalls, impurity peaks on gas chromatograms of such samples may be due to the presence of other anesthetics in the sample (24).

6.21 Gas Chromatography

Chapman et al (57) made an extensive study of impurities in halothane manufactured by high temperature bromination of 2-chloro-1,1,1-trifluoroethane. They used gas chromatography, mass spectrography, nuclear magnetic resonance, infrared absorption, and microchemical analyses to identify impurities, and took advantage of the larger impurity concentrations available either from distilling the crude halothane or from process streams. In this manner they were able to find sixteen impurities in the crude halothane, and devise gas chromatographic methods for analyzing for them at the p.p.m. level in the finished product. The sixteen impurities found are as follows:

Impurity Number	Identity
1	trans-2-Chloro-1,1,1,4,4,4-hexafluoro-2-butene
2	2-Chloro-1,1,1-trifluoroethane
3	cis-2-Chloro-1,1,1,4,4,4-hexafluoro-2-butene
4	trans-2-Bromo-1,1,1,4,4,4-hexafluoro-2-butene
5	2,2-Dichloro-1,1,1-trifluoroethane
6	2-Bromo-1,1,1-trifluoroethane.
7	trans-2,3-Dichloro-1,1,1,4,4,4-hexafluoro-2-butene
8	cis-2,3-Dichloro-1,1,1,4,4,4-hexafluoro-2-butene
9	1,1,2-Trichloro-1,2,2-trifluoroethane
10	Bromodichlorofluoromethane
11	2,2-Dibromo-1,1,1-trifluoroethane
12	Chloroform
13	2,2-Dibromo-2-chloro-1,1,1-trifluoroethane
14	**2-Bromo-2-chloro-1,1-difluoroethane**
15	2-Bromo-2,2-dichloro-1,1,1-trifluoroethane
16	1,2-Dichloro-1,1-difluoroethane

Table 2 lists the retention times of these impurities, relative to 1,1,2-trichloro-1,2,2-trifluoroethane (impurity number 9), on two gas chromatographic systems: (a) column 6 ft. long by 3/16 in. I.D., of 30 weight percent Aroclor 1254 (chlorinated diphenyl) on 60-80 mesh Chromosorb P, nitrogen carrier gas at 30 ml./min., column temperature 60°C., flame ionization detector; (b) column 6 ft. long by 3/16 in. I.D., of 30 weight percent polyethylene glycol 400 on 72-85 mesh Celite, joined to a column 3 ft. long by 3/16 in. I.D., of 30 weight percent dinonyl phthalate on 72-85 mesh Celite, nitrogen carrier gas at 40 ml./min., column temperature 60°C, flame ionization detector. These two systems separate all impurities except numbers 5 and 6. Only four of the impurities are usually detectable, numbers 2, 5, 9, and 14, at concentrations typically 1, 8, 12, and 1 p.p.m., respectively (57).

TABLE 2

Relative Retention Times of Some Possible Halothane Impurities (57)

Retention Times Relative to $CF_2ClCFCl_2$

Impurity Number	(a) Chlorinated Diphenyl Column	(b) Polyethylene Glycol-Dinonyl Phthalate Column
1	0.12	0.38
2	0.22	1.00
3	0.30	0.73
4	0.40	1.00
5	0.62	2.35
6	0.62	2.35
7	0.62	0.73
8	0.78	0.85
9	1.00	1.00
10	3.48	ND
11	4.85	ND
12	6.43	10.9
13	8.73	ND
14	M	1.50
15	M	3.28
16	M	4.22

ND - Not determined.
M - Peak masked by halothane.

Scipioni et al (58) report a somewhat different set of impurities for halothane made by high temperature bromination of 2-chloro-1,1,1-trifluoroethane on a laboratory scale, and they also investigated the changes in concentrations with changes in reaction temperature and contact time.

A number of other investigators have used gas chromatography to detect or analyze for trace impurities in halothane. Cohen et al (19, 20) first drew attention to the presence of cis and trans 2,3-dichloro-1,1,1,4,4,4-hexafluoro-2-butene in halothane made by high temperature processes; these materials were separated from halothane by gas chromatography, with identification by mass spectrometry. Albin et al (24) similarly confirmed the presence of these butenes and identified 2-bromo-1,1,1,4,4,4-hexafluoro-2-butene as an impurity as well. Gjaldbaek and Worm (25) reported 1,1,2-trichloro-1,2,2-trifluoroethane in halothane from both high and low temperature processes. The absence of the butenes in halothane of later manufacture has been reported (20, 57, 59). A trace of 1,2-dichlorohexafluorocyclobutane was found in halothane made by a low temperature process (25, 60). Two groups of investigators were unable to separate 1-bromo-2-chloro-1,1,2-trifluoroethane, the starting material of one low temperature process, from halothane by gas chromatography (3, 25) (see below).

6.22 Infrared Absorption

Although the concentrations of practically all the impurities in the market product are too low to detect by infrared absorption, it has been used to analyze for 1-bromo-2-chloro-1,1,2-trifluoroethane at the 0.1 percent level, in halothane made by isomerizing the former material (3, 25).

6.23 Mass Spectrometry

After separation by gas chromatography, impurities have been identified by mass spectrometry in some cases (19, 20, 24, 57).

6.3 Analysis for Thymol

The dibromoquinone-chlorimide colorimetric method is the official method for thymol analysis in U.S.P. XVIII (56). Thymol can also be analyzed for by gas chromatography; one technique uses an SE-30 column, flame ionization detection, helium carrier gas, and temperature programming from 100 to 150° (25).

7. Determination in Body Fluids and Tissues

7.1 Gas Chromatographic Methods

7.11 Methods Using Prior Extraction

Halothane in blood and tissues can be analyzed by gas chromatography after extraction into heptane (36, 43, 61-63). This method takes more time than the direct injection methods mentioned below, but it is not necessary to clean the injection port frequently, and column life may be extended.

Use of electron capture detectors increases the sensitivity to halothane and decreases the relative sensitivity to heptane, improving the overall sensitivity and speed of the method (64, 65).

Carbon tetrachloride containing chloroform as an internal standard can be used to extract halothane from blood; it clears SE-30 columns faster than heptane (66).

Table 3 lists gas chromatographic systems used with this technique.

7.12 Methods Using Prior Distillation or Gas Phase Partitioning

Halothane in blood or tissues can be analyzed after distillation into heptane from a sample-water mixture. The distillate and heptane are mixed by shaking, the heptane is separated and dried, and an aliquot is injected into the gas chromatographic apparatus (67).

TABLE 3

Gas Chromatographic Systems for Methods Using Extraction Prior to Chromatography

Reference Number	Column	Carrier Gas	Column Temp., °C	Detector
61, 62	12-30 in. long, 10 wt. % silicone M.S. 550 on 80-100 mesh Celite	H_2	30	Flame Ionization
43, 63	6.5 ft. long x 0.062 in. I.D., silicone M.S. 550 on Chromosorb P, 60-80 mesh	air	88	Flame Ionization
64	6 ft. long x 0.125 in. O.D., 35 wt. % silicone M.S. 550 on 44-60 mesh Celite	N_2	44	Electron Capture
36	6 ft. long x 0.25 in. O.D., Tide detergent	He	90	Thermal Conductivity
66	5 ft. long x 0.125 in., 5% SE-30 on 60/80 mesh Chromosorb W	N_2	55	Flame Ionization

Halothane in body fluids and tissue has also been determined after distillation from the sample with toluene. The distilled toluene, containing the halothane, is dried and analyzed by gas chromatography (68, 69).

Where the blood-gas partition coefficient is known, halothane can be determined on the gas or air space above blood samples equilibrated with the gas. Conversely, the blood-gas partition coefficient can be determined by gas chromatographic analysis of gas equilibrated with blood if the halothane content of the system and the blood and gas volumes are known (70-73). By using a syringe with stops set to definite volumes, Fink and Morikawa determined halothane partial pressures in blood by making two consecutive equilibrations with air, analyzing the air phase after each equilibration (74).

Table 4 lists systems used with this technique.

TABLE 4

Gas Chromatographic Systems for Methods Using Distillation or Partitioning Prior to Chromatography

Reference Number	Column	Carrier Gas	Column Temp., °C	Detector
67	Silicone oil (commercial Perkin-Elmer Type C column)	He	40	Thermal Conductivity
73	2 ft. long x 0.25 in., 20% SE-30 on Chromosorb	Argon	100	Flame Ionization
68, 69	12-24 ft., 10% di-2-ethylhexyl sebacate on 20-60 mesh firebrick	He	130	Thermal Conductivity
71	5 ft. long x 0.125 in., 5% silicone gum rubber on 60-80 mesh Chromosorb W	He	100	Flame Ionization
72	6 ft. long, 3.8 percent (w/w) silicone fluid on Chromosorb W	Not Stated	50	Flame Ionization

TABLE 4
(cont'd)

Reference Number	Column	Carrier Gas	Column Temp., °C	Detector
74	5 ft. long x 1/8 in., 5 percent SE-30 on DMCS Chromosorb W, 50-80 mesh	He	85	Flame Ionization

7.13 Direct Injection Methods

Halothane in blood and tissues can be determined by direct injection of such material into a gas chromatograph with a suitable injection port system. The injection port must provide for more or less frequent cleanout.

Cohen (75) described a special biopsy needle for sampling and injecting tissue. Lowe (38, 39) used glass capillaries which he loaded from a syringe and crushed in the injection port with a solid-sample injector. Yokota et al homogenized tissue with water ultrasonically; the homogenizing was done in a simple sealed system and the homogenized mixture was injected (77).

Blood samples may be injected with an ordinary microliter syringe (38, 39, 76, 77); Douglas et al (64) used a syringe with a needle which protuded from the tip to remove any dried blood.

When direct injection of samples is used, the water content of the samples may interfere with flame ionization detectors unless precautions are taken. This interference can be prevented by saturating the carrier gas with water, using a short section of column containing water (38, 39, 64, 76). Alternatively, a section of drying column may be used to remove water vaporized from the sample (38, 75, 77).

Lowe (38, 39, 76) points out that normal blood samples contain no volatile organic matter to interfere with flame ionization detectors. Therefore, little or no chromatographic separation is needed if only one anesthetic is present in blood samples.

TABLE 5

Gas Chromatographic Systems for Direct Injection

Reference Number	Column	Carrier Gas	Column Temp., °C	Detector
75	Pre-column of 0.5 x 0.25 in. 60-80 mesh Drierite treated with 12-15 wt. % Carbowax 400; partition column DC 550	Argon, 5% Methane	72	Electron Capture
38, 39, 76	Glass wool or Chromosorb P, 12 in. long x 0.25 in., alone or following 2 ft. Tide column	N_2	Room Temp.	Flame Ionization
64	18 in. x 0.125 in., 60-85 mesh Chromosorb P saturated with water	Not Stated	Room Temp.	Flame Ionization
77	75 cm. x 3 mm. I.D. molecular sieve 5A, 60-80 mesh, coated with 0.3% diethylene glycol succinate	N_2	120	Flame Ionization

7.2 Absorptiometric Methods

7.21 Turbidimetric Method

The first reported method of analysis for halothane in blood was by turbidimetric measurement. Halothane was extracted from blood with petroleum ether. The extract was heated with sodium amoxide in a sealed tube. The halide formed was then precipitated with silver nitrate and the absorption of the suspension measured at 520 nanometers. Standard halothane solutions were required because hydrolysis was incomplete (54).

This technique was modified by use of lithium aluminum hydride (53, 78) or sodium (79) ultrasonically dispersed in petroleum jelly instead of sodium amoxide. It was also used for tissue analysis (53).

7.22 Infrared Absorption

Halothane in blood was determined by extraction with an equal volume of carbon disulfide and measurement of the absorption at 7.90 microns in a 10 mm. microcell (80). The method was not suitable for tissue samples.

Larson et al used a commercial infrared halothane analyzer to analyze for halothane in gases equilibrated with blood and tissue, for determining the solubility of halothane in blood and tissue homogenates (49).

7.23 Ultraviolet Absorption

Halothane was determined in blood and various tissues after extraction into toluene and treatment with pyridine and sodium hydroxide. The absorption of the treated sample was measured at 367 nanometers (80).

7.3 X-ray Spectrography

Halothane in 1 to 2 g. tissue samples was determined by x-ray spectrographic (fluorescence) analysis for bromine in 10 ml. hexane extracts (81).

8. References

1. R. Theimer and J. R. Nielsen, J. Chem. Phys. 27, 887-90 (1957).
2. W. Kalow, Can. Anaesth. Soc. J. 4, 384-7 (1957).
3. B. Kakac and M. Hudlicky, Talanta 9, 530-3 (1962).
4. J. M. Pryce, personal communication.
5. C. W. Suckling, Brit. J. Anaesth. 29, 466-72 (1957).
6. G. A. Bottomley and G. H. F. Seiflow, J. Appl. Chem. (London) 13, 399-402 (1963).
7. C. W. Suckling and J. Raventos, Brit. Patent 767, 779 (1957); C.A. 51, 15546i.
8. J. Chapman and R. L. McGinty, Brit. Patent 805, 764 (1958); C.A. 53, 10035c.
9. H. Madai and R. Mueller, J. Prakt. Chem. 19, 83-7 (1962); C.A. 58, 11200g.
10. O. Scherer and H. Kuhn, U.S. Patent 2,959,624 (1960); C.A. 55, 8290e.
11. M. Hudlicky and I. Lejhancova, Collect. Czech. Chem. Commun. 28, 2455-61 (1963); C.A. 59, 11227a.
12. H. Madai, Brit. Patent 939,920 (1963); C.A. 60, 2751e.
13. J. Raventos, Brit. J. Pharmacol. 11, 394-409 (1956).
14. M. Hudlicky, Czech. Patent 107,890 (1963); C.A. 60, 5334d.
15. J. Chapman and R. L. McGinty, Brit. Patent 925,909 (1963); C.A. 59, 11246c.
16. R. L. McGinty, U.S. Patent 3,082,263 (1963); C.A. 59, 8592c.
17. C. W. Suckling and J. Raventos, U.S. Patent 2,921,098 (1960); C.A. 55, 4360h.
18. T. Satogawa and Y. Osaka, Japan. Patent 68 09,728; C.A. 70, 37152x.
19. E. N. Cohen, J. W. Bellville, H. Budzikiewicz, and D. H. Williams, Science 141, 899 (1963).
20. E. N. Cohen, H. W. Brewer, J. W. Belleville, and R. Sher, Anesthesiology 26, 140-53 (1965).
21. W. A. Sexton and W. G. Hendrickson, Science 142, 621-2 (1963).
22. D. S. Corrigan, G. V. McHattie, and J. Raventos, Brit. J. Anaesth. 35, 824-5 (1963).
23. R. A. Butler and H. W. Linde, Anesthesiology 25, 397-8 (1964).

24. M. S. Albin, L. A. Horrocks, and H. E. Kretchmer, Anesthesiology 25, 672-5 (1964); see also correspondence regarding this paper, ibid. 26, 236-7 (1965).
25. J. C. Gjaldbaek and K. Worm, Dansk Tidsskr. Farm. 39, 141-51 (1965).
26. A. Stier and H. Alter, Anaesthesist 15, 154-5 (1966).
27. K. Rehder, J. Forbes, H. Alter, O. Hessler, and A. Stier, Anesthesiology 28, 711-5 (1967).
28. A. Stier, H. Alter, O. Hessler, and K. Rehder, Anesth. Analg. 43, 723-8 (1964).
29. A. Stier, Biochem. Pharmacol. 13, 1544 (1964).
30. R. A. Van Dyke, M. B. Chenoweth, and A. Van Poznak, Biochem. Pharmacol. 13, 1239-47 (1964).
31. A. Stier, Naturwissenschaften 51, 65 (1964).
32. R. A. Van Dyke, M. B. Chenoweth, and E. R. Larsen, Nature 204, 471-2 (1964).
33. E. N. Cohen and N. Hood, Anesthesiology 31, 553-9 (1969).
34. E. N. Cohen, Anesthesiology 31, 560-5 (1969).
35. E. R. Adlard and D. W. Hill, Nature 186, 1045 (1960).
36. C. O. Rutledge, E. Seifen, M. H. Alper, and W. Flacke, Anesthesiology 24, 862-7 (1963).
37. R. A. Theye, Anesthesiology 25, 75-79 (1964).
38. H. J. Lowe, Anesthesiology 25, 808-14 (1964).
39. H. J. Lowe, J. Gas Chromatog. 2, 380-4 (1964).
40. S. Koudela and J. Lukaci, Soudni Lekarstvi 10, 8-11 [published in Cesk. Patol. 1 (1)] (1965); C.A. 63, 13859f.
41. K. Rehder, J. Forbes, O. Hessler, and K. Gossmann, **Anaesthesist** 15, 162-8 (1966).
42. S. P. Mirolyubova, Nauch. Tr. Aspir. Ordinatorov, 1-i Mosk. Med. Inst. 1967, 141-2 (from Ref. Zh., Khim. 1968, Abstr. No. 12N533); C.A. 70, 50505j.
43. D. J. Wortley, P. Herbert, J. A. Thornton, and D. Whelpton, Brit. J. Anaesth. 40, 624-8 (1968).
44. M. Tiengo, Ann. Ostet. Ginecol. 89, 543-57 (1967); C.A. 68, 58312y.
45. V. Patzelova, Chromatographia 4, 174-6 (1971).
46. B. Kakac and M. Hudlicky, Czech. Patent 112,162 (1964); C.A. 62, 8395f.

47. J. I. Davies, S. Bakerman, G. B. Gish, S. N. Angell, and E. L. Frederickson, Anesthesiology 23, 143-4 (1962).
48. P. H. Sechzer, H. W. Linde, R. D. Dripps, and H. L. Price, Anesthesiology 24, 779-83 (1963).
49. C. P. Larson, E. I. Eger, and J. W. Severinghaus, Anesthesiology 23, 349-55 (1962).
50. A. Robinson, J. S. Denson, and F. W. Summers, Anesthesiology 23, 391-4 (1962).
51. D. Langrehr, I. Kluge, and I. Riecken, Anaesthesist 19, 340-5 (1970).
52. B. Wolfson, Anesthesiology 29, 157-9 (1968).
53. W. A. M. Duncan, Brit. J. Anaesth. 31, 316-20 (1959).
54. R. R. Goodall, Brit. J. Pharmacol. 11, 409-10 (1956).
55. M. Luder, Anaesthesist 13, 360-4 (1964).
56. Halothane monograph, Pharmacopeia of the United States of America, Eighteenth Revision, Mack Printing Co., Easton, Pa., 1970, pp. 294-5.
57. J. Chapman, R. Hill, J. Muir, C. W. Suckling, and D. J. Viney, J. Pharm. Pharmac. 19, 231-9 (1967).
58. A. Scipioni, G. Gambaretto, G. Troilo, and C. Fraccaro, Chim. Ind. (Milan) 49, 577-82 (1967); C.A. 67, 108107n.
59. M. B. Chenoweth and H. W. Brewer, Toxicity Anesth., Proc. Res. Symp. 1967 (Pub. 1968), 65-76; C.A. 70, 60768m.
60. O. Scherer and W. Weigand, Anaesthesist 13, 313-4 (1964).
61. R. A. Butler and D. W. Hill, Nature 189, 488-9 (1961).
62. R. A. Butler and J. Freeman, Brit. J. Anaesth. 34, 440-4 (1962).
63. P. Herbert, J. Med. Lab. Technol. (London) 25(3), 233-7 (1968).
64. R. Douglas, D. W. Hill, and D. G. L. Wood, Brit. J. Anaesth. 42, 119-23 (1970).
65. D. D. Davies and J. A. Mathias, Br. J. Pharmacol. 40, 596P-7P (1970).
66. B. Wolfson, H. E. Ciccarelli, and E. S. Siker, Brit. J. Anaesth. 38, 591-5 (1966).
67. A. Dyfverman and J. Sjovall, Acta anaesthesiol. Scand. 6, 171-4 (1962).

68. R. H. Gadsden and W. M. McCord, J. Gas Chromatog. 2, 7-11 (1964).
69. R. H. Gadsden, K. B. H. Risinger, and E. E. Bagwell, Can. Anaesth. Soc. J. 12 (1), 90-8 (1965).
70. C. P. Larson, Jr., Uptake and Distribution of Anesthetic Agents, Conf., New York 1962, 5-19 (Pub. 1963); C.A. 60, 4360.
71. R. A. Butler, A. B. Kelly, and J. Zapp, Anesthesiology 28, 760-3 (1967).
72. L. H. Laasberg and J. Hedley-Whyte, Anesthesiology 32, 351-6 (1970).
73. I. F. H. Purchase, Nature 198, 895-6 (1963).
74. B. R. Fink and K. Morikawa, Anesthesiology 32, 451-5 (1970).
75. E. N. Cohen and H. W. Brewer, J. Gas Chromatog. 2, 261-2 (1964).
76. H. J. Lowe and L. M. Beckham, in "Biomedical Applications of Gas Chromatography," H. A. Szymanski, editor, Plenum Press, New York (1964).
77. T. Yokota, Y. Hitomi, K. Ohta, and F. Kosaka, Anesthesiology 28, 1064-73 (1967).
78. J. G. Robson and P. Welt, Can. Anaesth. Soc. J. 4, 388-93 (1957).
79. J. Burns and G. A. Snow, Brit. J. Anaesth. 33, 102-3 (1961).
80. Y. Maeda, Osaka City Med. J. 9 (1), 113-28 (1963); C.A. 60, 16197g.
81. M. B. Chenoweth, D. N. Robertson, D. S. Erley, and R. S. Gohlke, Anesthesiology 23, 101-6 (1962).

Acknowledgments

The writer wishes to thank Mr. A. Holbrook of Imperial Chemical Industries Ltd., and Dr. B. T. Kho of Ayerst Laboratories Inc., for their careful reading of the manuscript and helpful suggestions for improvements, as well as Dr. J. M. Pryce of Imperial Chemical Industries Ltd., who provided the mass spectral and nuclear magnetic resonance information for this profile.

LEVARTERENOL BITARTRATE

Charles F. Schwender

Reviewed by E. L. Pratt *et al.*[1]

CONTENTS

Analytical Profile - Levarterenol Bitartrate

1. Description
    1.1 Nomenclature
    1.2 Formula
    1.3 Molecular Weight
    1.4 Structure
    1.5 Appearance
2. Physical Properties
    2.1 Melting Point
    2.2 Solubility
    2.3 Crystal Properties
    2.4 Optical Rotation
    2.5 Optical Rotatory Dispersion
    2.6 Configuration
    2.7 Spectrophotometry
        2.71 Ultraviolet
        2.72 Determination of Dissociation Constants (pK'a)
        2.73 Infrared
        2.74 Mass Spectrum
3. Synthesis and Resolution
    3.1 Synthesis
    3.2 Resolution
4. Stability
    4.1 Solutions
    4.2 Racemization
    4.3 Oxidation
5. Metabolism
    5.1 Methylation
    5.2 Deamination
    5.3 Excretion Products
    5.4 Conjugates
    5.5 Minor Pathways
    5.6 Distribution
6. Methods of Analysis
    6.1 Elemental
    6.2 Colorimetric
        6.21 1,2-Naphthoquinone-4-sodium sulfonate
        6.22 Oxidation with Iodine

CONTENTS (cont'd)

      6.23   Arsenomolybdic Acid Reduction
      6.24   Color Reactions
  6.3 Fluorometric
      6.31   Ethylenediamine
      6.32   Trihydroxyindole
  6.4 Specific Rotation
7. Chromatography
  7.1 Electrophoresis
  7.2 Paper, Thin layer
  7.3 Gas-liquid
8. References

1. Description

    1.1 Nomenclature
    Levarterenol Bitartrate
    (-) α-(Aminomethyl)-3,4-dihydroxybenzylalcohol Bitartrate Monohydrate
    (-) 1-(3,4-Dihydroxyphenyl)-2-aminoethanol Bitartrate Monohydrate
    (-) -2-Amino-1-(3,4-dihydroxyphenyl)ethanol Bitartrate Monohydrate
    Norepinephrine, Noradrenaline Tartrate

    1.2 Formula
    $C_8H_{11}NO_3 \cdot C_4H_6O_6 \cdot H_2O$

    1.3 Molecular Weight 337.29

    1.4 Structure

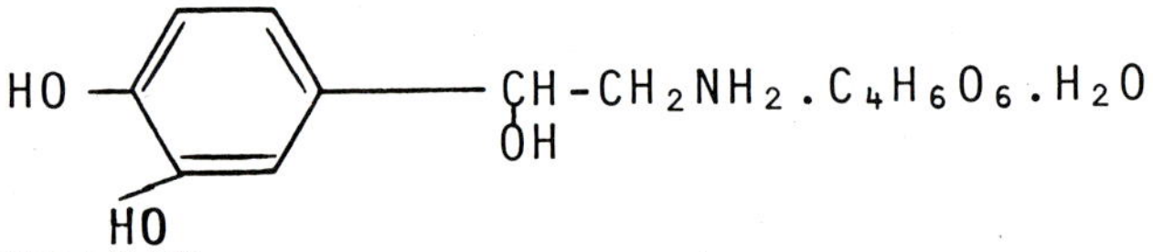

    HO—⟨benzene ring⟩—$CH-CH_2NH_2 \cdot C_4H_6O_6 \cdot H_2O$
                              $|$
                              OH
    HO—

    1.5 Appearance
    White or faint gray, odorless crystalline powder.

2. Physical Properties

    2.1 Melting Point
    Levarterenol bitartrate monohydrate
    mp   100-106°[2]
    mp   102-104°[3]
    mp   102-104.5°[7]
    Levarterenol bitartrate, anhydrous
    mp   147-150° dec.[3]
    mp   158-159° dec.[3]
    mp   160°[7]
    Levarterenol Hydrochloride
    mp   146.5-147.5°[7]
    Levarterenol
    mp   216.5-218°[7]

2.2 Solubility[2]
Freely soluble in water (greater than 10%)

Slightly soluble in alcohol (0.1-1.0%)
Practically insoluble in chloroform, ether (less than 0.0001%)

2.3 Crystal Properties[3]
Lone orthorhombic prisms
optically (+)
$2V = 85° \pm 2°$ (calcd.)
elongation (-)
$n\alpha = 1.620 \pm 0.002$
$n\beta = 1.577 \pm 0.002$
$n\gamma = 1.531 \pm 0.002$
Twinning common, end view under polarized light.

2.4 Optical Rotation

Levarterenol Form	$[\alpha]_D^T$	T°C	Solution	Reference
Bitartrate·H_2O	-11.3°	18	---	12
	-10 to -12°	25	c=5,H_2O	2
	-10.7°	25-29	c=1.6,H_2O	7
Tartrate, anhydrous	-12.5°	24	c=2,H_2O	3
	-11.0°	24	c=1.6,H_2O	3
Base	-37.3°	25-29	c=5,dil.HCl	7
	-39°	20	c=5,H_2O	8
	-39.5°	20	dilute HCl	8
	-41.7°	20	---	9
	-44.0°	20	c=2,H_2O	4
Hydrochloride	-39°	25-29	c=6,H_2O	7
	-40°	25	c=5.0,1N HCl	7
	-45.7°	20	c=2.0,5M HCl	4
	-40.6°	20	c=3,H_2O	11
	-47.5°	25	c=4,dil.HCl	11
	-46.7°	20	c=4,dil.HCl	11
$O^{3,4}$,N-Triacetyl	-81.3°	25	c=1,$CHCl_3$	13

2.5 Optical Rotatory Dispersion

$[\alpha]^{25}_{333} = -333°$ (peak)

$[\alpha]^{25}_{276.5} = -2055°$ (trough)

$[\alpha]^{25}_{260} = -945°$ (peak)

$[\alpha]^{25}_{238} = +2955°$ (shoulder)

$[\alpha]^{25}_{210.5} = +10,460°$ (peak)

C = 0.09, EtOH/5% 0.1N NaOH, ± 5% error of $[\alpha]$.

2.6 Configuration of Levarterenol

Rotatory dispersion curves indicated the configuration to be D through its relation to D-mandelic acid and D-lactic acid by their negative Cotton effects.[5] The assignments were confirmed through an independent chemical transformation.[14]

2.7 Spectrophotometry
2.71 Ultraviolet[15,16]

Solution	λ max, mμ	$\varepsilon \times 10^{-3}$
0.1N HCl	285 sh.	2.45
	280	2.75
	208	7.00
	295	4.50
	243	7.10
pH 7	280	2.80
	220	7.30
pH 10.5	293	4.90
	243	7.60
Isosbestic points	281	2.75
	266	1.40
	232	4.80

2.72 Determination of Dissociation Constants (pK'a)

First phenol	Amine	Second phenol	Reference
8.72	9.72		15
8.90 ± 0.06	9.78 ± 0.09	>12	18

2.73 Infrared absorption of Levarterenol Bitartrate[17,20]

ν in cm^{-1} (KBr) :3450 (OH); 2700-2200 (acidic hydrogen); 1740 (COOH); 1650, 1630, 1590 and 1400 (C=C, COO$^-$, NH$_3^+$).

2.74 Mass Spectrum

Mass	m/e (rel. int., %)
169	169 (3); 153 (7); 151 (8); 139 (30); 137 (7); 124 (4); 123 (6); 111 (5); 93 (30); 77 (7). (The spectrum was taken on a Hitachi-Perkin Elmer RMU-6D with electron beam energy of 70eV using the direct heated inlet system.)[19]
169	169 (25); 139 (100); 111 (21); 93 (75); 62 (69); 32 (52); 30 (97). (JEACO JMS-01, photo plate-Mattauch-Herzog).[20]

3. Synthesis and Resolution

3.1 The synthesis of arterenol (IV) has been reported[21] with utility in large scale preparation. The route employed a phosphorous oxychloride mediated acylation of catechol (I) by α-chloroacetic acid. The crude 4-chloroacetylcatechol (II) obtained was further reacted with ammonia in alcohol and arterenone (III) was obtained. The purified arterenone was hydrogenated as its hydrochloride salt over palladium on charcoal as catalyst. Racemic arterenol (IV) was obtained by precipitation with ammonia. A similar preparation of dl-arterenol utilizing the protected 3,4-diacetoxy-α-(bromo) acetophenone intermediate has been described.[22]

3.2 The first sucessful resolution of racemic arterenol was described by Tullar.[6,7] An industrial scale resolution was described using d-tartaric acid and methanol.[21] The levarterenol d-bitartrate fraction obtained by crystallization was converted to its free base form by precipitation from a methanol solution with ammonia. Repeated precipitations of l-enriched arterenol from d-tartrate solutions with ammonia gave the levarterenol with the required optical purity. The resolved base was also crystallized as its d-bitartrate salt. Arterenol of lesser optical purity was recovered, racemized with hydrochloric acid and recycled through the resolution procedure. Tullar[6,7] modified the procedure through a recrystallization of the bitartrate salt from water. Ruschig and Stern have described the resolution of arterenol utilizing l-mandelic acid.[8]

4. Stability of Levarterenol

4.1 Solutions

Levarterenol bitartrate may be stored at pH 3.6 in a well-filled ampul in the presence of 0.1% $NaHSO_3$. Exposure to air, in an alkaline or neutral pH resulted in deterioration of the sample accompanied by a darkening of the solution to a brown color. Dilution of levarterenol in plasma, 5% dextrose, or saline containing ascorbic acid resulted in no significant loss of activity after 9 hours at room temperature. Saline diluent without ascorbic acid allowed loss of activity. Levarterenol as the bitartrate salt or in the free base form behaved similarly. Solutions of levarterenol bitartrate in the presence of sodium bisulfite could be sterilized at 115°C for 30 minutes with apparent negligible loss of activity.[23]

Infusion solutions of levarterenol in isotonic glucose, sodium chloride or sodium bicarbonate were stable at least four hours at room temperature.[24]

4.2 Racemization

Levarterenol bitartrate monohydrate will become racemized upon heating for 3 minutes at 120°. Concentrated hydrochloric acid will cause complete racemization of levarterenol solutions after 2 hours at 80-90°.[4]

4.3 Oxidation

Levarterenol undergoes an autoxidation to noradrenochrome in the presence of oxygen and divalent metal ions such as Cu^{+2}, Mn^{+2} or Ni^{+2}. Since chelating agents prevent the oxidation, a levarterenol-metal ion complex probably is necessary for the oxidation to occur. The reaction may proceed at 37° in either sodium bicarbonate or sodium phosphate buffer and may be followed spectrally by the appearance of the oxidation product at 400 mµ.[16]

Noradrenochrome

5. Metabolism

Levarterenol (I) is metabolized in man by the action of two enzymes, monoamine oxidase (MAO) and catechol O-methyl transferase (COMT). Deamination and O-methylation leads to a number of urine metabolites.

5.1 Methylation

Levarterenol (I) is methylated by catechol O-methyl transferase giving normetanephrine (II).[27,28] Deamination of II to an aldehyde intermediate and oxidation or reduction of the aldehyde leads to 4-hydroxy-3-methoxy-phenylglycol (III), or 4-hydroxy-3-methoxy-mandelic acid (IV).[25,26,29]

5.2 Deamination

Deamination of levarterenol by the monoamine oxidase gives an aldehyde intermediate which leads to the glycol (V) or 3,4-dihydroxy-mandelic acid (VI). Catechol O-methylation of V and VI leads to III and IV.

5.3 Excretion Products

Human urinary metabolites isolated from parenterally administered, isotopically labelled, levarterenol indicated that the major metabolites were II and IV which were each present in quantities of 20-40% of the injected dose.[25,31] Another 5% appeared as the sulfate conjugate of III.[30] Also appearing in the urine was about 4% of I and 10% of VI.[31] The ratios of the excretion products changed when the rate of infusion varied or endogenous levarterenol was studied. Endogenous levarterenol was metabolized mainly to IV while II occurred in only one-tenth the amount of IV. Other catechols were excreted in only minor quantities.[3] The methylation pathway was found to account for two-thirds of an injected dose while the deaminase route was the initial route for only one-fourth of the dose.[30]

5.4 Conjugates

In man, IV is excreted as the free acid while II is a glucuronide and III is the sulfate form. The metabolism of levarterenol in the rat leads to the same urinary products. However, the glycol sulfate (III) replaces IV as the major metabolite. The rat also excretes IV as the glucuronide while the free acid is excreted in man.[30]

LEVARTERENOL BITARTRATE

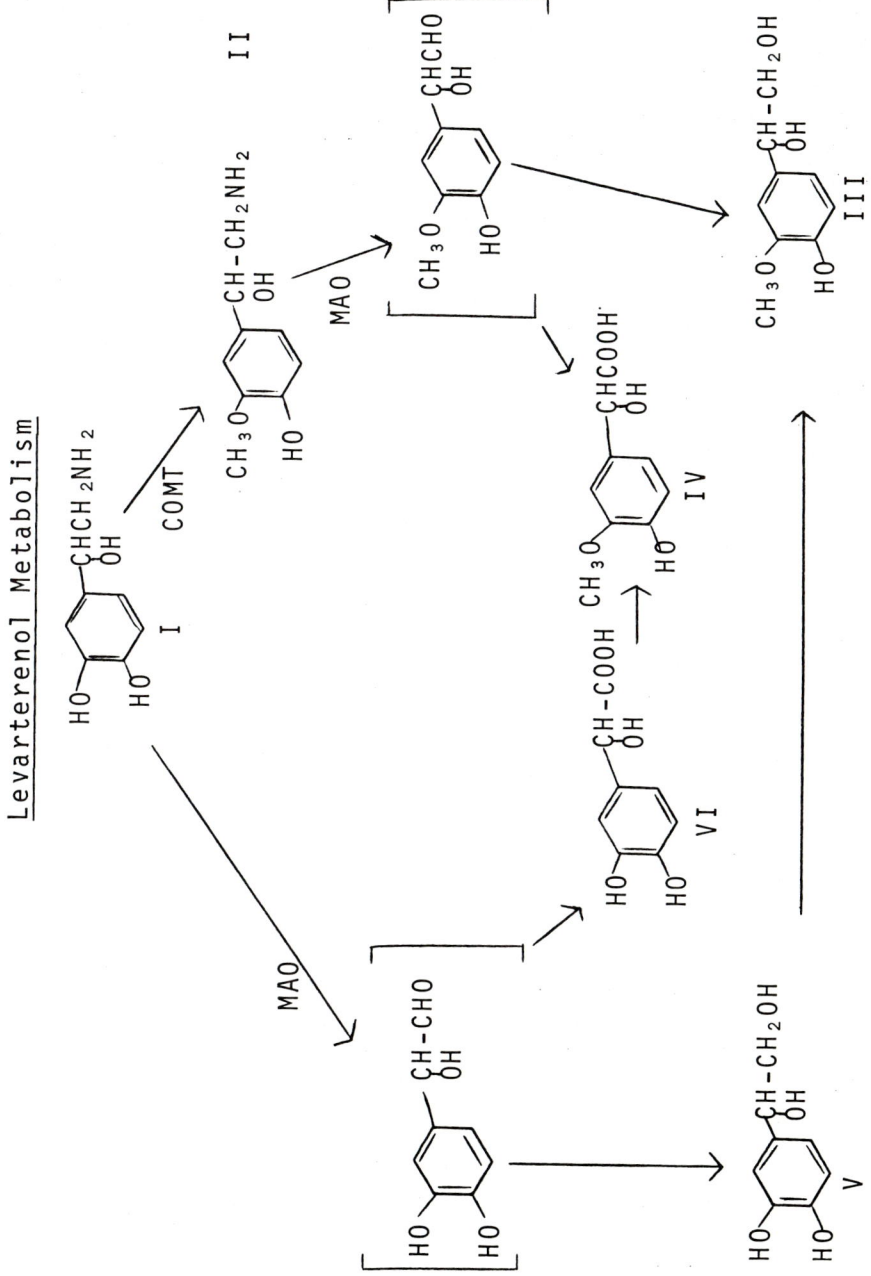

5.5 Minor Metabolic Pathways

The instability of levarterenol *in vitro* is related to oxidative cyclization to noradrenochrome. The possibility exists that a minor route of metabolism *in vivo* may involve such an oxidation. However, such a metabolite has not been detected. Therefore, it was concluded that the oxidation product is not formed in significant amounts if such a pathway exists.

5.6 Distribution

Selective uptake and retention of levarterenol by the adrenal gland, heart and spleen followed intravenous injections in cats and mice. All tissues studied appeared to bind small amounts of levarterenol.[33]

6. Methods of Analysis

6.1 Elemental Analysis

Calcd. for $C_8H_{11}NO_3 \cdot C_4H_6O_6 \cdot H_2O$: C, 42.73; H, 5.68; N, 4.15; O, 47.44.
Calcd. for $C_8H_{11}NO_3 \cdot C_4H_6O_6$: C, 45.20; H, 5.33; N, 4.40: O, 45.07.

6.2 Colorimetric Analysis

Color complexes with catecholamines are generally the least specific and least accurate methods for determining concentrations. However, highly colored complexes are often useful in the visualization of chromatograms.

6.21 The condensation of 1,2-naphthoquinone-4-sodium sulfonate with arterenol generates a purple color which can be quantitated at 540 mµ. This procedure is selective for the primary amine in the presence of epinephrine and can be applied to pharmaceutical quality control.[34,35] The sensitivity is limited to 50 µg.

6.22 Oxidation of arterenol by iodine in an acetate buffer pH 6 forms a color due to the production of noradrenochrome measured at

529 mμ.[35] The sensitivity is limited to 20 μg.

6.23 The reduction of arsenomolybdic acid by catecholamines[35] can be measured quantitatively at 690 mμ. by the appearance of a blue color. The reduction is believed to proceed through the alkaline oxidation product of catecholamines. This procedure is sensitive to 40 ng.

6.24 Color Reactions

Reagent	λ max. (color)	Reference
sodium molybdate	400 mμ	39
blue tetrazolium	530 mμ	40
p-nitrobenzene diazonium chloride	640 mμ (yellow-green)	41
Hg(OAc)$_2$,HOAc,NaNO$_2$	530 mμ (purple)	43
Ninhydrin	mustard	43
K$_3$Fe(CN)$_6$	lavender-brown	43
K$_3$Fe(CN)$_6$-Fe$_2$(SO$_4$)$_3$	blue	43
FeCl$_3$	green	2

6.3 Fluorometric Analysis

Quantitative analysis of catechloamines derived from biological sources most often has been performed utilizing fluorometric procedures. The fluorescence is measured on products derived from the oxidation of the catechol nucleus of arterenol to noradrenochrome. Noradrenochrome is stabilized as noradrenolutin with alkali or converted to a tricyclic condensation product by ethylenediamine.

arterenol → noradrenochrome

ethylenediamine condensate   noradrenolutin

6.31 The fluorometric analysis of levarterenol by the ethylenediamine method[35,36] involves heating levarterenol at acidic pH and 50° in the presence of ethylenediamine. The resulting condensate is measured at an excitation wavelength of about 420 and fluorescence at wavelengths of 510 and about 600 mµ. The readings are compared with standard solutions. The sensitivity limit is 2 ng.

6.32 The trihydroxyindole method[37] of arterenol analysis utilizes the oxidation of arterenol to noradrenochrome at pH 6 by ferricyanide. The conversion of noradrenochrome to the fluorescent noradrenolutin is accomplished by the addition of alkali. The addition of ascorbic acid or 2-mercaptoethanol stabilizes noradrenolutin from further oxidation and the subsequent loss of fluorescence, λ ex 400-λ em. 500 mµ.

6.4 Specific Rotation

The official U.S. Pharmacopeia assay[1] for levarterenol involves the preparation of the triacetyl derivative of levarterenol from an aqueous solution. The specific rotation of the isolated triacetyl derivative of levarterenol is utilized to calculate the weight of levarterenol present in the original aqueous solution by the formula 1.1423W(0.5+0.5R/80) when W is the weight in mg. and R is the specific rotation of the isolated triacetyl levarterenol.

7. Chromatography

7.1 Electrophoresis

Levarterenol was applied near the anode of Whatman 3MM paper. An EEL apparatus (Evans Electroselenium, Ltd.) was used with 0.04 M sodium acetate buffer pH 5.6 with a potential gradient of 10 V/cm. Levarterenol migrated 8-10 cm. toward the cathode in 2 hours. The spot was visualized by its fluorescence with ethylenediamine.[5]

7.2 Paper and Thin Layer Chromatographic Analysis

Solvent System[a]	Type Support	R_f	Visualization[c]	Ref.
Phenol	paper[b]	0.22	1, 2	46
88% Phenol (saturated with 1N HCl)[d]	paper (Whatman #1)	0.27-0.33	1, 2	43, 52
BuOH (saturated with 1N HCl)	paper	0.30	6	47
BuOH-HOAc-H_2O (4:1:5)	paper	0.28	2	45, 48, 49
BuOH-HOAc-H_2O (12:3:5)[e]	paper (Whatman #4)	0.58	1	50
BuOH-C_5H_5N-HOAc (7:2:1)	paper	0.35	1, 5	51
PrOH-EtOAc-NH_4OH (7:9:1)	paper	0.13	1, 5	51
BuOH-C_5H_5N-0.2M NaOAc (1:1:1)	paper (Whatman #1)	0.75	1, 4	52
BuOH-C_5H_5N-H_2O (1:1:1)	paper (Whatman #1)	0.74	1, 4	52
C_6H_6-MeOH-BuOH-C_5H_5N-H_2O (1:2:1:1:1)	paper (Whatman #1)	0.63	1, 4	52
$PhCH_3$-EtOAc-C_5H_5N-H_2O-MeOH (1:1:1:1:1)	paper (Whatman #1)	0.82	1, 4	52
H_2O (saturated with MEK)	paper[b]	0.16	1, 4	52
BuOH-EtOH-H_2O (2:1:1)	paper (Whatman #1)	0.40	1, 4	52
$PhCH_3$-EtOAc-MeOH-0.1 N HCl (1:1:1:1)	paper (Whatman #1)	0.76	1, 4	52
t-BuOH-acetone-C_2H_5COOH-H_2O (160:160:1:39)	paper (Whatman #1)	0.34	1, 4	52
C_6H_6-C_2H_5COOH-H_2O (2:1:1)	paper (Whatman #1)	0.90	1, 4	52

Solvent System[a]	Type Support	R_f	Visualization[c]	Ref.
PrOH-HOAc-H_2O (70:5:25)	paper (Whatman #1)	0.51	1, 2	63
BuOH-C_5H_5N-H_2O (4:1:1)	glass fiber paper	0.57	1, 3	55
BuOH (saturated with 3N HCl)	cellulose thin layer	0.02	2	53
Cyclohexane-C_6H_6-$NH(CH_3)_2$ (15:3:2)	silica gel thin layer-0.1M KOH	0.02	6	54
MeOH	same	0.21	6	54
Acetone	same	0.75	6	54
MeOH	silica gel thin layer-0.1M $NaHSO_4$	0.66	6	54
EtOH	same	0.50	6	54
BuOH-HCOOH-H_2O (12:1:7)	silica gel thin layer-circular	0.5	1	56
2-PrOH-0.1N HCl (5:1)	cellulose thin layer	0.19	5, 7, 9	57
PrOH-H_2O (65:35)	cellulose thin layer	0.56	1, 4, 5, 8	58
Heptane-CCl_4-MeOH (7:4:3)	cellulose thin layer	0.08	1, 4, 5, 8	58

7.2 Footnotes

a. Abbreviations used: HOAc (acetic acid), C_6H_6 (benzene) BuOH (butanol, CCl_4 (carbon tetrachloride), $CHCl_3$ (chloroform), EtOH (ethanol), EtOAc (ethyl acetate), HCOOH (formic acid), MeOH (methanol), MEK (methyl ethyl ketone), PrOH (propanol), C_2H_5COOH (propionic acid), C_5H_5N (pyridine), NaOAc (sodium acetate), $PhCH_3$ (toluene).

b. Ascending development

c. Reagents: 1 (ninhydrin), 2 ($K_3Fe[CN]_6$), 3 (ethylenediamine-fluorescence), 4 (diazotized sulfanilic acid), 5 (diazotized p-nitroaniline), 6 (Iodine-methanol), 7 (ferrocitrate[57]), 8 (p-dimethylaminobenzaldehyde), 9 (Ferric chloride).

d. 8-Hydroxyquinoline added to prevent tailing.

e. 5% trichloroacetic acid added to prevent tailing and secondary spots due to salt mixtures.

7.3 Gas-Liquid Chromatographic Analysis of Levarterenol

Derivative	Column	Temperature, Flow Rate	Retention (min.)	Ref.
Trifluoroacetyl	Gaschrom P/20% GE-XF1105	190°, 80ml/min, N_2	1.92	59
	Gaschrom P/12% DC1107	170°, 80ml/min, N_2	0.92	
	Gaschrom P/7% DC560	150°, 80ml/min, N_2	0.86	
Trimethylsilyl	Gaschrom P (acid wash)/0.5% QF-I, 0.05% ethyl-eneglycol succinate	115° - 125° 1.5 Kg/cm^2, A	14	60
	Chromosorb W/6% SE-30	200°, 38ml/min, N_2	2.04	61
		220°, 38ml/min, N_2	1.84	
Trimethylsilyl ether-2-pentyl-imine	Gaschrom P/3.5% SE-30	180°, 120ml/min, N_2	12.5	62

8. **References**

1. Reviewed by E. L. Pratt, M. A. Borisenok, R. S. Browning, J. P. Dulin, T. G. Gerding, W. G. Gorman, W. W. Houghtaling, R. K. Kullnig, F. C. Nachod, G. A. Portmann, L. D. Shargel, J. U. Shepardson, I. S. Shupe and B. F. Tullar of Sterling-Winthrop Research Institute and Winthrop Laboratories.
2. The United States Pharmacopeia, Eighteenth Revision, Mack Publishing Co., Easton, Pa., 1970, p. 362.
3. R. L. Clarke, *J. Am. Pharm. Assoc. 43*, 681 (1954).
4. H. Hellberg, *J. Pharm. Pharmacol. 7*, 191 (1955).
5. J. C. Craig and S. K. Roy, *Tetrahedron 21*, 1847 (1965).
6. B. F. Tullar, *J. Am. Chem. Soc. 70*, 2067 (1948).
7. B. F. Tullar, U.S. 2,774,789 (Dec 18, 1956).
8. H. Ruschig and L. Stein, U.S. 2,820,827 (Jan. 21, 1958).
9. F. Fabian, Brit. 816,857 (July 22, 1959).
10. Lucius and Bruning, Brit. 790,920 (Feb. 19, 1958).
11. C. B. Friedmann, C. W. Picard and F. Fabian, Brit. 747,768 (April 11, 1956).
12. W. Langenbeck, *Pharmazie 5*, 56 (1950).
13. L. H. Welsh, *J. Am. Pharm. Assoc., Sci. Ed. 44*, 507 (1955).
14. P. Pratesi, A. LaManna, A. Campiglio and V. Ghislandi, *J. Chem. Soc. 1958*, (2069).
15. T. Kappe and M. D. Armstrong, *J. Med. Chem. 8*, 368 (1965).
16. E. Waalas, O. Walaas and S. Haavaldsen, *Arch. Biochem. Biophys. 100*, 97 (1963).
17. O. R. Sammul, W. L. Brannon and A. L. Hayden, *J. Assoc. Off. Anal. Chem. 47*, 918 (1964).

18. G. P. Lewis, *Brit. J. Pharmacol. 9*, 488 (1954).
19. J. Reisch, R. Pagnucco, H. Alfes, N. Jantos and H. Mollmann, *J. Pharm. Pharmacol. 20*, 81 (1968).
20. R. W. Kullnig, Sterling-Winthrop Research Institute, personal communication.
21. K. R. Payne, *Ind. Chemist 37*, 523 (1961).
22. D. R. Howton, J. F. Mead and W. H. Clark, *J. Am. Chem. Soc. 77*, 2896 (1955).
23. G. B. West, *J. Pharm. Pharmacol. 4*, 560 (1952).
24. J. Haggendal and G. Johnsson, *Acta Pharmacol. Toxicol. 25*, 461 (1967).
25. M. Sandler and C. R. J. Ruthven, *Prog. Med. Chem. 6*, 200 (1969).
26. J. Axelrod, *Adrenergic Mechanisms*, Ciba Foundation Symposium, p. 28, 1960. Little, Brown and Company. Boston. (Eds. J. R. Vane, G. E. W. Wolstenholme and M. O'Connor).
27. J. Axelrod, S. Senoh and B. Witkop, *J. Biol. Chem. 233*, 697 (1958).
28. J. Axelrod, *Sci. 127*, 754 (1958).
29. J. Axelrod, *Pharmacol. Rev. 11*, 402 (1959).
30. J. Axelrod, I. J. Kopin and J. D. Mann, *Biochim. Biophys. Acta 36*, 576 (1959).
31. McC. Goodall, *Pharmacol. Rev. 11*, 416 (1959).
32. R. W. Schayer and R. L. Smiley, *J. Biol. Chem. 202*, 425 (1953).
33. L. G. Whitby, J. Azelrod and H. Weil-Malherbe, *J. Pharmacal. Exp. Ther. 132*, 193 (1961).
34. M. E. Auerbach, *Drug Standards 20*, 165 (1952).
35. H. Persky, *Methods Biochem. Anal. 2*, 57 (1954).
36. H. Weil-Malherbe and A. D. Bone, *Biochem. J. 51*, 311 (1952).

37. H. Weil-Malherbe, *Methods Biochem. Anal. 16*, 293 (1968).
38. T. Canback and J. G. L. Harthon, *J. Pharm. Pharmacol. 11*, 764 (1959).
39. J. Holmekoski and A. Kivinen, *Farm. Aikak 75*, 223 (1966), *Chem. Abstrs. 66*, 5796g (1967).
40. E. F. Salim, P. E. Manni and J. E. Sinsheimer, *J. Pharm. Sci. 53*, 391 (1964).
41. K. H. Beyer, *J. Am. Chem. Soc. 64*, 1318 (1942).
42. H. Jensen, J. P. Gillet and E. Neuzil, *Ann. Biol. Clin. 26*, 73 (1968), *Chem. Abstrs. 68*, 102195 p (1968).
43. M. Goldenberg, M. Faber, E. J. Alston and E. C. Chargaft, *Sci. 109*, 534 (1949).
44. C. Valori, C. A. Brunori, V. Renzini and L. Corea, *Anal. Biochem. 33*, 158 (1970).
45. H. Weil-Maherbe and A. D. Bone, *Biochem. J. 67*, 65 (1957).
46. W. O. James, *Nature 161*, 851 (1948).
47. C. Romano, *Bull. Soc. ital. biol. sper. 26*, 1230 (1950) *Chem. Abstrs. 45*, 7622e (1951).
48. D. M. Shepherd and G. B. West, *Nature 171*, 1160 (1953).
49. H. Weil-Malherbe and A. D. Bone, *J. Clin Pathol. 10*, 138 (1957).
50. B. Robinson and D. M. Shepherd, *J. Pharm. Pharmacol. 13*, 374 (1961).
51. J. Krautheim and J. Blumberg, *Med. Exp. 7*, 8 (1962).
52. E. G. McGeer and W. H. Clark, *J. Chromatog. 14*, 107 (1967).
53. W. P. dePotter and R. F. Vochten, *Experientia 21*, 482 (1965).
54. W. W. Fike, *Anal. Chem. 38*, 1967 (1966).
55. J. A. Stern, M. J. Franklin and J. Mayer, *J. Chromatog. 30*, 632 (1967).
56. A. Alessandro and F. Mari, *G. Med. Mil. 117*, 281 (1967). *Chem. Abstrs. 68*, 16190b (1968).

57. L. Chafetz, A. I. Kay and H. Schriftman, *J. Chromatog. 35,* 567 (1968).
58. J. Dittmann, *J. Chromatog. 32,* 764 (1968).
59. S. Kawai and Z. Tamura, *Chem. Pharm. Bull. 16,* 699 (1968).
60. S. Lindstedt, *Clin. Chim. Acta 9,* 309 (1964).
61. N. P. Sen and P. L. McGeer, *Biochem. Biophys. Res. Commun. 13,* 390 (1963).
62. S. Kawai, T. Nagatsu, T. Imanari and Z. Tamura, *Chem. Pharm. Bull. 14,* 618 (1966).
63. W. Drell, *J. Am. Chem. Soc. 77,* 5429 (1955).

MEPERIDINE HYDROCHLORIDE

Nancy P. Fish and Nicholas J. DeAngelis

CONTENTS

1. Description
    1.1 Name, Formula, Molecular Weight
    1.2 Appearance, Color, Odor
2. Physical Properties
    2.1 Infrared Spectra
    2.2 Nuclear Magnetic Resonance Spectra
    2.3 Ultraviolet Spectra
    2.4 Mass Spectra
    2.5 Optical Rotation
    2.6 Melting Range
    2.7 Differential Thermal Analysis
    2.8 Solubility
    2.9 Crystal Properties
    2.10 Polymorphism
3. Synthesis
4. Stability - Degradation
5. Drug Metabolic Products
6. Identification
7. Methods of Analysis
    7.1 Elemental Analysis
    7.2 Spectrophotometric Assay
    7.3 Titrimetric Assay
    7.4 Colorimetric Assay
        7.41 Acid Dye
        7.42 Ammonium Reineckate
    7.5 Fluorometric Assay
    7.6 Miscellaneous
    7.7 Chromatography
        7.71 Paper and Thin Layer Chromatography
        7.72 Column Chromatography
        7.73 Gas Chromatography
8. References

1. Description

1.1 Name, Formula, Molecular Weight

Meperidine hydrochloride is designated by the following chemical names: ethyl 1-methyl-4-phenylisonipecotate hydrochloride[1], N-methyl-4-phenyl-4-carbethoxypiperidine hydrochloride[2], ethyl 1-methyl-4-phenylpiperidine-4-carboxylate hydrochloride[2], and the hydrochloride of the ethyl ester of 1-methyl-4-phenylisonipecotic acid[2]. In Chemical Abstracts the compound is listed under the heading-isonipecotic acid: 1-methyl-4-phenyl, ethyl ester and/or meperidine hydrochloride. The most commonly used trade or trivial names are: Demerol Hydrochloride, Pethidine Hydrochloride, and Dolantin. However, the Merck Index, 8th Edition[2], lists twenty-two others.

$$\text{structure} \cdot \text{HCl} \qquad M.W. = 283.80$$

1.2 Appearance, Color, Odor

Meperidine hydrochloride is a fine, white, odorless, crystalline powder.[1]

2. Physical Properties

2.1 Infrared Spectra

The infrared (I.R.) spectrum of meperidine hydrochloride, U.S.P. (Wyeth Lot No. F-665901; I.R. spectrum #6728) is given in Figure 1[3]. The I.R. spectrum was obtained on a KBr pellet and is identical to that presented by Manning[4]. Levi, Hubley, and Hinge[5] published the I.R. spectrum of meperidine HCl taken in a mineral oil (Nujol) mull and it is essentially identical to that shown in Figure 1. Assignment of some of the absorptions is given in Table I.

I.R. spectra of meperidine base in a mineral oil (Nujol) mull have been published by Levi, Hubley and Hinge[5]; in carbon disulfide by Carol[6]; and in chloroform by Pro and Nelson[7] and Levi, Hubley and Hinge[5].

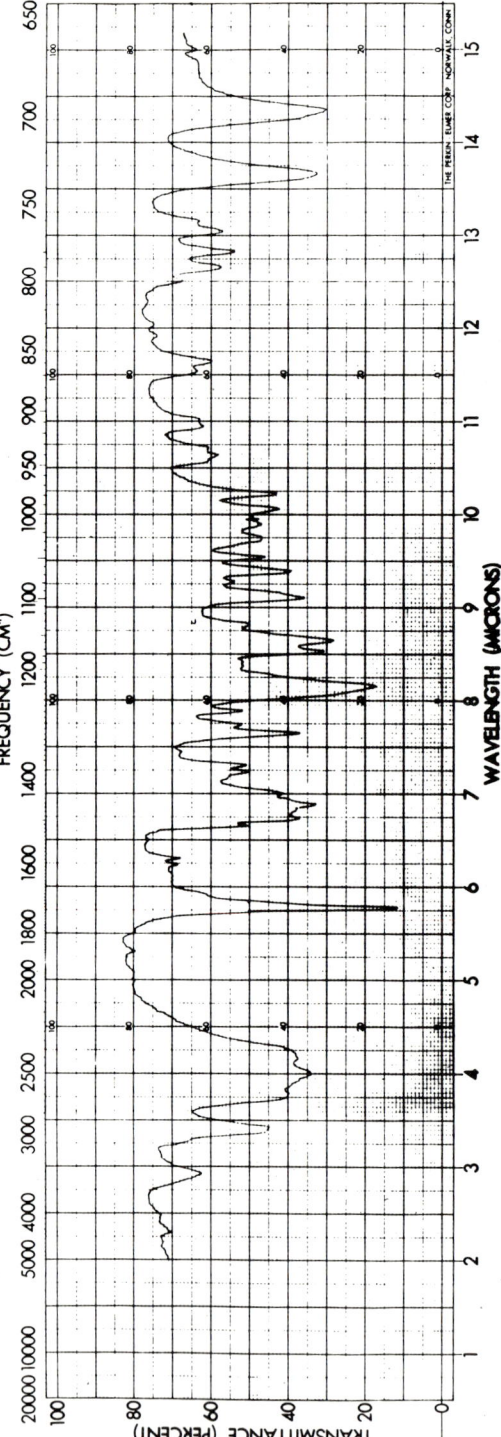

Fig. 1 I.R. Spectrum of Meperidine Hydrochloride, U.S.P., Wyeth Lot No. F-665901. 1% KBr Pellet - Instrument: Perkin Elmer Model 21.

TABLE I

IR Spectral Assignments of Meperidine HCl

Wavelength of Absorption (cm.$^{-1}$)	Vibration Mode
near 2900	CH_3, CH_2 stretch
1730	C = O stretch (ester)
near 1450	$\{$ -CH_2 - deformation CH_3 - N deformation
1230	C-O stretch (ester)
700 and 730	aromatic CH out of plane deformation (monosubstituted benzene ring)

TABLE II

NMR Spectral Assignments of Meperidine HCl

Chemical Shift (ppm.)	Protons	
1.16	CH_3-$\underline{CH_2}$-O	triplet
2.73 and 2.9 to 3.8	ring CH_2 groups	---
2.85	CH_3-N	singlet
4.18	$\underline{CH_3}$-CH_2-O	quartet
7.3	aromatic	singlet
11.3	HCl	singlet

2.2 Nuclear Magnetic Resonance Spectra

The nuclear magnetic resonance (NMR) spectrum (Fig. 2) was obtained by preparing a saturated solution of meperidine hydrochloride, U.S.P. (Wyeth Lot No.: F-665901) in deutero chloroform containing tetramethylsilane as internal reference[8]. The only exchangeable proton is the hydrogen associated with HCl. The NMR proton spectral assignments[8] are given in Table II.

2.3 Ultraviolet Spectra

Oestreicher, Farmilo, and Levi[9] reported λ max. at 251-252 mμ (ε -176), 257 mμ (ε -217) and 263 mμ (ε -174) for meperidine hydrochloride in water. Pro and Nelson[7] published a U.V. spectrum with identical λ max. Meperidine hydrochloride, U.S.P. (Wyeth Lot No. F-665901) when scanned[10]

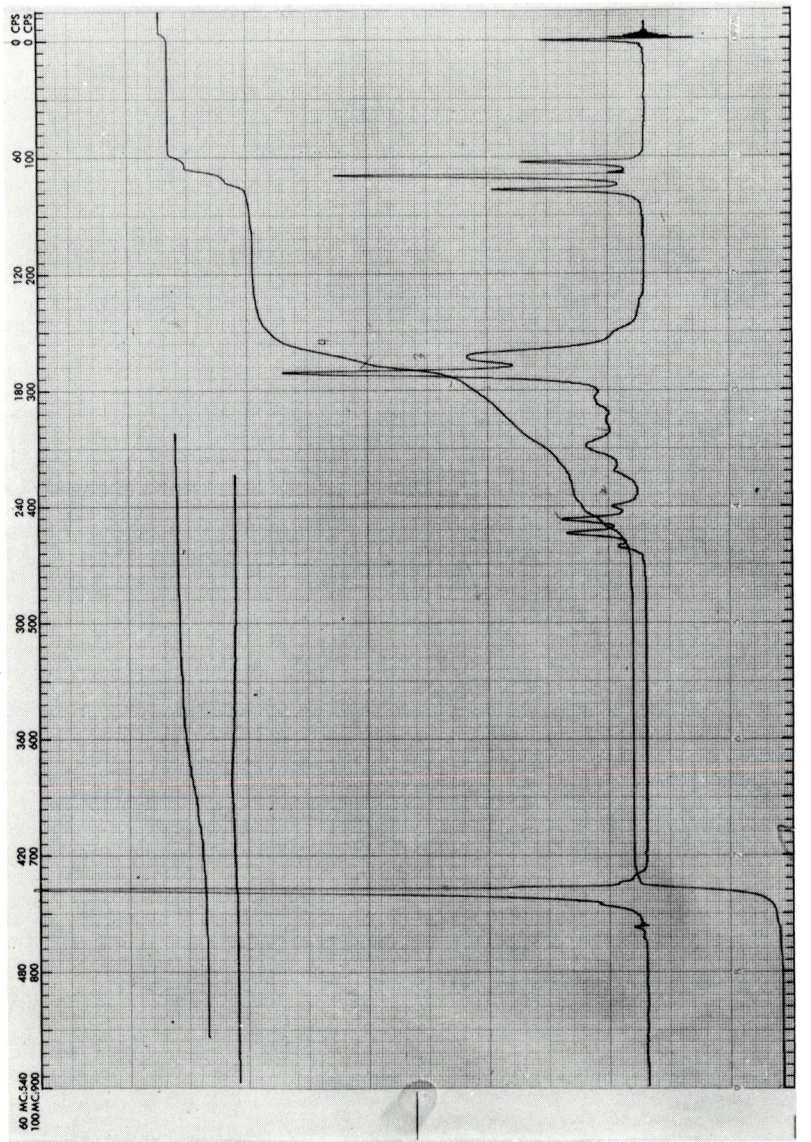

Fig. 2 NMR Spectrum of Meperidine Hydrochloride, U.S.P., Wyeth Lot No. F-66590l. Solvent: deuterochloroform, internal standard: tetramethylsilane. Instrument: Jelco Model C-60 HL.

MEPERIDINE HYDROCHLORIDE

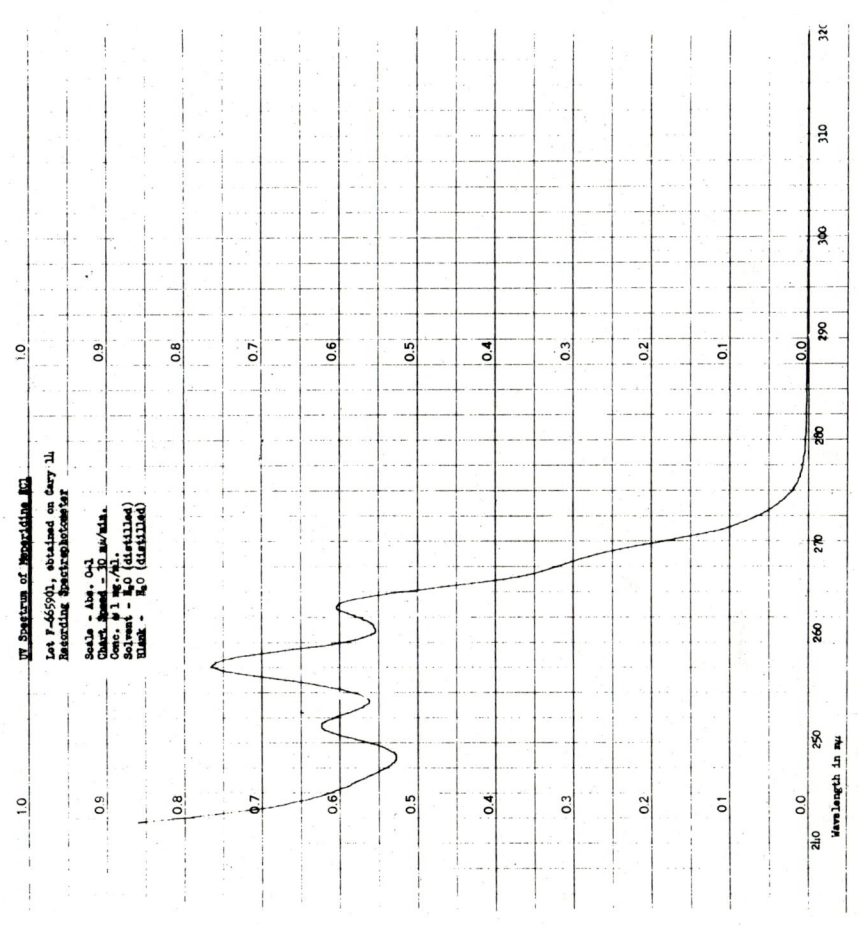

Fig. 3 UV Spectrum of Meperidine Hydrochloride, U.S.P., Wyeth Lot No. F-665901. Solvent: distilled water. Instrument: Cary Model 14

between 350 mμ and 240 mμ exhibited λ max. at 251.5 mμ (ϵ -174), 257.5 mμ (ϵ -214), and 263 mμ (ϵ -169). This spectrum is shown in Fig. 3.

2.4 Mass Spectra

The mass spectrum of meperidine HCl, U.S.P. (Wyeth Lot No.: F-665901) was obtained[11] by direct insertion of the sample into an MS-902 double focusing mass spectrometer. Results are summarized as a bar graph (Figure 4). The ion source temperature was 140°C. and the ionizing electron beam energy was 70 eV. Data were compiled by Kuhlman and Shrader[11] with the aid of an on-line PDP 8 Digital Computer.

Meperidine HCl gave a molecular ion of the free base at m/e 247.1563 and there is an intense M-1 peak. The first prominent fragment is at mass 232.1345 which corresponds to the loss of $\cdot CH_3$. A peak at m/e 218.1178 corresponds to the loss of $\cdot C_2H_5$ from the molecular ion. Fragments appear at masses 202 and 174 which are formed by a carbonyl cleavage with the loss of $\cdot OCH_2CH_3$ and $\cdot \overset{O}{\underset{\|}{C}}OCH_2CH_3$ respectively from the molecular ion. The nitrogen containing ions C_2H_4N, C_3H_7N, and C_4H_9N are present. The peak at m/e 42 is the base peak.

The high resolution mass spectrum assignments of the prominent ions are given in Table III[11].

TABLE III
High Resolution Mass Spectrum Assignments of Meperidine HCl

Measured Mass	Calculated Mass	Formula
247.1563	247.1571	$C_{15}H_{21}O_2N_1$
246.1490	246.1493	$C_{15}H_{20}O_2N_1$
232.1345	232.1337	$C_{14}H_{18}O_2N_1$
218.1178	218.1180	$C_{13}H_{16}O_2N_1$
202.1225	202.1231	$C_{13}H_{16}O_1N_1$
174.1292	174.1282	$C_{12}H_{16}N_1$
172.1123	172.1125	$C_{12}H_{14}N_1$
140.0706	140.0711	$C_7H_{10}O_2N$
131.0860	131.0860	$C_{10}H_{11}$
71.0723	71.0734	C_4H_9N
57.0579	57.0578	C_3H_7N

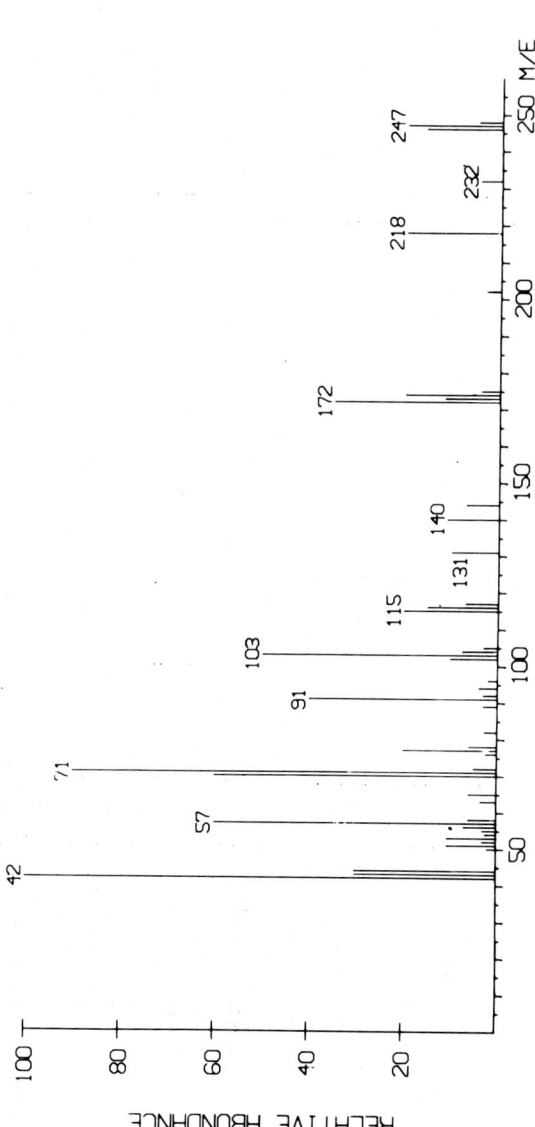

Fig. 4 Low Resolution Mass Spectrum of Meperidine Hydrochloride, U.S.P., Wyeth Lot No. F-665901. Instrument: AEI - Model MS 902

2.5 Optical Rotation
Meperidine hydrochloride is not optically active.

2.6 Melting Range
Meperidine hydrochloride exhibits a sharp melting point in the temperature range of 186°C. to 189°C.[1] The melting temperature range does not change significantly with variations in heating rate of from 1 to 10°C./min. Melting points outside the range stated above have not been reported for pure forms of meperidine hydrochloride.

2.7 Differential Thermal Analysis
The differential thermal analysis (DTA) curve of meperidine hydrochloride run from room temperature to the melting point exhibits no endotherms or exotherms other than that associated with the melt. The DTA curve[10] of meperidine hydrochloride, U.S.P. (Wyeth Lot No.: F-665901) run on a Dupont 900 DTA using a micro cell and a heating rate of 5°C./min. is shown in Figure 5.

2.8 Solubility
The following solubilities have been determined for meperidine hydrochloride at room temperature.

	U.S.P. XVIII[1]	Authors[10]
water	very soluble	greater than 1 g./ml.
95% ethanol	soluble	400 mg./ml.
chloroform		500 mg./ml.
ether	sparingly soluble	less than 0.1 mg./ml.
benzene		less than 1 mg./ml.
acetone		25 mg./ml.

2.9 Crystal Properties
a. Keenan[12] reported the following optical crystallographic properties for meperidine hydrochloride:
elongated six sided prisms; extinction: parallel; elongation sign: positive.

n_D^{20} : α = 1.545; β = 1.581; γ = 1.618 ($\pm$ 0.002)

In the first supplement to the N.F. XIII[13] identical refractive indexes, elongation and extinction data are given with the additional information that the optic sign

MEPERIDINE HYDROCHLORIDE

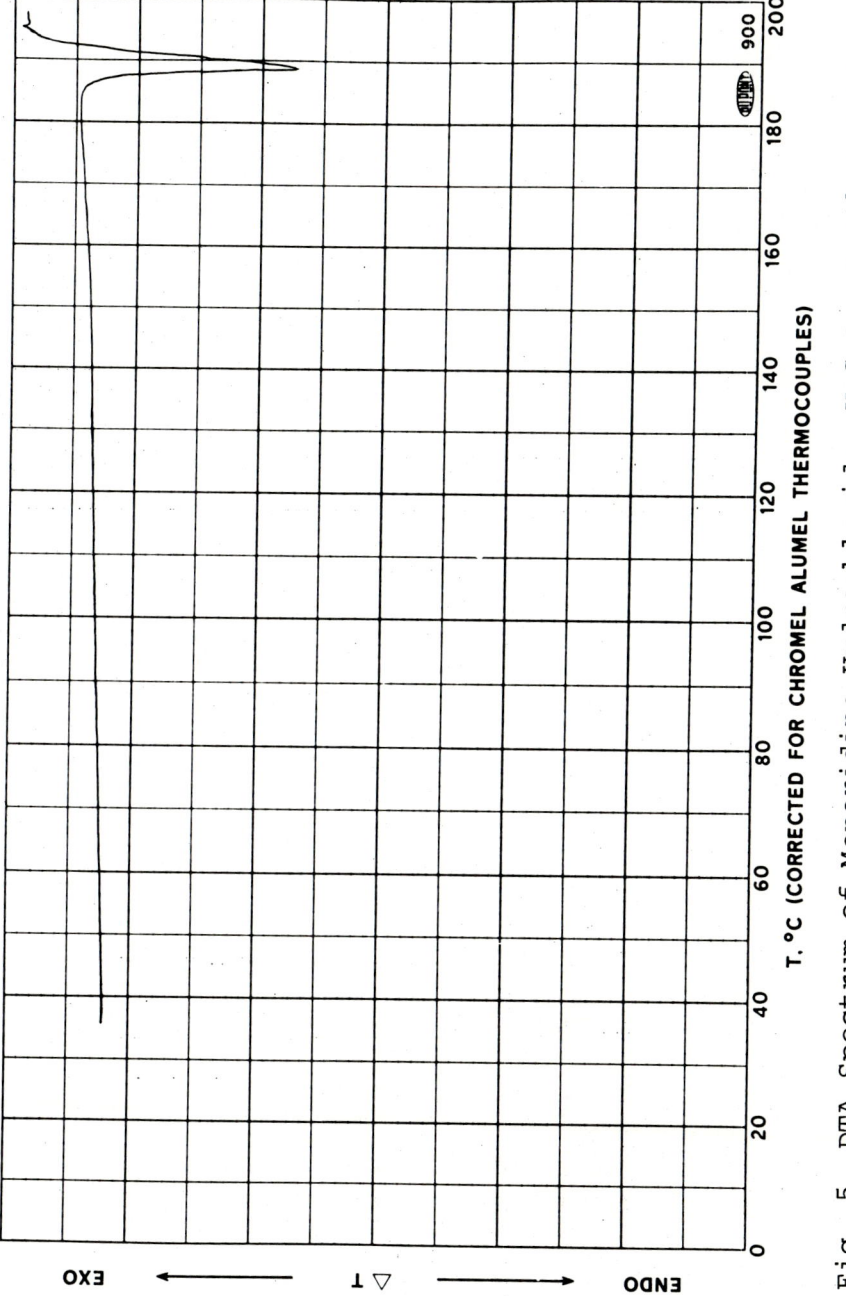

Fig. 5 DTA Spectrum of Meperidine Hydrochloride, U.S.P., Wyeth Lot No. F-665901, scan speed 5°/min. Instrument: DuPont 500 DTA.

is positive.

b. The X-ray powder diffraction pattern of meperidine hydrochloride, U.S.P. (Wyeth Lot No. F-665901) obtained[10] with a Philips diffractometer[14] using CuK$_\alpha$ radiation is shown in Figure 6.

The calculated d spacings[10] for the diffraction pattern shown in Fig. 6 as well as the d spacings published by Barnes and Sheppard[15] and Gross and Oberst[16] are given in Table IV. The Barnes and Sheppard data were obtained from a powder photograph using a cylindrical camera of 114.6 mm. diameter and Cobalt K$_\alpha$ radiation. The data of Gross and Oberst were obtained from a powder photograph using a flat film mounted perpendicular to the X-ray beam and Cu K$_\alpha$ radiation.

2.10 Polymorphism

Brandstätter and Grimm[17] reported three polymorphic forms of meperidine hydrochloride. In addition to the stable Form I (m.p. 187 - 191°), they prepared Forms II (m.p. 163-165°C.) and III (m.p. 154 - 156°C.) by sublimation, and described the appearance and melting behavior of the crystals they obtained.

3. Synthesis

In the original synthesis of meperidine HCl by Otto Eisleb[18] diethanol-methylamine was converted to methyl di-(β-chloroethyl)-amine by means of thionyl chloride. This was then condensed with benzyl cyanide in the presence of sodium amide to 1-methyl-4-phenylpiperidine-4-carboxylic acid nitrile which was then hydrolyzed, esterified and converted to the hydrochloride (see Figure 7). Numerous modifications of this method employing different starting materials and/or condensing agents have been reported.[19,20,21,22] Another published method uses a piperidine derivative as the starting material[23].

An alternate synthesis was developed by E. Smissman and G. Hite[24], using a modified Favorski rearrangement. Isonicotinic acid was methylated, reduced to 1-methyl-4-piperidinecarboxylic acid hydrochloride, and converted to the acid chloride hydrochloride. This was then condensed with benzene, chlorinated, treated with alkali, esterified, and converted to the hydrochloride (See Figure 8).

TABLE IV (concluded)

1. Authors Data
2. Data of Barnes and Sheppard (see reference 15).
3. Data of Gross and Oberst (see reference 16).
(B) Broad
I_1 = Intensity of the strongest maximum
I = Intensity of the maximum corresponding to the indicated d value
d = (interplanar distance) $\frac{n\lambda}{2\sin\theta}$

4. Stability - Degradation

Meperidine hydrochloride in the solid state is very stable[25]; however, as is common for esters it will undergo hydrolysis to the corresponding acid in aqueous solution. Patel et al[26] have studied the hydrolysis of meperidine hydrochloride in aqueous acid solution and have reached the following conclusions.

Hydrolysis of meperidine hydrochloride in dilute hydrochloric acid is a specific hydronium ion catalyzed reaction which is first order with respect to hydrogen ion concentration. A positive, primary salt effect was noted in dilute acid solution. The hydrolysis of the protonated form of meperidine hydrochloride was found to be first order with respect to meperidine hydrochloride over a wide range of hydrogen ion concentration (pH 1-7). A catalytic effect with respect to meperidine hydrolysis was exhibited by dihydrogen phosphate ion, the only phosphate ion studied in the investigation, and the possibility of similar effects with respect to other buffer systems should be recognized. The rate of hydrolysis was found to be very slow in the pH range of 3.5 to 5 with maximum stability at pH 4.01. The calculated half-life for meperidine hydrochloride at pH 4.9 is 23.8 years at 25°C.

Mariani Marelli[27] investigated the stability of meperidine hydrochloride in buffered aqueous solutions during sterilization by steam, and determined the maximum stability to be at pH 7 - 7.5. However, temperature, ionic strength, and specific ion effects were not considered.

The stability of meperidine hydrochloride solutions in polyethylene containers when exposed to high temperature, oxygen, and UV light is reported on by Kempa et al.[28]

5. Drug Metabolic Products

Meperidine hydrochloride has been found to undergo biotransformation rapidly and almost completely in man[29,30]. In a study by Burns et al[29] in which the plasma levels of meperidine hydrochloride were measured after intravenous injection in man, it was inferred that metabolism occurs at a rate varying from 10 to 20% per hour (17% average) in ten different subjects. No accumulation of the drug after repeated administration was found in this study and only 5% of the drug was found to be excreted unchanged in the urine. Plotnikoff et al[31] determined that the biotransformation products of meperidine hydrochloride excreted in human urine are: normeperidine (I), meperidinic acid (II), normeperidinic acid (III), and bound forms of meperidinic acid (IV) and normeperidinic acid (V). The following metabolic pathways were proposed:

R = unidentified conjugate

Asatoor et al[32] determined that in highly acidic urine the main route of removal of meperidine hydrochloride from the body is by excretion of meperidine and normeperidine. If the urine is alkaline, excretion of the hydrolysis products meperidinic and normeperidinic acids, both as free acids and as conjugates, is the more important means of elimination.

Bernheim and Bernheim[33] reported that meperidine hydrochloride is hydrolyzed by livers of various animals but not by brain, blood, kidney, spleen, or heart tissue. Clark[34] reported on the N-demethylation of meperidine hydrochloride and the in vitro inhibition of this demethylation.

6. <u>Identification</u>

Meperidine hydrochloride can be identified by virtue of its characteristic IR, UV, and X-ray spectra (see 2.1, 2.3 and 2.9b). The N.F. XII[35] describes a test (based on the formation of an orange red color when meperidine hydrochloride is reacted with sulfuric acid-formaldehyde test solution) which distinguishes meperidine hydrochloride from morphine and hydromorphone. The characteristic melting point (188 - 191°C.) of the picrate salt[1] of meperidine is also useful as an identity test.

Ternikova[36] and Bonino[37] describe several color identification tests and Levine[38] gives a crystal formation identity test.

7. <u>Methods of Analysis</u>

   7.1 <u>Elemental Analysis</u>*

Element	% Theory	Determined[8]
C	63.48	63.71
H	7.81	8.00
N	4.94	4.95
Cl	12.49	12.33

*Meperidine HCl, U.S.P. (Wyeth Lot No.: F-655901).

   7.2 <u>Spectrophotometric Assay</u>

The UV absorption maximum of meperidine hydrochloride at 257 mμ has been extensively utilized for assay purposes. Determination of the absorptivity value is useful for assay of the raw material and also for formulations after the meperidine has been isolated. In regard to the isolation of meperidine it should be borne in mind that meperidine base can be readily steam distilled. Furthermore, extraction of the base from aqueous basic solution separates intact meperidine from its hydrolytic degradation products.

Pro and Nelson[7] quantitatively determined meperidine in a mixture with acetophenetidin (phenacetin) by comparing its absorbance to that of a standard after separating the meperidine from the phenacetin by steam distillation.

Kaysak[39] isolated meperidine from biological material by ethylene dichloride extraction of an alkaline homogenate of the sample. The ethylene dichloride extract was washed with pH 5.8 phosphate buffer, and then meperidine was extracted with 0.5N H_2SO_4 and quantitatively determined by measuring its UV absorbance.

Machek and Lorenz[40] describe an indirect spectrophotometric assay for a mixture of meperidine hydrochloride with benzyl-2-δ-ethylpiperidine cyclohexanone-2-carboxylate hydrochloride (Cetran). Marozzi and Falzi[41] separated meperidine from other compounds by paper chromatography, cut out the appropriate zones, eluted them with HCl, and determined meperidine by its UV absorbance.

Jonas Carol[6] developed a quantitative IR method for meperidine in pharmaceutical preparations in which it is the only active using a CS_2 solution of the drug.

7.3 Assay - Titrimetric

The tertiary amine group in meperidine is directly titratable in non-aqueous media. Meperidine hydrochloride is likewise titratable if mercuric acetate is added to tie up the chloride ion. Standard non-aqueous titration techniques are applicable using either visual indicators or potentiometric end point detection. Several such methods for both meperidine and its hydrochloride salt have been published[42-48].

Pellerin, Gautier, and Demay[49] developed a semimicro two-phase ($CHCl_3$ and acidic H_2O) titration of organic bases including meperidine using 0.01M sodium lauryl sulfate as the titrant and methyl yellow as the indicator. Johnson and King[50] describe a similar titration technique for meperidine using sodium dioctylsulphosuccinate as the titrant. Mundell[51] determined meperidine after steam distilling it from a weakly alkaline solution by collecting the base in a standard sulfuric acid solution and then titrating the excess acid.

7.4 Assay - Colorimetric

7.41 Acid dye Extraction

Modifications of the well-known acid dye method of Brodie[52] have been applied to the determination of meperidine successfully. Organic base-sulfonic acid molecular complexes are extracted into organic solvents, and the concentration of the organic base is determined indirectly by measurement of the sulfonic acid in the organic solvent. Gettler and Sunshine[53] applied this method to the determination of microgram quantities of meperidine in human tissues. The tissue sample was extracted with hot water containing tartaric acid, and a modification of Brodie's method using methyl orange and chloroform was used to determine meperidine. Lehman and Aitken[54] used bromothymol blue and extracted the yellow meperidine-dye complex into benzene from pH 7.5 buffer. The sodium salt of the dye was then determined colorimetrically after extraction into aqueous alkali. Oberst[55] used a similar technique for meperidine in urine, but determined the benzene solution colorimetrically. He did not get color formation with hydrolysis products of meperidine.

7.42 Ammonium Reineckate

Kum-Tatt[56] identified meperidine by preparing the ammonium reineckate derivative and determining several physical properties of the reineckate including its ultraviolet and visible absorption spectra in ethanol. Basu and Dutta[57] quantitatively determined meperidine by dissolving the ammonium reineckate in acetone and measuring the absorbance at 525 mμ.

7.5 Fluorometric Assay

Although meperidine has no native fluorescence, Kozuki and Kawase[58] report a determination based on induced fluorescence after the addition of Marquis reagent to meperidine in solution.

7.6 Miscellaneous

Wasilewska[59] developed a method for the determination of tertiary and quaternary organic bases in various pharmaceutical preparations. He added standardized tungstosilicic acid in excess of the amount required to react with the organic base, and then quantitatively precipitated the excess acid with bismuth nitrate. The solution was filtered and the tungstosilicic acid in the filtrate was titrated with di-sodium Versenate after the addition of Xylenol Orange and potassium nitrate.

7.7 Chromatography

7.71 Paper and Thin Layer Chromatography

Many thin layer and paper chromatographic methods have been found suitable for the isolation and identification of meperidine and meperidine HCl. A number of these are summarized in Tables V and VI. Some of the specific references cited give sample preparation techniques, and methods for eluting the drug from the developed chromatogram which will enable one to obtain quantitative chromatographic results by spectrophotometry or some other suitable means if desired.

Maruyama[60] used a ring oven paper chromatographic method to separate meperidine and normeperidine from the free acids.

7.72 Column Chromatography

McMartin, Simpson and Thorpe[71] developed an automated assay procedure suitable for separating meperidine from urine and other basic compounds. They used two 1.2 x 6 cm. CM-cellulose columns and an eluent of pH 8.5, 0.05M borate buffer at a flow rate of 0.33 ml./min. and report a retention time of 87 min. The procedure is sensitive enough that the authors suggest its applicability to the determination of meperidine in blood plasma also.

Elvidge, Proctor, and Baines[72] found that oxycellulose used as a carboxylic cation-exchange column would separate phenol and chlorocresol from various alkaloids, including meperidine before determination by ultraviolet spectroscopy. 1N H_2SO_4 followed by water was used to elute the phenol and chlorocresol and additional H_2SO_4 eluted the meperidine. This method was devised for the assay of injection formulations.

S. L. Tompsett[73] used a Dowex 50X-12 column to separate meperidine from urine after the urine had been acidified with HCl. The meperidine was eluted with strong acid (5-8N HCl) and identified by paper chromatography after

Table V
Paper Chromatographic Systems for Meperidine HCl

Solvent	Paper	Paper Treatment	Rf	Visualization Technique	Reference
A	Whatman No. 2	Buffered, pH 3.0	0.50	A	61
A	Whatman No. 2	Buffered, pH 5.0	0.61	A	61
A	Whatman No. 2	Buffered, pH 6.5	0.72	A	61
A	Whatman No. 2	Buffered, pH 7.5	0.90	A	61
A	Whatman No. 1	Buffered, pH 5.0	0.57	A	62
B	Schleicher and Schull 2034b	Buffered, pH 5.7	0.58	B	63
C	Schleicher and Schull 2034b	Buffered, pH 5.7	0.67	B	63
D	Schleicher and Schull 2034b	Buffered, pH 5.7	0.54	B	63
E	Schleicher and Schull 2034b	Buffered, pH 5.7	0.73	B	63
F	Whatman No. 3 MM	Buffered, pH 4.2	0.91	A	64
G	Whatman No. 3 MM	Buffered, pH 4.2	0.63	A	64
H	Whatman No. 3 MM	Impregnated with light paraffin (descending)	0.95	A	64
I	Whatman No. 3 MM	Descending paper impregnated with sec. octanol	0.60	A	64
J	Whatman No. 3 MM	Impregnated with formamide	0.78	A	64

continued.....

Table V (concluded)
Paper Chromatographic Systems for Meperidine HCl

Solvent	Paper	Paper Treatment	R_f	Visualization Technique	Reference
K	Whatman No. 1	NONE	0.64	A,C,D,E,	65
L	Whatman No. 1	NONE	0.74	A,C,D,E	65
M	Whatman No. 1	NONE	0.94	A,C,D,E	65
N	Whatman No. 1	Buffered, pH 5.5	0.60	A,C,D,E	65

Solvents
A n-butanol, saturated with buffer
B n-butanol – H_2O (80–30)
C n-butanol–acetic acid (10–30) saturated with H_2O
D butyl acetate–butanol–acetic acid, H_2O (85–15–40–22)
E butyl acetate–butanol–isobutanol–acetic acid–H_2O (50:25:25:50:75)
F isobutanol–glacial acetic acid–water (100:10:24)
G butyl acetate–glacial acetic acid–water (35:10:3)
H propyl alcohol – water – diethyl amine (1:8:1)
I ammonium formate in water (10%) saturated with sec.–octanol
J Light paraffin – diethylamine (9:1)
K Isoamyl alcohol – acetic acid (10:1) saturated with H_2O
L Butanol – acetic acid (10:1) saturated with H_2O
M Isoamyl alcohol – NH_4OH (10:1) saturated with H_2O
N Isoamyl alcohol saturated with H_2O

Visualization
A Potassium iodoplatinate spray
B $Bi(NO_3)_3$ – KI Reagent
C Dragendorff's Reagent
D Thallium Iodide
E Iodine

Table VI
Thin Layer Chromatographic Systems for Meperidine HCl

Solvent System	Adsorbent	Visualization Technique	R_f	Reference
A	Silica Gel G	A	0.63	66
B	Silica Gel G	A	0.70	66
C	Silica Gel G	A	0.70	66
D	Silica Gel G	A	0.27	66
E	Silica Gel G	A	0.28	66
F_1	Silica Gel G	A	0.00	67
F_2	Silica Gel G	A	0.04	67
F_3	Silica Gel G	A	0.06	67
F_4	Silica Gel G	A	0.14	67
F_5	Silica Gel G	A	0.17	67
F_6	Silica Gel G	A	0.32	67
F_7	Silica Gel G	A	0.43	67
F_8	Silica Gel G	A	0.23	67
F_9	Silica Gel G	A	0.40	67
F_{10}	Silica Gel G	A	0.54	67
F_{11}	Silica Gel G	A	0.54	67
F_{12}	Silica Gel G	A	0.61	67
F_{13}	Silica Gel G	A	0.76	67
G	Silica Gel G	A	0.64	68
H	Silica Gel with Fluorescent Indicator	A-E	0.34	69
I	Silica Gel with Fluorescent Indicator	A-E	0.47	69
J	Silica Gel with Fluorescent Indicator	A-E	0.45	69
K	Silica Gel G	C	0.55	70

Solvent Systems
A Ethyl alcohol - dioxane-benzene-ammonium hydroxide (5:40:50:5)
B Chloroform-dioxane-ethyl acetate-ammonium hydroxide (25:60:10:5)
C Ethyl alcohol-chloroform-dioxane-petroleum ether (30-60°) benzene-ammonium hydroxide-ethyl acetate (5:10:50:15:10:5:5)
D Ethyl acetate-benzene-ammonium hydroxide (60:35:5)

continued....

Table VI (concluded)
Thin Layer Chromatographic Systems for Meperidine HCl

Solvent Systems: (concluded)
E Ethyl acetate-n-butyl ether-ammonium hydroxide (60:35:5)
$F_1 - F_{13}$ Developing tank with 110 ml. solvent plus a beaker containing 10 ml. 28% ammonia

F_1 Petroleum ether (30-60 °C.)
F_2 Carbon tetrachloride
F_3 Isopropyl ether
F_4 Benzene
F_5 Ethylene dichloride
F_6 Methylene chloride
F_7 Chloroform
F_8 Ethyl ether
F_9 Ethyl acetate
F_{10} n-butyl alcohol
F_{11} Isopropyl alcohol
F_{12} Acetone
F_{13} Methyl alcohol

G Benzene:dioxane:ethanol:ammonia-25% (50:40:5:5)
H Acetone:Ammonia-25% (99:1)
I Methanol
J Methanol:Ammonia-25% (99:1)
K Methanol:Acetone:Triethanolamine (1:1:0.03)

Visualization Methods:
A Potassium Iodoplatinate Reagent
B Ultraviolet Light
C Dragendorff Spray
D Potassium Permanganate Spray
E p-Dimethylaminobenzaldehyde Spray

the eluent was made basic and extracted with a mixture of chloroform and isopropyl alcohol.

Brioch, DeMayo, and Dal Cortivo[74] describe a semi-automated method for the separation and determination of narcotics mixed with mannitol and lactose. A SE-Sephadex C-25 cation exchange column buffered with pH 4.6 phosphate buffer is used to effect separation, followed by ultraviolet monitoring of the column effluent for the detection of the narcotics, including meperidine. A retention volume of 71.1 ml. was reported for meperidine when eluted with phosphate buffer from a 50 x 0.9 cm. column packed 40 cm. high.

Gunderson, Heiz and Klevstrand[75] used Dowex 2 resin, converted to the OH- form, to form meperidine base from meperidine HCl in tablet preparations. Elution with 70% ethanol was followed by titration of the base with 0.1N HCl standardized by titration of Borax in 70% ethanol with bromothymol blue.

7.73 Gas Chromatography

Stainier and Gloesener[76] hydrolyzed meperidine with 3.5% KOH and determined the ethanol evolved using a 2 meter Carbowax 1500 column, operated at 70°C. Alcohol recoveries of 97% to 102% were reported.

Using a 1/8", 5', 5% SE-30 Chromosorb W column operated at 230°C. and at 250°C., Parker, Fontan and Kirk[77] separated meperidine from other alkaloids, barbiturates, sympathomimetic amines and tranquilizers. This work was directed towards development of a screening process rather than quantitative determination.

Kazyak and Knoblock[78] chromatographed meperidine and other drugs using a 4 mm. glass, 6 ft. 1% SE-30 liquid phase on Chromosorb W column at temperatures of 150°, 165°C., 180°C. and 200°C.

Boon and Mace[79] applied gas chromatography to the ion-pair complex of meperidine with bromothymol blue after it was extracted from pH 5 solution with chloroform. A glass, 4 ft. 1% Carbowax 20M plus 2% KOH on Gas Chrom P column at 150°C. was used.

8. References

1. "United States Pharmacopeia", 18th Ed., Mack Printing Co., Easton, Pa.
2. "Merck Index", 8th Edition, Merck and Co., Inc., Rahway, N.J.
3. Peter Rulon, Wyeth Laboratories, personal communication.
4. James J. Manning, Bull. Narcotics 7 (1), 85-100 (1955).
5. L. Levi, C. E. Hubley, and R. A. Hinge, Bull. Narcotics 7 (1), 42-84 (1955).
6. J. Carol, J. Ass. Off. Agr. Chem. 37, 692 (1954).
7. M. J. Pro and R. A. Nelson, J. Ass. Off. Agr. Chem. 40 (4), (1957).
8. Bruce Hofmann, Wyeth Laboratories, personal communication.
9. P. M. Oestreicher, C. G. Farmilo, and L. Levi, Bull. Narcotics 6 (3), 42-70 (1954).
10. N. Fish and N. DeAngelis, Wyeth Laboratories, unpublished results.
11. C. Kuhlman and S. Shrader, Wyeth Laboratories, personal communication.
12. G. J. Keenan, J. Am. Pharm. Assoc. Sci. Ed. 35 (11), 338-9 (1946).
13. "National Formulary", 13th Ed., Mack Printing Co., Easton, Pa.
14. North American Philips Company, Mount Vernon, New York.
15. W. H. Barnes and H. M. Sheppard, Bull. Narcotics 6 (2), 27-68 (1954).
16. S. T. Gross and F. W. Oberst, J. Lab. Clin. Med. 32, 94-101 (1947).
17. M. Brandstätter and H. Grimm, Mikrochim. Acta 2, 1175-1182 (1956).
18. Otto Eisleb, Ber. Deut. Chem. Ges. 74B, 1433-50 (1941); U.S. Patent 2,167,351 - 7/25/39 ; C.A. 35, 5646; C.A. 36, 5465.
19. Keyung Ho Kim, Taehan Naekwa Hakhoe Chapchi 6 (4), 201-3 (1963); C.A. 65, 16934a.
20. F. Bergel, A. L. Morrison, H. Rinderknecht, J. W. Haworth, and N. C. Hindley, J. Chem. Soc. 6, 261-272 (1944) U.S. Patent 2,418,289, 4/1/42.
21. K. Meischer and H. Kaegi; U.S. Patent 2,486,792, 11/1/49; C.A. 44 7887a.

22. Tanabe Drug Co., Japan Patent 153,615, 10/10/42; C.A. 43, 3471 h.
23. Welcome Foundation, Ltd. and E. Walton, Brit. Patent 592,016, 9/4/47; C.A. 42, 1610h.
24. E. E. Smissman and G. Hite, J. Am. Chem. Soc. 81, 1201 (1959).
25. R. M. Patel, Dissertation Abstract 26 (10), 5735-6 (1966), Order No. 66-3477, C.A. 65, 3684e.
26. R. M. Patel, T.-F. Chin, and J. L. Lach, Amer. J. Hosp. Pharm. 25, 256-61 (1968).
27. O. Mariani Marelli, Rend. Ist. Super. Sanita (Rome) 14, 282-86 (1951).
28. H. Kempa, G. Brockelt, R. Pohloudek-Fabini, Gyogyszereszet 13 (10), 389 (1969).
29. J. J. Burns, B. L. Berger, P. A. Lief, A. Wollack, E. M. Papper and B. B. Brodie, J. Pharmacol. Exp. Ther., 114 (3), 289-98 (1955).
30. E. Leong Way, A. I. Gimble, W. P. McKelway, H. Ross, C.-Y. Sung and H. Ellsworth, J. Pharmacol. Exp. Ther., 96 (4), 477-84 (1949).
31. N. P. Plotnikoff, E. Leong Way and H. W. Elliott, J. Pharmacol. Exp. Ther. 117, 414-19 (1956).
32. A. M. Asatoor, D. R. London, M. D. Milne and M. L. Simenhoff, Brit. J. Pharmacol. and Chemother., 20 (2), 285-98 (1963).
33. F. Bernheim and M. L. C. Bernheim, J. Pharmacol. Exp. Ther., 85 (1), 74-77 (1945).
34. B. Clark, Biochem. Pharmacol., 16 (12), 2369-85 (1967)
35. "National Formulary", 13th Ed., Mack Printing Co., Easton, Pa.
36. R. M. Ternikova, (Pharm. Inst. Moscow); Aptechnoe Delo 6 (2), 38-43 (1957); C.A. 52, 5751b.
37. R. C. D'Alessio de Carnevale Bonino, Semana méd. (Buenos Aires) 1943 I, 289-93; C.A. 37, 3024⁴.
38. J. Levine, Ind. Eng. Chem., Anal. Ed. 16, 408-10 (1044); C.A. 38 4381⁶.
39. L. Kaysak, J. Forensic Sci. 4, 264-75 (1959); C.A. 53, 11501b.
40. G. Machek and F. Lorenz, Sci. Pharm. 31, 17-26 (1963); C.A. 59, 1440e; G. Machek and F. Lorenz, Anales Farm. Hosp. 10 (22), 237-249 (1967).
41. E. Marozzi and G. Falzi, Med. Leg. Assicurazioni, 13 (3-4), 239-58 (1965); C.A. 64, 20181a.

42. R. Vasiliev, J. Fruchter and M. Jecu, Rev. Chim. 16 (5), 293-4 (1965); C. A. 63, 8123g.
43. A. Anastasi, U. Gallo and E. Mecarelli, Il Farmaco (Pavia), Ed. prat. 10, 604-11 (1955); C.A. 50, 5242b.
44. Poey Seng Bouw, Suara Pharm. Madj. 5, 1-5 (1960) (English); C.A. 58, 13719d.
45. R. Pohloudek-Fabini and K. König, Pharmazie 13, 752-6 (1959); C.A. 53, 20693h.
46. M. Rink and R. Lux, Deut. Apoth - Ztg. 101, 911-18 (1961); C.A. 55, 25167d.
47. J. B. Schute, Kongr. Pharm. Wiss. Vortr. Originalmitt. 23, 695-701 (1963); C.A. 62, 6347; C.A. 62, 7588d; Schute, Pharm. Weekblad 99 (39), 1053-70 (1964).
48. T. Espersen, Dan. Tidsskr. Farm. 33, 113-24 (1959); C.A. 53, 22735c.
49. F. Pellerin, J. A. Gautier, and D. Demay, Ann. Pharm. Fr. 22 (8-9), 495-504 (1964); C.A. 62, 12979g.
50. C. A. Johnson and R. E. King, J. Pharm. Pharmacol. 15 (9), 584-8 (1963).
51. M. Mundell, J. Ass. Offic. Agr. Chem. 28, 711-14 (1945); C.A. 40, 985⁸.
52. B. B. Brodie, S. Udenfriend, J. E. Baer, T. Chenkin, and W. Dill, J. Biol. Chem. 158, 705-14 (1945); C.A. 39, 3881¹.
53. A. O. Gettler and I. Sunshine, Anal. Chem. 23, 779-81 (1951).
54. R. A. Lehman and T. Aitken, J. Lab. Clin. Med. 28, 787-93 (1943).
55. F. W. Oberst, J. Pharmacol. 79, 10-15 (1943).
56. L. Kum-Tatt, J. Pharm. Pharmacol. 12, 666-76 (1960).
57. K. Basu and B. N. Dutta, J. Proc. Inst. Chemists (Calcutta) 34, 142-59 (1962); C.A. 58, 4378e.
58. H. Kozuki and S. Kawase, Kagaku Keisatsu Kenkyusho Hokoku 20 (4), 252-56 (1967); C.A. 69, 89728p.
59. L. Wasilewska, Diss. Pharm. Pharmacol. 18 (3), 305-13 (1966); C.A. 66, 5805f.
60. Yuzi Maruyama, Igaku To Seibutsugaku, 75 (4), 123-4 (1967); C.A. 68, 84846v.
61. L. R. Goldbaum and L. Kazyak, Anal. Chem. 28, 1289-90 (1956).
62. L. A. Dal Cortivo, C. H. Willumsen, S. B. Weinberg and W. Matusiak, Anal. Chem. 33 (9), 1218 (1961).
63. R. Fischer and N. Otterbeck, Sci. Pharm. 25, 242-8 (1957); C.A. 52, 13192i.

64. K. Genest and C. G. Farmilo, J. Chromatogr. 6, 343-349 (1961).

65. G. J. Mannering, A. C. Dixon, N. V. Carroll, and O. B. Cope, Jour. Lab. and Clin. Med. 44 (2), 292-300(1954).

66. J. A. Steele, J. Chromatogr. 19, 300-303 (1965).

67. J. L. Emmerson and R. C. Anderson, J. Chromatogr. 17, 495-500 (1965).

68. P. Schweda, Anal. Chem. 39 (8), 1019 (1967).

69. A. Noirfailise, J. Chromotogr. 20, 61-77 (1965).

70. K. Randerath, "Thin Layer Chromatography", p. 195, Academic Press, 2nd Ed., New York.

71. C. McMartin, P. Simpson and N. Thorpe, J. Chromatogr. 43, 72-83 (1969).

72. D. A. Elvidge, K. A. Proctor and C. B. Baines, Analyst (London), 82, 367-72 (1957); C.A. 51, 12437c.

73. S. L. Tompsett, Acta Pharmacol. et Toxicol. 17, 295-303 (1960).

74. J. R. Broich, M. M. DeMayo and L. A. Dal Cortivo, J. Chromatogr. 33, 526-529 (1968).

75. F. O. Gunderson, R. Heiz and R. Klevstrand, J. Pharm. Pharmacol 6, 608-14 (1953); C.A. 47, 12754h.

76. C. Stainier and E. Gloesener, Farmaco, Ed. Prat. 15, 721-31 (1960); C.A. 55, 18011.

77. K. D. Parker, C. R. Fontan and P. L. Kirk, Anal. Chem. 35 (3), 356-59 (1963).

78. L. Kazyak and E. O. Knoblock, Anal. Chem. 35 (10), 1448-52 (1963).

79. P. F. G. Boon and A. W. Mace, J. Chromatogr. 41 (1), 105-6 (1969).

MEPROBAMATE

C. Shearer and P. Rulon

CONTENTS

1. Description
   1.1 Name, Formula, Molecular Weight
   1.2 Appearance, Color, Odor
2. Physical Properties
   2.1 Infrared Spectra
   2.2 Nuclear Magnetic Resonance Spectra
   2.3 Mass Spectrum
   2.4 Melting Range
   2.5 Differential Thermal Analysis
   2.6 Solubility
   2.7 Crystal Properties
3. Synthesis
4. Stability
5. Drug Metabolic Products
6. Methods of Analysis
   6.1 Elemental Analysis
   6.2 Direct Spectrophotometric Analysis
      6.21 Nuclear Magnetic Resonance Analysis
      6.22 Infrared Spectrophotometric Analysis
   6.3 Colorimetric Analysis
   6.4 Oscillopolarographic Analysis
   6.5 Titrimetric Analysis
   6.6 Chromatographic Analysis
      6.61 Paper
      6.62 Thin Layer
      6.63 Gas
      6.64 Column
   6.7 Electrophoretic Analysis
   6.8 Refractometric Analysis
7. References

MEPROBAMATE

1. Description

   1.1 Name, Formula, Molecular Weight
   The name used by Chemical Abstracts and U.S.P.XVIII for meprobamate is 2-methyl-2-propyl-1,3 propanediol dicarbamate. It can also be called 2,2-di(carbamoyloxymethyl)pentane or 2-methyl-2-propyltrimethylene carbamate. The empirical formula is $C_9H_{18}N_2O_4$ with a molecular weight of 218.25.

$$\begin{array}{c} CH_2\,O_2\,CNH_2 \\ | \\ H_3CCH_2\,CH_2\,CCH_3 \\ | \\ CH_2\,O_2\,CNH_2 \end{array}$$

   1.2 Appearance, Color, Odor
   Meprobamate is a white powder having a characteristic odor.

2. Physical Properties

   2.1 Infrared Spectra
   An infrared spectrum of meprobamate (N.F. Reference Standard material) has been run in a KBr pellet. The spectrum is enclosed as Figure 1. This spectrum agrees with published spectra[1,2,3]. The following band assignments have been made.

Wavelength, μ	Characteristic of	Reference
2.9 and 3.0	NH stretching	5
3.4	CH stretching	4
5.8	Amide I band	5
6.25	Amide II band	5
7.2	CH_2 symmetrical deformation	4
7.5	Amide III	5

   2.2 Nuclear Magnetic Resonance Spectra
   Meprobamate (N.F. reference sample) was dissolved in dimethyl sulfoxide with tetramethylsilane as the internal standard[6]. The spectrum obtained on a 60 MHz NMR spectrophotometer is enclosed as Figure 2. This spectrum conforms with a published spectrum[7]. The following peak

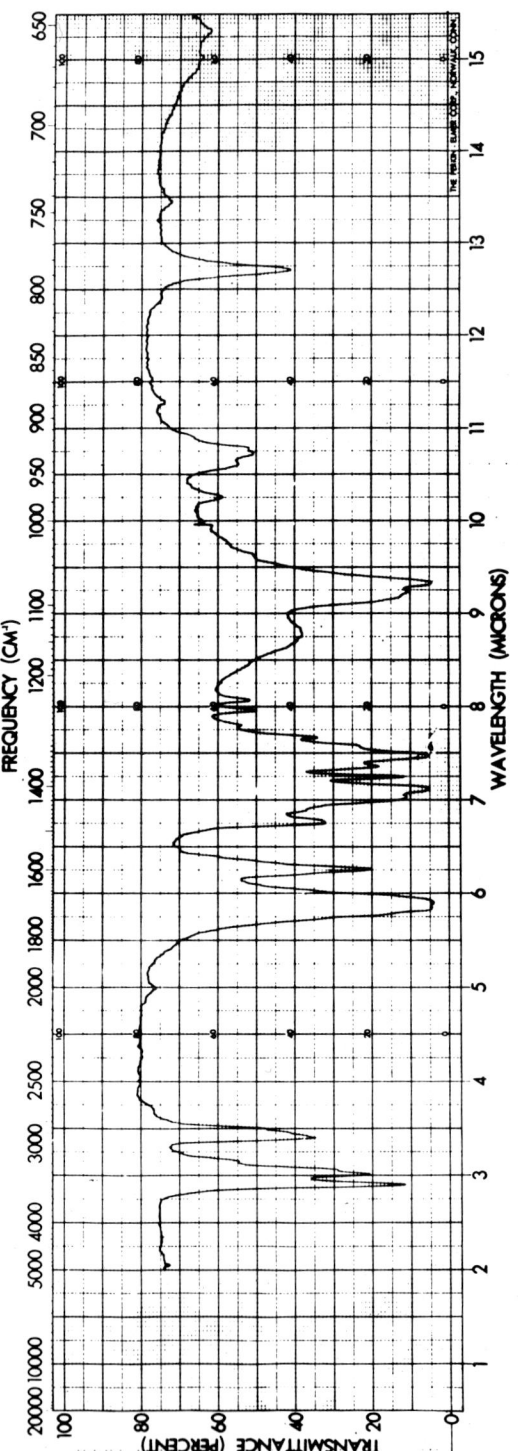

Fig. 1 Infrared Spectrum - Meprobamate N.F. Reference Lot 65075.

MEPROBAMATE

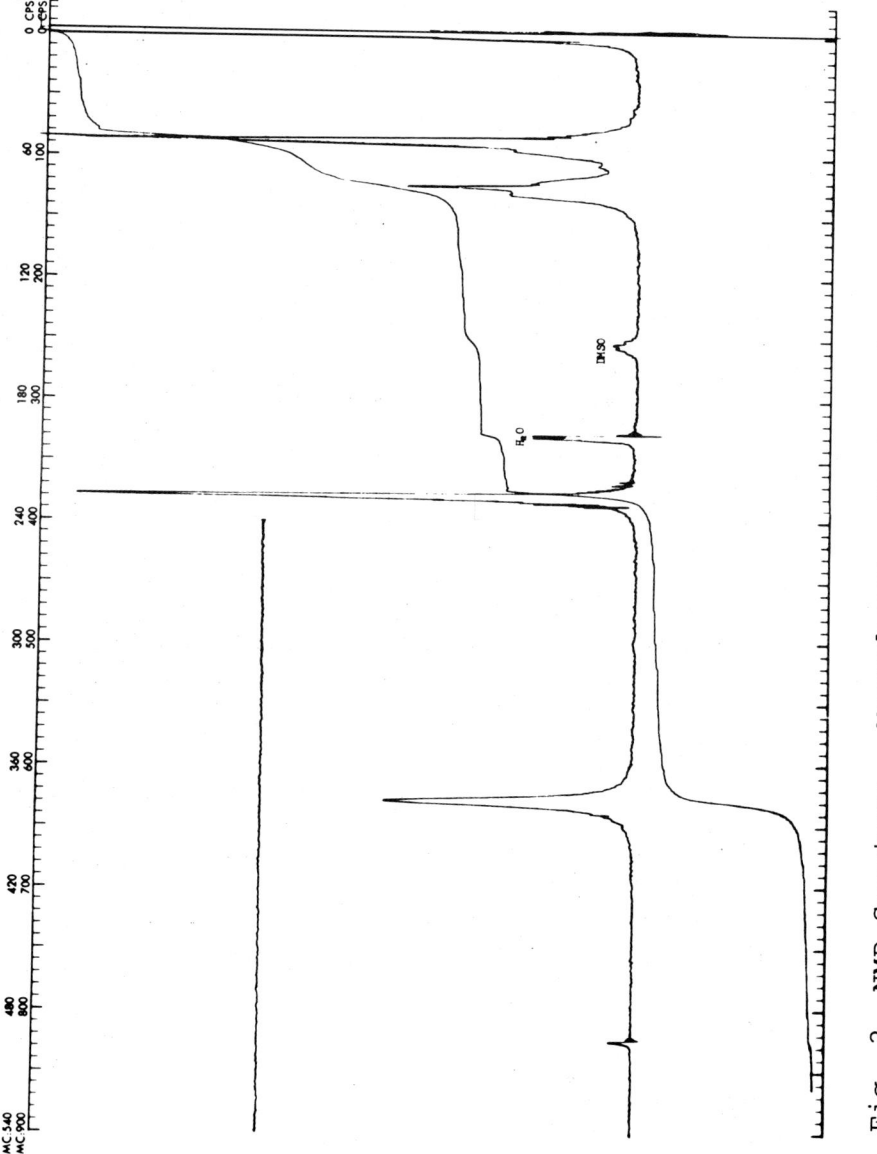

Fig. 2 NMR Spectrum - Meprobamate, N. F. Reference Lot 65075.

assignments have been made.

Chemical Shift (δ)	Proton
0.9	$H_3C - C$
1.3	$C - CH_2 - C$
3.8	$C - CH_2 - O$
6.3	$C - NH_2$

2.3 Mass Spectrum

The mass spectrum of meprobamate (N.F. reference standard) was obtained by direct insertion of the sample into the MS-9 double focusing, high resolution mass spectrometer[8]. The sample was run at 100°C. and 1.6 x 10^{-6} torr. The high resolution data was compiled and tabulated with the aid of the PDP-8 Digital Computer. The data is given in Table I and Figure 3.

Meprobamate (molecular weight 218) gave no molecular ion. We did obtain however, a P + 1 peak at mass 219.1361. The first prominent fragment is at mass 144.1007 which corresponds to cleavage at the tertiary carbon atom with the loss of CH_2OOCNH_2.

An ion of mass 135.0406 with the formula $C_3H_7O_4N_2$ is observed. This fragment is of interest because it cannot originate by way of simple cleavage or simple rearrangement but must involve rearrangement of the carbamate portion of the molecule and loss of a neutral C_6H_{11}.

The most abundant peak in the spectrum (m/e 83) is the charged ion C_6H_{11} formed at least partially from the loss of $NH_2CO_2H_2$ from the fragment at m/e 144. This transition is supported by the presence of a metastable ion at m/e 47.8.

There are metastables also showing m/e 96 to m/e 81 (M* 68.2) and m/e 43 to m/e 41 (M* 39.1).

The intense fragment at mass 62.0242 ($C_1H_4O_2N$) is $[NH_2C\begin{smallmatrix}CH\\CH\end{smallmatrix}]^+$ and is probably formed by the documented double hydrogen rearrangement[9].

2.4 Melting Range

The following melting point temperatures (°C.) have been reported: 104-106[11,12]
104-105[13]
105-106[14]

MEPROBAMATE

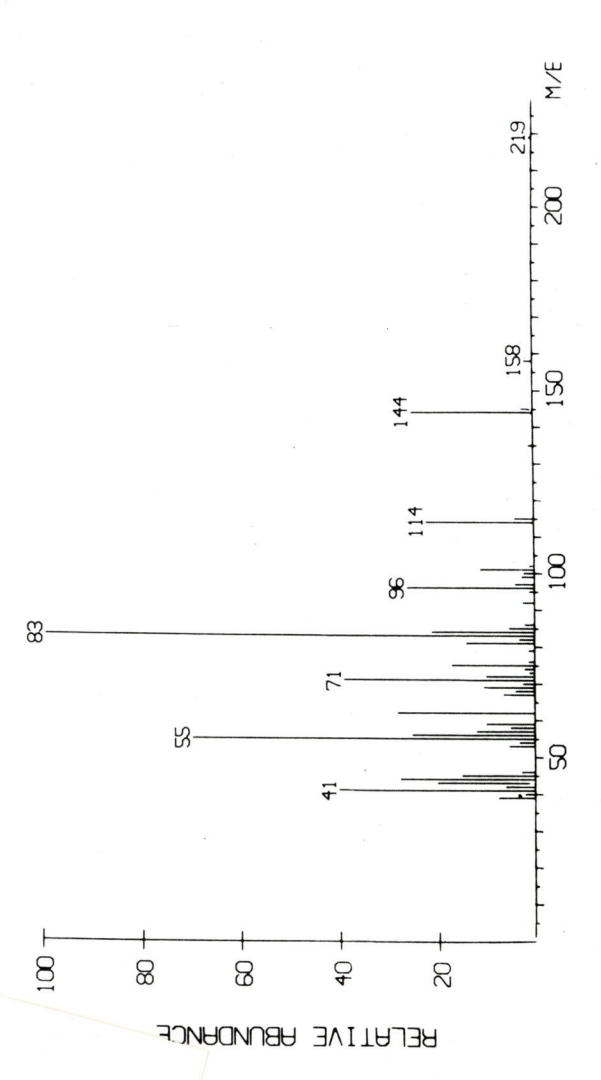

Fig. 3 Mass Spectrum - Meprobamate, N. F. Reference Lot 65075.

TABLE I

Mass Spectrum of Meprobamate

Measured Mass	Calculated Mass	Formula
219.1361	219.1344	$C_9H_{19}O_4N_2$
158.1174	158.1180	$C_8H_{16}O_2N$
144.1007	144.1024	$C_7H_{14}O_2N$
135.0406	135.0405	$C_3H_7O_4N_2$
114.1032	114.1044	$C_7H_{14}O$
101.0958	101.0966	$C_6H_{13}O$
96.0931	96.0938	C_7H_{12}
83.0844	83.0860	C_6H_{11}
81.0695	81.0704	C_6H_9
75.0316	75.0320	$C_2H_5O_2N$
71.0496	71.0485	C_4H_7O
62.0242	62.0241	CH_4O_2N
57.0704	57.0704	C_4H_9
57.0340	57.0347	C_3H_5O
55.0547	55.0552	C_4H_7
44.0134	44.0136	CH_2ON
41.0384	41.0391	C_3H_5

2.5 Differential Thermal Analysis

A differential thermal analysis was performed on meprobamate (N.F. Reference Standard material)[15]. A melting endotherm at 106°C. was observed. The endotherm did not appear to shift with a change in heating rate, from 5°C./min. to 20°C./min.

2.6 Solubility

The following solubility data[16] were obtained at room temperature:

    250 mg./ml. in 95% ethanol
    160 mg./ml. in isopropyl alcohol
    110 mg./ml. in acetone
    140 mg./ml. in chloroform
    650 mg./ml. in dimethylformamide

The solubilities in water at 20°C. and 37°C. were determined as 3.4 mg./ml. and 7.9 mg./ml., respectively. The solubility of meprobamate in isotonic sodium chloride solution at room temperature was found to be 3.7 mg./ml.[17]

2.7 Crystal Properties

The X-ray powder diffraction pattern of meprobamate (N.F. Reference Standard) is presented in Table II[15]. It is in agreement with two published patterns[10,18].

TABLE II

X-Ray Powder Diffraction Pattern

Sample: Meprobamate N.F. Reference Standard Lot 65075
Source: Cu K

°2θ	d	
6.4	13.8*	
10.85	8.15	
12.95	6.82*	
17.1	5.18	
18.0	4.93	
18.8	4.72	
20.2	4.395	
22.6	3.93*	
24.15	3.685	
25.3	3.52	
28.55	3.125	*= most
32.75	2.735	intense peaks

3. Synthesis

Meprobamate has been synthesized by several procedures (See Figure 4). The first step is to synthesize 2-methyl-2-propyl-1,3-propanediol. This has been done[14] by reacting 2-methyl pentanal and formaldehyde in presence of KOH. The diol has also been prepared by reducing 2-methyl-2-n-propyl malonic acid, diethyl ester with lithium aluminum hydride[19].

The diol can then be reacted[13,14,19,20] with phosgene in the presence of sodium hydroxide or an organic base and then with ammonia to give meprobamate. Other procedures which have been used, include reacting the diol with (a) urea in the presence of lead acetate or other salts[21,22] and (b) ethyl urethane in the presence of aluminum isopropoxide[23,24] or sodamide[25] to give meprobamate.

4. Stability

Meprobamate is very stable as a solid. Meprobamate is stable in dilute acid and dilute alkali and is not broken down in gastric or intestinal fluid[26]. It is recognized that heating solutions of meprobamate in strong acid will cause hydrolysis of the material[26]. It has been proposed that during treatment of meprobamate with alkaline alcohol the sodium salt of carbamic acid[27] is formed, which may undergo a dehydration to form a cyanate[28]. The amino moieties of the carbamate can form its diacetyl derivative with acetic anhydride[26,27], and will condense with aldehydes[27].

5. Drug Metabolic Products

Five metabolic products of meprobamate have been identified. Ludwig[29] characterized the major metabolic product as 2-methyl-2-(β-hydroxypropyl)-1,3-propanediol dicarbamate (Compound I, Figure 5). Other studies[12,30,31,32] as well as Ludwig's showed the presence of a glucuronide of meprobamate. Four metabolites in rabbits and/or dogs were characterized by Yamamoto and coworkers[33]. The structures are given as II, III, IV and V in Figure 5.

6. Methods of Analysis

The assay procedure listed under the following sections have been applied to meprobamate as a raw material and/or in a pharmaceutical formulation: 6.1, 6.21, 6.22, 6.31, 6.32, 6.33, 6.35, 6.36, 6.371, 6.7321, 6.38, 6.4, 6.5, 6.7

MEPROBAMATE

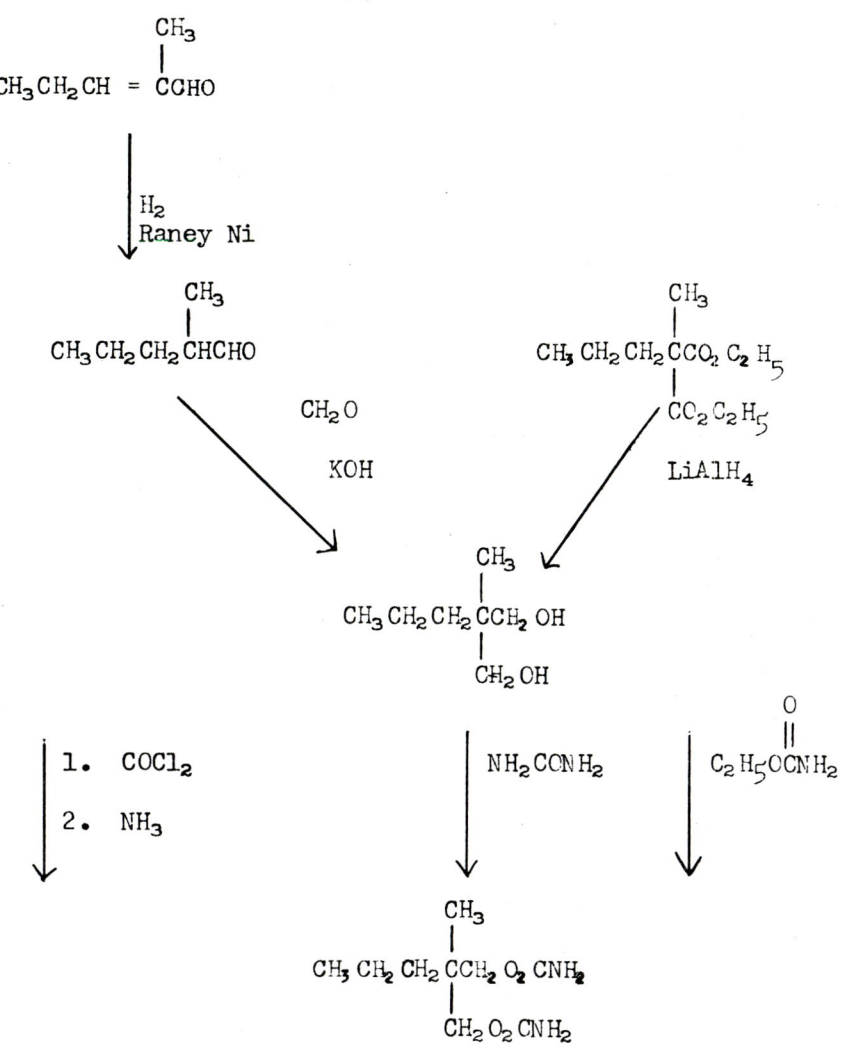

Figure 4 -- Synthesis Routes for Meprobamate

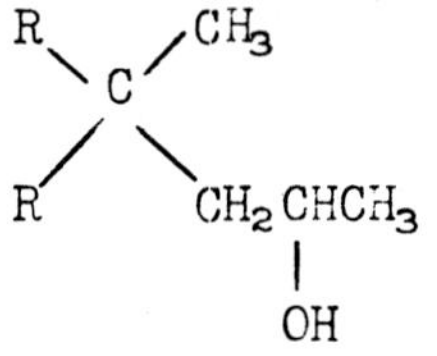

Figure 5 -- Metabolic Products of Meprobamate

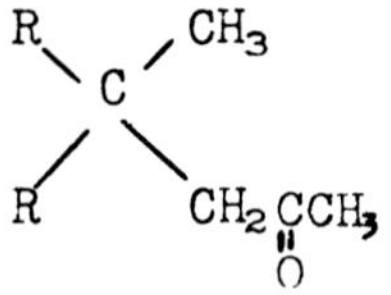

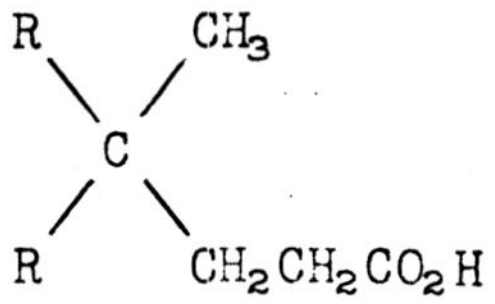

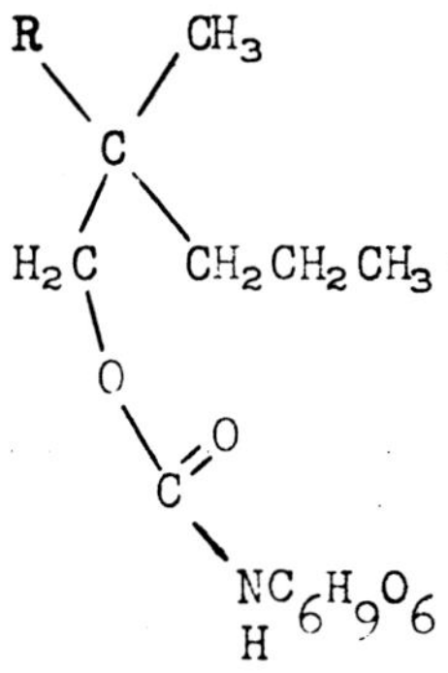

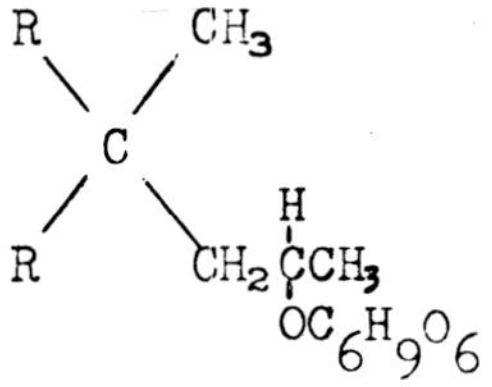

$R = H_2NCO_2CH_2-$

and 6.8. The assay procedure listed under the following sections have been applied to determining meprobamate in biological materials: 6.31, 6.32, 6.33, 6.34, 6.35, 6.36, 6.372, 6.38 and 6.7. The absence in the list of an assay procedure does not necessarily mean it will not be applicable to either situation. The chromatographic procedures were not included in this listing. Some assay procedures are not included in this listing since the article or abstract did not indicate to what material the test was applied.

6.1 Elemental Analysis

Data obtained on N.F. Reference Sample No.: 65075

Element	% Theory	Reported[6]
C	49.53	49.38
H	8.31	8.30
N	12.84	12.99

6.2 Direct Spectrophotometric Analysis

6.21 Nuclear Magnetic Resonance Analysis

An NMR procedure was developed for the analysis of meprobamate tablets using malonic acid as an internal standard[7]. The integrated area of meprobamate's peak at 3.8 p.p.m. is compared to the integrated area of the malonic acid peak at 3.3 p.p.m.

6.22 Infrared Spectrophotometric Analysis

Meprobamate can be assayed using infrared spectrophotometry. This technique compares the relative intensity of the amide band at 6.32μ of the sample dissolved in chloroform to that of a standard meprobamate solution[34,35]. The N-H stretching bands at 2.820μ and 2.914μ have also been used[36].

6.3 Colorimetric Analysis

A variety of colorimetric methods have been used to assay meprobamate.

6.31
Meprobamate is hydrolyzed in an alkaline alcoholic medium to yield a cyanate which forms a blue colored complex with cobalt ions. For analyses[28,37,38,39] the intensity of the color has been determined at 605 mμ.

6.32 Meprobamate forms a colored complex, absorption maximum at 408 mμ, with hydroquinone in concentrated sulfuric acid[40,41].

6.33 A hypochlorite solution at pH 10.5 reacts with meprobamate to form an "active" chlorine derivative. After decomposing the excess hypochlorite with phenol in dilute hydrochloric acid, the "active" derivative is reacted with excess potassium iodide. The color produced by the iodine liberated has been measured at 350 mμ. The "active" derivative will also react with potassium iodide-starch solution to give a color with a maximum absorbance of 625 mμ[42].

6.34 Meprobamate when heated in 1N NaOH is hydrolyzed to ammonia and carbon dioxide. The concentration of the ammonia formed was measured by the use of Nessler reagent[43].

6.35 The red color produced when meprobamate is heated with concentrated sulfuric acid has been used to assay the material[30,44,45].

6.36 Meprobamate has been assayed by reacting it with Zimmerman's reagent[46,47]. The color produced has a maximum absorbance of 395 mμ.

6.37 Meprobamate condenses with several different aldehydes to form colored compounds.

6.371 Furfuraldehyde has been reacted with meprobamate in an acid medium to give an absorbance maximum at 570 mμ when in 10% sulfuric acid in acetic acid[48], or at 550 mμ when in 20% hydrochloric acid in ethanol[49]. When meprobamate was heated in acetic acid and then reacted with furfural in the presence of antimony trichloride a colored product with a maximum absorbance at 591 mμ was formed[50].

6.372 p-Dimethylaminobenzaldehyde has been used extensively as the color forming reagent for meprobamate.

6.3721 The intensity of the yellow color was determined at 416 mμ when meprobamate was reacted with p-dimethylaminobenzaldehyde in hydrochloric acid[51]. When meprobamate was heated with p-dimethylaminobenzaldehyde in a sulfuric acid medium and then diluted with ice water, a violet color was formed which can be used as a quantitative measure of the meprobamate[52].

6.3722 Meprobamate reacts with p-dimethylaminobenzaldehyde and antimony trichloride in acetic anhydride to give a red-violet colored species having absorbance maximum at 550 mμ[53,54,55].

6.3723 Meprobamate reacts with p-dimethylaminobenzaldehyde in anisole in the presence of a mixed acetone - $AlCl_3$ catalyst to give a color with a peak at 540 mμ[56]

6.373 p-Dimethylaminocinnamaldehyde can replace the p-dimethylaminobenzaldehyde and in the presence of 3% concentrated sulfuric acid in acetic acid a red color was formed with a maximum absorbance at 520 mμ[57].

6.4 Oscillopolarographic Analysis

Meprobamate in N KOH and at a concentration of 10^{-3}M produces a peak of Q value 0.555[58]. For analytical purposes this peak was compared to that of a standard[59]. Meprobamate in N NaOH has been observed to give a response to single sweep A.C. oscillographic polarography. Kalvoda[60] indicates the possibility for the determination of substances in concentrations from 10^{-5} to 10^{-7} mol/l.

6.5 Titrimetric

Meprobamate, after hydrolysis, has been titrated in a variety of ways:

6.51 Meprobamate was hydrolyzed in conc. HCl. The solution was neutralized to methyl red end point and then formaldehyde added to form hexamethyltetramine, which is then titrated with the sodium hydroxide to a phenolphthalein end point[27,61].

6.52 Meprobamate was refluxed in pyridine sodium methoxide solution. The excess sodium methoxide was titrated with 0.1N benzoic acid in benzene using thymol blue[62,63] as the end point indicator.

6.53 Meprobamate was refluxed in hydrochloric acid and then taken to dryness. The residue was taken up in acetic acid and mercuric acetate added. The non-aqueous titration is made with perchloric acid in acetic acid using methylrosaniline as an indicator[64,65].

6.54 Meprobamate was boiled with a measured amount of KOH[66] or NaOH[67]. The excess base was then titrated with hydrochloric acid using a combination of phenolphthalein and alizarin yellow as indicators.

6.6 Chromatographic Analysis
All of the more common chromatographic techniques have been applied to meprobamate.

6.61 Paper Chromatographic Analysis
Meprobamate has been chromatographed on paper in many solvent systems. The pertinent information is given in Table III. The various spray reagents used for the detection of meprobamate on paper chromatograms are given in Table IV.

6.62 Thin Layer Chromatographic Analysis
The various eluant and adsorbent systems used for thin layer chromatography of meprobamate are given in Table V. Table VI gives spray reagents used for the detection of meprobamate on thin layer chromatographs.

6.63 Gas Chromatography
Gas phase chromatography has been used to analyze meprobamate. The necessary data of the various methods is enclosed as Table VII.

6.64 Column Chromatography
Meprobamate has been separated from the hydroxy-and keto meprobamate (see Figure 5, Compound I and II) on an alumina column using stepwise gradient elution. The eluants used were benzene, ethylacetate, acetone and methanol, both alone and in mixtures[33].

Meprobamate and pentaerythritol tetranitrate have been separated by selective elution from a dual layer column of Celite-sodium hydroxide and Celite-phosphoric

TABLE III
Paper Chromatography of Meprobamate

Rf	Paper	Direction	Eluant	Ref.
0.00	S&S 2043b	asc.	benzene:butanol:water (100:3:1)	68
0.13	S&S 2043b	asc.	petroleum ether: dioxane (5:2)	68
0.19	Whatman #1	des.	di-n-butyl ether	69
0.32	S&S 2043b	asc.	methylal:decalin:chloroform:hexane (11:3:3:8)	70
0.35	Whatman #1	des.	carbon tetrachloride:acetic acid:water (1:2:1)	71
0.35	n.a.	asc.	toluene:n-butanol:water (19:1:20)	72
0.74	Whatman #1	des.	chloroform	69
0.75	n.a.	n.a.	isoamyl alcohol:10% acetic acid	73
0.77	Whatman #1	asc.	butanol:water (100:15)	40
0.80	Whatman #1	des.	diethyl ether	69
0.81	Toyo Roshi #50	asc.	butanol containing 3% conc. ammonium hydroxide	33
0.85	Toyo Roshi #50	asc.	butanol:acetic acid:water (4:1:5)	33
0.85	Whatman #1	asc.	butanol:acetic acid:water (4:1:5)	74
0.85	n.a.	asc.	butanol saturated with water	75
0.85	Whatman #1	des.	benzene:butanol:acetic acid:water (3:1:1:5)	74
0.88	n.a.	asc.	butanol saturated with 5N NH$_4$OH	75
0.93	n.a.	asc.	butanol:acetic acid (10:3) saturated with water	75
0.95	Whatman #1	asc.	benzene:acetic acid:water (2:2:1)	74
0.95	Whatman #1	des.	chloroform:acetic acid:water (100:2:5)	74

asc. - ascending
des. - descending
n.a. - information not available

TABLE IV

Spray Reagents for Paper Chromatography

Reagent	Color	Ref.
A. Dimethylaminobenzaldehyde-$SbCl_3$ spray	n.a.	75
B. Ehrlich reagent	n.a.	76
C. Dinitrophenol	yellow	68
D. Iodine atmosphere	yellow	68
E. 10% H_2SO_4 then 2.5% Dimethylaminobenzaldehyde with heat	yellow	68
F. 1% $Co(NO_3)_2$ in absolute ethanol	red-yellow	68
G. Dragendorff Reagent	yellow	68
H. 10% Sulfuric acid in ethanol	yellow to brown	68, 70
I. Bromphenol blue	gold	68
J. Chlorine atmosphere then benzidine plus KI	bright blue	70
K. Chlorine atmosphere then starch plus KI	blue	71, 77
L. Furfural then conc. HCl	n.a.	78
M. Chlorine atmosphere then fluorescein sodium	n.a.	79
N. Sodium hypochlorite then ethanol then KI and starch	blue	80

n.a. - information not available

---

acid, using chloroform and benzene, respectively, as eluting solvents[98]. Meprobamate was detected and quantitatively measured by IR spectrophotometry as described in Section 6.22.

6.7 Electrophoretic Analysis

Meprobamate was separated from the carboxymeprobamate and the glucuronides, which are formed as metabolites, but not from the hydroxy or keto-meprobamate by electrophoresis[33]. A horizontal open strip type apparatus with a 1% borax (electrolyte) solution was used. A constant voltage of 400 V was applied to a filter paper strip (24 x 12 cm.) for 1 hour.

MEPROBAMATE

TABLE V
Thin Layer Chromatography Systems for Meprobamate

Rf	Adsorbent	Eluant	Ref.
0.00	Special prep 1	Benzene:chloroform (1:4) saturated with formamide	81
0.03	Special prep 2	Cyclohexane:diethylamine:benzene (75:20:15)	82
0.05	Silica gel	Chloroform:diethyl ether (85:15)	83
0.08	Silica gel G	Benzene:12N ammonium hydroxide:ethanol (95:5:15)	82
0.18	Silica gel G	Benzene:dioxane:28% ammonium hydroxide (75:20:5)	84,85
0.22	Kieselgel G	Methanol:acetic acid:ether:benzene (1:9:30:60)	86
0.23	Silica gel	Benzene:acetone (4:1)	85
0.30	Silica gel	Cyclohexane:ethanol (85:15)	87
0.35	Special prep 1	Carbon tetrachloride saturated with formamide	81
0.36	Silica gel G	Acetone:chloroform (1:1)	48
0.37	Silica G	Acetic acid:carbon tetrachloride:chloroform:water (100:60:90:50)	81
0.42	Special prep 2	Chloroform:methanol (90:10)	82
0.51	Silica gel G	Dioxane:benzene:25% ammonium hydroxide (40:50:10)	88
0.53	Silica gel	Chloroform:ethanol (90:10)	83
0.65	Silica gel	Chloroform:acetic acid (20:1)	74
0.71	Silica G	Methanol:12N ammonium hydroxide (100:1.5)	82
0.75	Silica gel	Chloroform:acetone (4:1)	74
0.76	Special prep 2	Acetone	82
0.80	Silica	Benzene:acetone (2:1)	74
0.80	Silica	Diethyl ether	74
0.89	Silica gel	Acetone:cyclohexane:ethanol (4:4:2)	85
n.a.	Silica gel	Chloroform:acetone:ammonium hydroxide (80:20:1)	89
n.a.	Kieselguhr G	Chloroform:acetone (1:1)	90

n.a. = information not available, Special prep 1 - silica gel impregnated with formamide, Special prep 2 - silica gel plates prepared using 0.1N NaOH

225

TABLE VI
TLC Spray Reagents for the Detection of Meprobamate

	Reagent	Color	Ref.
1.	Conc. H_2SO_4 with heat	yellow	87
2.	Cl_2 atmosphere then spray with KI-benzidine acetate	n.a.	83,90
3.	Cl_2 atmosphere then spray with KI-o-tolidine - HCl	n.a.	83
4.	1% furfural:5% H_2SO_4 in acetone	blue-black	81,83
5.	I_2 vapor	brown	89
6.	2.5% p-dimethylaminobenzaldehyde in conc. H_2SO_4	n.a.	86
7.	5% Vanillin in conc. H_2SO_4	yellow, with heat - blue	81
8.	2% $HgCl_2$ and 0.05% diphenylcarbazone in diethylamine	pink	91
9.	Ehrlich reagent then Dragendorff reagent	n.a.	85
10.	1% $HgNO_3$	n.a.	84

n.a. = information not available

6.8 Refractometric Analysis

Meprobamate tablets have been assayed by taking advantage of the refractive index of meprobamate. Meprobamate (3 to 10% w/w) was dissolved in ethanol and the refractive index is measured at 20°C. A formula was then used to determine the concentration of the meprobamate[99].

TABLE VII
Gas Chromatography of Meprobamate

Form of Meprobamate	Column	Temp.	Retention Time (minutes)	Internal Standard	Ref.
Alkaline Hydrolysis Product	2% methyl silicone SE 30 on Gas-Chrom P	n.a.	n.a.	2-methyl 2-sec-butyl 1,3 propanediol	92
	20% SE 30 on Chromosorb W	190°C.	n.a.		93
Extracted from Biological Material	3% SE 30 on Chromosorb W	165°C.	7.4		2
	3.8% Me silicone on Diatoport S	180°C.	n.a.	di-butyl phthalate	97
	Hi-Eff. 8B 1% on Gas-Chrom P	220°C.	4		96
	1% SE 30 on Chromosorb W	165°C.	10.4		97
	1% SE 30 on Chromosorb W	180°C.	4.4		97
	1% SE 30 on Chromosorb W	220°C.	1.8		97
p-dimethyl-aminobenzaldehyde derivative	2.5% SE 30 on Chromosorb G	195°C.	n.a.		94

n.a. = information not available

7. References
1. O. R. Samuel, W. L. Brannen and O. L. Hayden, *J. Ass. Offic. Anal. Chem.*, 47, 918 (1964).
2. R. K. Maddock, Jr., and H. A. Bloomer, *Clin. Chem.*, 13, 333 (1967).
3. I. Sunshine and S. R. Gerber, "Spectrophotometric Analysis of Drugs Including Atlas of Spectra", Charles C. Thomas Co., Springfield, Illinois, 1963, p. 167.
4. L. J. Bellamy, "The Infrared Spectra of Complex Molecules", 2nd ed., John Wiley and Sons, Inc., New York, N.Y., 1964, Ch. 2.
5. Ibid, Ch. 12.
6. Personal communication - B. Hofmann, Wyeth Laboratories, Inc.
7. J. W. Turczan and T. C. Kram, *J. Pharm. Sci.*, 56, 1643 (1967).
8. Personal communication - S. Shrader and C. Kuhlman, Wyeth Laboratories, Inc.
9. F. W. McLafferty, "Interpretation of Mass Spectra", W. A. Benjamin, Inc., New York, N.Y., (1969), pp 137-138.
10. W. G. Penprase and J. A. Biles, *J. Am. Pharm. Ass. Sci. Ed.*, 47, 523 (1958).
11. P. G. Stecher, Ed., "The Merck Index," Eighth Edition, Merck and Co., Inc., Rahway, New Jersey, (1968), p 657.
12. F. M. Berger, *J. Pharmacol. Exp. Ther.*, 112, 413 (1954).
13. F. M. Berger and B. J. Ludwig, U.S. Patent 2,724,720 (1955).
14. B. J. Ludwig and E. C. Piech, *J. Am. Chem. Soc.*, 73, 5779 (1951).
15. Personal communication - N. DeAngelis, Wyeth Laboratories, Inc.
16. Unpublished results, P. Rulon, Wyeth Laboratories, Inc.
17. Personal communication - S. Young, Wyeth Laboratories, Inc.
18. P. Rajeswaren and P. L. Kirk, *Bull. Narcotics, U.N., Dept. Social Affairs*, 14, 19 (1962).
19. S. Lepitit, Brit. Patent 797,494 (1958); *Chem. Abstr.*, 53, 4131i (1959).
20. P. Nantka-Namirsik, et. al., Pol. Patent 42,591 (1960); *Chem. Abstr.*, 55, 20964g (1961).
21. G. Ferrari, Chim. Ind., (Milan), 40, 13 (1958); *Chem. Abstr.*, 52, 10875c (1958).

22. S. Raymahasay, Indian Patent 67,983 (1960); Chem. Abstr., 55, 15356a (1961).
23. G. Ghielmetti, Farmaco Ed. Sci., 11, 1014 (1956); Chem. Abstr., 56, 1284b (1962).
24. S. Bienfest, P. Adams and J. Halpern, U.S. Patent 2,837,560 (1958); Chem. Abstr., 52, 17112a (1958).
25. Fabrica Espanola de Productos Quimicos y Farmaceutics S.A., Span. Patent 233,310 (1957); Chem. Abstr., 52, 1556g (1958).
26. P. E. Carlo in "Psychotropic Drugs", S. Garattini and V. Ghetti, Ed., Elsevier Press, Amsterdam, (1957), p 392.
27. K. A. Connors in "Pharmaceutical Analysis", T. Higuchi and E. Brochmann-Hanssen, Ed., Interscience Publishers, New York, N.Y., (1961), Ch. 6.
28. G. Devaux, P. Mesnard and J. Cren, Prod. Pharm., 18, 221 (1963).
29. B. J. Ludwig, J. F. Douglas, L. S. Dowell, M. Meyer, and F. M. Berger, J. Med. Pharm. Chem., 3, 53 (1961).
30. B. W. Agranoff, R. M. Bradley and J. Axelrod, Proc. Soc. Exp. Biol. Med., 96, 261 (1957).
31. H. Tsukamato, H. Yoshimura and K. Tatsumi, Chem. Pharm. Bull., (Tokyo), 11, 421 (1963).
32. H. Tsukamato, H. Yoshimuro and K. Tatsumi, Life Sciences, 2, 382 (1963); Chem. Abstr., 60, 4638g (1964).
33. A. Yamamoto, H. Yoshimura and H. Tsukamato, Chem. Pharm. Bull., (Tokyo), 10, 522 (1962).
34. P. M. Castilla and Ch. B. Leon, Anales Real Acad. Farm., 24, 31 (1963); Chem. Abstr., 60, 1541a (1964).
35. Wm. R. Maynard, Jr., J. Ass. Offic. Agr. Chem., 43, 791 (1960).
36. S. Sherken, J. Ass. Offic. Agr. Chem., 51, 616 (1968)
37. G. Devaux, P. Mesnard, and J. Cren, Bull. Soc. Pharm. Bordeaux, 100, 231 (1961); Chem. Abstr., 61, 17131f (1964).
38. G. Bors, L. Armasescu, I. Ionescu, and N. Ioanid, Farmacia, (Bucharest), 12, 607 (1964); Chem. Abstr., 62, 4312g (1965).
39. L. F. Cullen, L. J. Heckman and G. J. Papariello, J. Pharm. Sci., 58, 1537 (1969).
40. S. Walkenstein, C. Knebel, J. A. Macmullen, J. Pharmacol. Exp. Ther., 123, 254 (1958).
41. M. Nedergaard, Farm. Revy, 65, 622 (1966); Chem. Abstr., 66, 22286n (1967).

42. G. H. Ellis and C. A. Hetzel, Anal. Chem., 31, 1090 (1959).
43. E. S. Harris and J. J. Reik, Clin. Chem., 4, 241 (1958).
44. G. Cau, A. Boucherle, M. Yacoub, and J. Faure, Ann. Biol. Clin. (Paris), 22, 1133 (1964); Chem. Abstr., 62, 11000a. (1965).
45. H. S. Bedson, Lancet, 1, 288 (1959).
46. S. Salveson and R. Nissen-Meyer, Acta Endocrinol. (Copenhagen), 29, 224 (1958).
47. S. Salvesen and R. Nissen-Mayer, J. Clin. Endocrinol Metab. 17, 914 (1957).
48. A. Heyndrickx, M. Schauvliege, and A. Blomme, J. Pharm. Belg., 20, 117 (1965); Chem. Abstr., 63, 11256b (1965).
49. G. Lagrange and J. J. Thomas, J. Pharm. Belg., 13, 402 (1958); Chem. Abstr., 53, 10663b (1959).
50. B. J. Ludwig and A. J. Hoffman, Arch. Biochem. Biophys., 72, 234 (1957).
51. M. Shimizu and S. Ichimura, Yakugaku Zasshi, 78, 1183 (1958); Chem. Abstr., 53, 3602f (1959).
52. M. Chambon, Therapie, 14, 771 (1959): Chem. Abstr., 55, 24901c (1961).
53. A. J. Hoffman and B. J. Ludwig, J. Am. Pharm. Assoc. Sci. Ed., 48, 740 (1959).
54. O. D. Madsen, Clin. Chim. Acta, 7, 481 (1962).
55. G. A. Ponomarev and A. I. Terekhina, Farmakol. Toksikol., 27, 238 (1964); Chem. Abstr., 61, 15190a (1964).
56. R. Bourdon and A. M. Nicaise, Ann. Biol. Clin. (Paris), 26, 897 (1968); Chem. Abstr., 70, 10072j (1969).
57. R. Tulus and Y. Aydogan, Istanbul Univ. Eczacilik Fak. Mecmuasi, 3, 181 (1967); Chem. Abstr., 69, 5264y (1968)
58. I. Hynie and J. Prokes, Chem. Zvesti, 18, 425 (1964) Chem. Abstr., 61, 15190a (1964).
59. I. Hynie, J. Prokes, and K. Kacl, Casopis Lekaru Ceskych, 103, 412 (1964); Chem. Abstr., 63, 4823c (1965).
60. R. Kalvoda, Collect. Czeck. Chem. Commun., 34, 1076 (1969).
61. "Pharmacopeia of the United States of America," 18th Revision, Mack Publishing Co., Easton, Penna. (1970), p 401.
62. O. Cerri and A. Spialtini, Boll. Chim. Farm., 97, 259 (1958); Chem. Abstr., 52, 17619f (1958).

63. O. Cerri, A. Spialtini, and U. Ga lo, Pharm. Acta Helv., 34, 13 (1959); Chem. Abstr., 53, 14413h (1959).
64. O. Laurent, J. Pharm. Belg., 21, 589 (1966); Chem. Abstr., 66, 7965n (1967).
65. B. Salvesen and O. Solli, Med. Norsk. Farm. Selskap 21, 85 (1959); Chem. Abstr., 53, 22732a (1959).
66. I. Solomen-Ionesch, D. Popescu and H. Nicola, Farmacia (Bucharest), 10, 627 (1962); Chem. Abstr., 58, 6648c (1962).
67. K. Boichinov and G. Cholakova, Farmatsiya (Sofia), 16, 11 (1966); Chem. Abstr., 65, 18429g (1966).
68. H. P. Kloecking, Pharmazie, 16, 604 (1961).
69. A. C. Maehly and M. K. Linturi, Acta Chem. Scand., 16, 283 (1962).
70. A. Dressler, Deut. Z. ges. gerichtl. Med., 50, 457 (1960); Chem. Abstr., 55, 11761c (1961).
71. J. F. Douglas and A. Schlosser, J. Chromatogr., 6, 540 (1961).
72. L. J. Roth, K. E. Wilzbach, A. Heller, and L. Kaplan, J. of Amer. Pharm. Assoc. Sci. Ed., 48, 415 (1959).
73. A. Viala and R. Monnet, Trav. Soc. Pharm. Montpellier, 18, 79 (1958); Chem. Abstr., 53, 8534h (1959).
74. I. Hynie, J. Konig. and K. Kacl, J. Chromatogr., 19, 192 (1965).
75. L. Marques de Sa and A. Flori, Anais Farm. Quim. Sao Paulo 14, 155 (1963); Chem. Abstr., 61, 7332a (1964).
76. K. Sato, Kagaku Keisatsu Kenkyusho Hokoku, 15, 237 (1962); Chem. Abstr., 57, 16986d (1962).
77. H. N. Rydon and P. W. G. Smith, Nature, 169, 922 (1952).
78. M. S. Moss and J. V. Jackson, J. Pharm. Pharmacol., 13, 361 (1961).
79. J. L. Emmerson and T. S. Miya, Anal. Chem., 31, 2104 (1959).
80. S. C. Pan and J. D. Dutcher, Anal. Chem., 28, 836 (1956).
81. T. W. McConnell Davis, J. Chromatogr., 29, 283 (1967).
82. I. Zingales, J. Chromatogr., 31, 405 (1967).
83. R. Lindfors, Ann. Med. Exp. Biol. Fenniae, (Helsinki), 41, 355 (1963); Chem. Abstr., 60, 12346b. (1964).
84. K. Tsutsui, T. Sakai, and M. Takahashi, Yakuzaigaku, 25, 53 (1965); Chem. Abstr., 63, 17802a (1965).

85. M. Tomoda, Kyritsu Yokka Daigaku Kenkyu Nempo, 10, 18 (1965); Chem. Abstr., 66, 98525k (1967).
86. O. V. Olesen, Acta Pharmacol. Toxicol., 24, 183 (1966); Chem. Abstr., 64, 4103e (1966).
87. M. Marigo, Arch. Kriminol., 128, 99 (1961); Chem. Abstr., 56, 5068i (1962).
88. J. C. Morrison and J. M. Orr, J. Pharm. Sci., 35, 36 (1966).
89. T. Fuwa, T. Kido and H. Tanaka, Yakuzaigaku, 25, 138 (1965); Chem. Abstr., 64, 6405f (1966).
90. J. Koenig, B. Chundela and M. Sulcova, Bratislav. Lek. Listy, 48, 655 (1967); Chem. Abstr., 68, 48046a (1968).
91. G. LeMoan, R. Rouillac, C. Jolivet and A. Nidrecourt, Ann. Fals. Expert Chim., 62, 207 (1969); Chem. Abstr., 72, 125090x (1970).
92. O. Cerri, Boll. Chim. Farm., 108, 217 (1969); Chem. Abstr., 71, 5363k (1969).
93. R. F. Skinner, J. Forensic Sci., 12, 230 (1967); Chem. Abstr., 67, 105848u (1967).
94. B. S. Finkle, J. Forensic Sci., 12, 509 (1967); Chem. Abstr., 68, 36574h (1968).
95. J. F. Douglas, T. K. Kelley, N. B. Smith and J. A. Stockage, Anal. Chem., 39, 956 (1967).
96. N. C. Jain and P. L. Kirk, Microchem. J., 12, 256 (1967).
97. L. Kazyak and E. C. Knoblock, Anal. Chem., 35, 1448 (1963).
98. J. L. Hamilton, J. Ass. Offic. Anal. Chem., 53, 594 (1970).
99. E. Kalinowska and L. Majumia, Farm. Pol., 22, 406 (1966); Chem. Abstr., 66, 31989w (1967).

NORTRIPTYLINE HYDROCHLORIDE

J. L. Hale

CONTENTS

1. Description
   1.1 Name, Formula, Molecular Weight
   1.2 Appearance, Color, Odor
2. Physical Properties
   2.1 Infrared Spectrum
   2.2 Nuclear Magnetic Resonance Spectrum
   2.3 Ultraviolet Spectrum
   2.4 Mass Spectrum
   2.5 X-ray Powder Diffraction
   2.6 Differential Thermal Analysis
   2.7 Thermogravimetric Analysis
   2.8 Melting Range
   2.9 Solubility
3. Synthesis
4. Stability - Degradation
5. Drug Metabolic Products
6. Methods of Analysis
   6.1 Elemental Analysis
   6.2 Ultraviolet Spectrophotometric Analysis
   6.3 Colorimetric Analysis
   6.4 Qualitative Analyses
   6.5 Isotope Derivative Analysis
   6.6 Chromatographic Methods of Analysis
      6.61 Thin Layer
      6.62 Paper
      6.63 Gas-Liquid
7. Pharmacokinetics
8. References

1. Description

1.1 Name, Formula, Molecular Weight

Nortriptyline Hydrochloride is 10,11-dihydro-N-methyl-5H-dibenzo [a,d]-cycloheptene- Δ^5, γ-propylamine hydrochloride[1]. It is also known as 5H-dibenzo [a,d] cycloheptene-Δ^5, γ-propylamine, 10,11-dihydro-N-methyl hydrochloride[2]. It is known as 5-(α-methylamino propylidene)-dibenzo [a,d] cyclohepta[1,4] diene hydrochloride and 3-(10,11-dihydro-5H-dibenzo[a,d] cyclophetene-5-ylidene)-N-methyl propylamine hydrochloride[3]. Common names for the compound are desitriptylina and desmethylamitriptyline hydrochloride[3].

$C_{19}H_{21}N \cdot HCl$ Mol. Wt.: 299.85

1.2 Appearance, Color, Odor

The compound is a white to off-white powder having a slight characteristic odor[1].

2. Physical Properties

2.1 Infrared Spectrum

The spectrum in Figure No. 1 was obtained using a Perkin-Elmer 221, Infrared Spectrophotometer[4]. A 13 mm. KBr pellet was used. The max. at 2930 cm^{-1} is typical of C-H stretch (CH_2, CH_3). The band at 2440 cm^{-1} is assigned to the secondary amine hydrochloride group. The band at 1493 cm^{-1} is due to CH_2 bend, and the bands at 720-770 cm^{-1} are a mixture of ethylene and aromatic deformations.

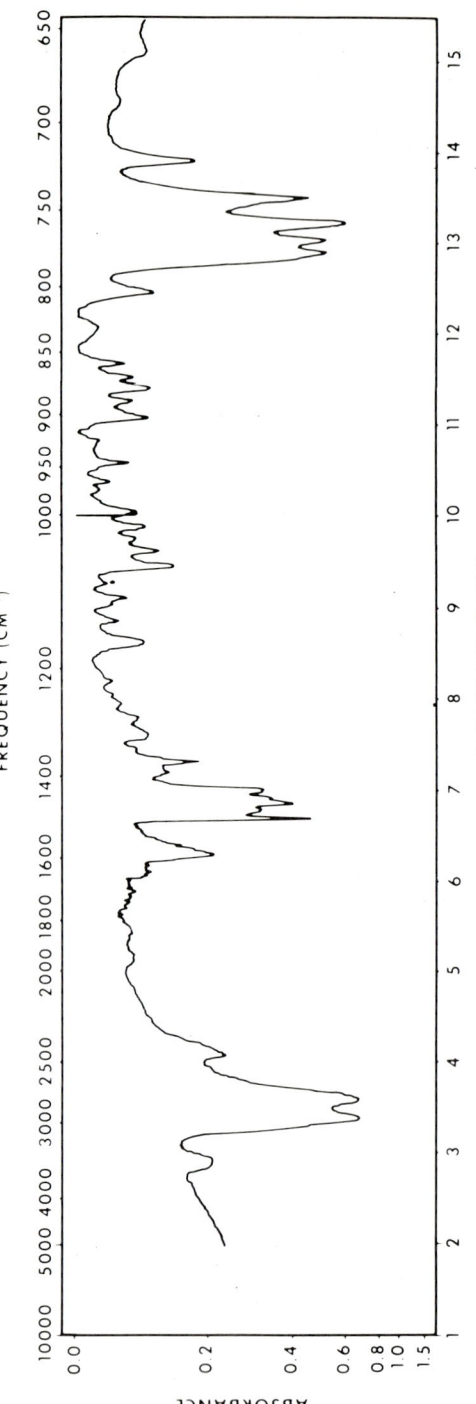

Fig. 1. Infrared spectrum of nortriptyline hydrochloride; instrument: Perkin-Elmer 221

2.2 Nuclear Magnetic Resonance Spectrum

The NMR spectrum in Figure No. 2 was obtained using a Varian A-60 instrument[4]. The solvent was deuterated chloroform at ambient temperature. The singlet at 2.46δ is typical of -NCH$_3$, the multiplet at 2.5-3.3δ is due to methylene multiplets, the triplet at 5.80δ is due to the vinyl hydrogen, and the multiplet at 7.0-7.5δ is typical of aromatic hydrogens.

2.3 Ultraviolet Spectrum

The λ max. at 239 nm is typical of styrene chromophores which have undergone a wavelength shift and loss of intensity due to lack of coplanarity among the components of the chromophore[4]. The solvent used was 95% alcohol, and a molar absorptivity value of 12,300 was obtained. The instrument used was a Cary 14 Spectrophotometer.

2.4 Mass Spectrum

The low resolution mass spectrum when run on a Hitachi, Perkin-Elmer RMU6 Mass Spectrometer gives two characteristic m/e peaks: 263, the M$^+$ of the free base, and 220, which is due to [M-(CH$_2$=NCH$_3$)][7,4].

2.5 X-ray Powder Diffraction

The pattern was obtained using Cu with a Ni filter at λ = 1.5405. Readings and relative intensities are tabulated in Table I taken from ASTM 17-1197.

2.6 Differential Thermal Analysis

The DTA was done using a DuPont 900 Different Thermal Analyzer. A sharp melting endotherm was observed at 217°C.[4]

2.7 Thermogravimetric Analysis

The TGA when done with a DuPont 950 instrument resulted in 1% weight loss at 207°C.[4]

2.8 Melting Range

Early reports of the melting range gave values of 210-212°C.[5] Later reports were given as 216-219°C.[6]

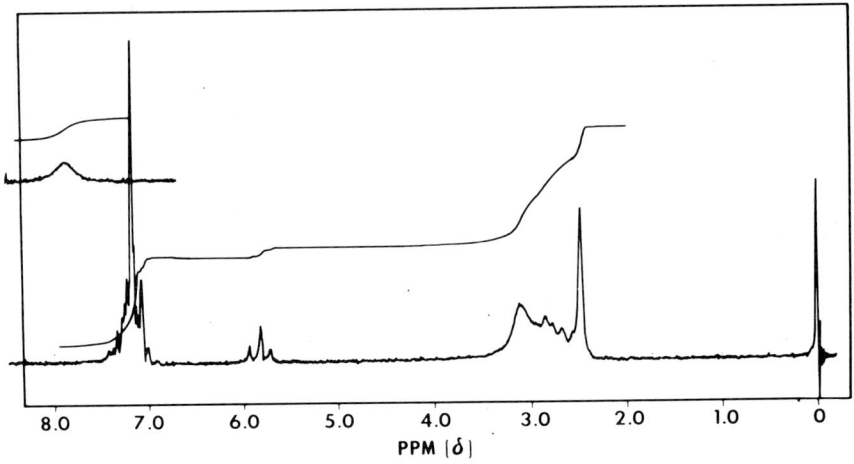

Fig. 2. NMR spectrum of nortriptyline hydrochloride; instrument: Varian A-60

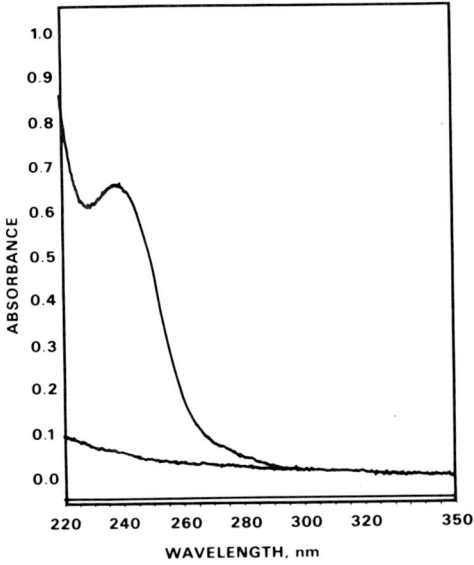

Fig. 3. Ultraviolet spectrum of nortriptyline hydrochloride; instrument: Cary 14

TABLE I

d	I/I_0	d	I/I_0
10.21*	5	2.73	10
9.20	40	2.64	5
8.14	30	2.49	15
6.00	5	2.41	10
5.67	50	2.26	5b
5.13	20	2.19	5
4.94	20	2.16	10
4.60	10b	2.11	5
4.34	70	2.08	5
4.12	15	2.01	5
3.93	15	1.93	5
3.68	10		
3.58	00		
3.51	15		
3.29	15		
3.19	10		
3.04	15		
2.90	5		
2.81	5		

*This value is erroneously reported as 12.9 in ASTM.

2.9 <u>Solubility</u>
The following solubility data[7] were obtained at room temperature:
>water - 11 mg./ml.
>95% ethanol - 33 mg./ml.
>chloroform - 50 mg./ml.

methanol - 100 mg./ml.
Practically insoluble in ether, acetone, and benzene[8].

3. <u>Synthesis</u>

Northriptyline Hydrochloride is prepared by the reaction of 5-oxodibenzo [a,d] cyclohepta-1,4-diene with the sodium derivate of N-methylpropargylamine, followed by hydrogenation and then dehydration[8]. A general scheme[5] follows:

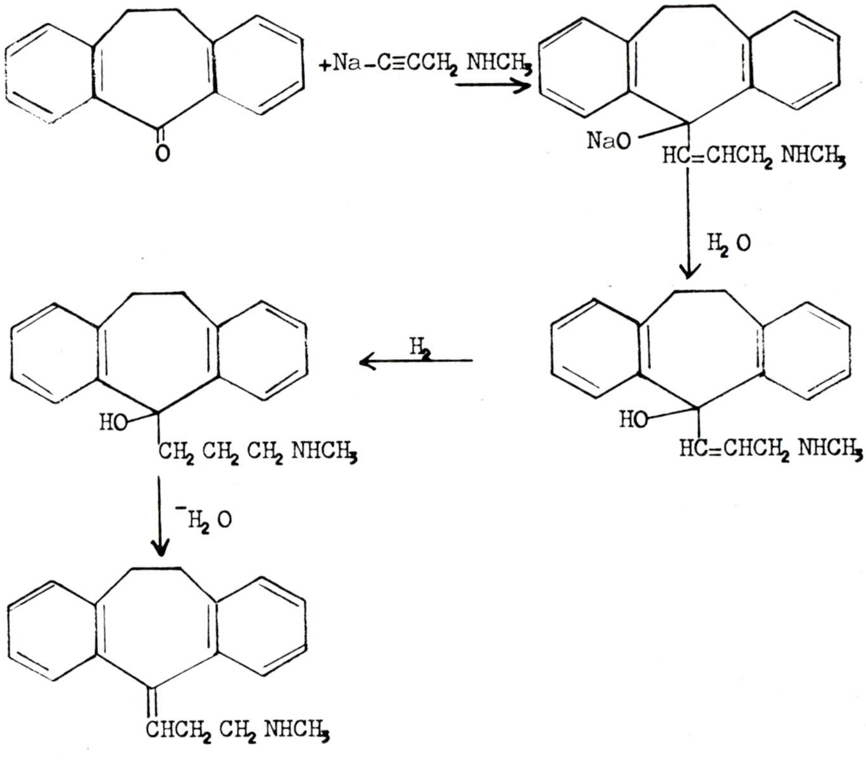

Peters and Hennion reported the method using propargylic intermediates[6]. An alternate synthesis of this type of compound is by hydroboration of 5-allylidene-5H-dibenzo [a,d]-10,11-dihydrocycloheptene[9].

4. Stability - Degradation

Nortriptyline hydrochloride is quite stable as a solid and in solution. Basic, aqueous solutions may be heated for as long as 16 hours on a steam bath without appreciably degrading the base[10]. Degradation takes place in the presence of strong oxidizing agents such as hydrogen peroxide and is evident by the increasing absorbance at 239nm[10]. Irradiation under ultraviolet light in aqueous solutuion results in approximately 50% degradation after 16 hours, and is evidenced by the appearance of a brownish-yellow color of the solution and precipitate[10]. Degradation products in the foregoing have not been identified.

5. Drug Metabolic Products

Four metabolites have been found in the urine of man in addition to small amounts of the unchanged drug. Amundson and Manthey separated the metabolites by extraction of the urine and then thin layer chromatography[11]. They identified two of the metabolites as 10-hydroxy nortriptyline and the conjugated derivative formed when the dimethylene bridge is replaced by a double bond. McMahon, et al, investigated the metabolism of nortriptyline-N-methyl[14]C in rats and found that N-demethylation and hydroxylation of one of the bridgehead carbons were the major metabolic changes observed[12]. They found that tissue levels of radiocarbon in rats one-half day after administration of radionortriptyline were highest in liver followed in order by kidney, lung, plasma, blood, and brain. It is of interest to note that nortriptyline is a metabolic product of amitriptyline, the N,N-dimethyl analog[13].

6. Methods Of Analysis

6.1 Elemental Analysis (as $C_{19}H_{22}NCl$)

Element	%Theory	Reported[9]
C	76.10	75.61
H	7.39	7.05
N	4.67	4.38

6.2 Ultraviolet Spectrophotometric Analysis

The λ max. for nortriptyline hydrochloride is at approximately 239 nm, and an absorbance of 0.482 is reported for a 10 mcg./ml. solution of the drug in 0.1N hydrochloric acid[11]. The UV absorbance of the compound is utilized in the N.F. XIII assay for capsule formulations. The UV spectrum is also useful as a means of differentiating spectrophotometrically between nortriptyline and its metabolites[11].

6.3 Colorimetric Analysis

The copper sulfate-carbon disulfide method has been used for the quantitation of nortriptyline, an amitriptyline metabolite, in rabbit's urine[14]. The assay is advantageous in that it is specific for secondary amines. The analate is reacted with $CuSO_4$ and CS_2 giving a yellow to dark-brown colored salt which is then extracted into benzene. The linear portion of the Beer's Law curve is in the 1 to 8 mcg./ml. range.

6.4 Qualitative Analyses (Other than spectrophotometric)

6.41 Precipitate the amine base from a solution of 25 mg. of nortriptyline hydrochloride in 5 ml. of water by addition of dilute ammonia water and then filter. Acidify the filtrate with nitric acid and add excess dilute silver nitrate. The precipitate is insoluble in excess nitric acid and soluble in diluted ammonia solution[8]. This confirms the presence of chloride.

6.42 A solution of 5 mg. of nortriptyline hydrochloride in 2 ml. of sulfuric acid produces a reddish-orange color. The color disappears upon addition of 10 ml. of water[8].

6.5 Isotope Derivative Analysis

This method was developed for secondary amines by Hammer and Brodie[15] and was applied to nortriptyline by Sjoqvist et al.[16] The drug is extracted initially into hexane which excludes interference from polar metabolites; it is then acetylated with H^3-acetic anhydride. The hexane is evaporated, and the excess acetic anhydride is hydrolyzed with 0.1N NaOH. The radioactive amide is extracted into heptane and quantitated by scintillation spectrometry. Specificity is checked by thin-layer chromatography.

6.6 Chromatographic Methods of Analysis
6.61 Thin Layer
(See Attached Chart)

6.62 Paper

Developer	Detection	Rf	Ref.
15% BuOH-HC$_{0\,2}$H-H$_2$O(12:1:7)	---	0.74	19
BuOH satd. with NH$_3$	---	0.94	19

6.63 Gas-Liquid

Street carried out the gas-liquid phase chromatography using a Perkin-Elmer Model 800 or an F & M Model 810^{20}. Injector temperature was about 50° above the column temperature of 235° and detector temperature was about the same as the column temperature. Oxygen-free nitrogen was the carrier gas at flow rates of 50-60ml./min. The column was a 6 ft. length of stainless steel tubing 1/8 in. o.d. and 0.085 in. i.d. Column packing was SE30 on Chromosorb W(100-120 mesh). Retention time at 235° is given as 3.7 minutes, and the minimum detectable amount at attenuation X20 is 0.2 µg in 1 µl of ethanol.

7. Pharmacokinetics

Recent studies by Sjoqvist et al. indicate that there are large individual differences in the steady state plasma levels of nortriptyline in human beings[21]. The time to reach steady state has been found to vary from

Thin Layer Chromatographic Analyses

Adsorbent	Developer	Visualization	Rf	Ref.
Silica Gel G	Ethyl Ac-MeOH-NH_4OH (170:20:10)	Iodoplatinate-Dragendorff's reagent	0.75	17
Silica Gel G	MeOH-12N NH_4OH (100:1.5)	Folin-Ciocalteu reagent	0.31	18
Silica Gel G (0.1N NaOH Slurry)	Cyclohexane-diethylamine-benzene (75:20:15)	Folin-Ciocalteu reagent	0.50	18
Silica Gel G (0.1N NaOH Slurry)	$CHCl_3$-MeOH (90:10)	Folin-Ciocalteu reagent	0.19	18
Silica Gel G	Benzene-EtOH-12N NH_4OH (95:15:5)	Folin-Ciocalteu reagent	0.51	18
Silica Gel G	MeOH-$CHCl_3$ (1:2)	----	0.15	19

Thin Layer Chromatographic Analyses (Continued)

Adsorbent	Developer	Visualization	Rf	Ref.
Silica Gel G	Isopropanol-H_2O (1/3 sat. with NaCl) (88:12)	EtOH-H_2SO_4 (1:1) Heat	----	11
Silica Gel G	$CHCl_3$-Isopropanol-5%NH_4OH (74.4:25:0.6)	Dragendorff's Reagent	----	14

245

5 to 8 days[22]. The steady state was found to be reproducible in the same individual, and the drug level was found to be directly proportional to the administered daily dose which is an indication of a constant apparent volume over the studied dose range[21]. Only small amounts of the unchanged drug were found in the feces (less than 10 mcg./g), and renal elimination was negligible even at high plasma concentrations[21]. Relatively small amounts were found in the gastric juice of dogs after intravenous injection[16]. Half-life of the drug apparently depends again on the individual, i.e., individuals with highest plasma levels had the longest half-life[21]. One interesting point was that females maintained a higher plasma level than males[21]. Nortriptyline, an amitriptyline metabolite, was found in relatively high concentrations in liver tissue after four fatal, human poisonings with amitriptyline[23]. Rat studies also indicate that liver tissue absorbs a large portion of the drug, and that in order of decreasing concentration of nortriptyline or its metabolites, the various absorbing tissues are liver, adrenal, kidney, lungs and brain[12]. Little work has been done on the construction of mathematical models of the pharmacokinetic parameters of nortriptyline; however, some work has been done by Wolfgang and Brodie on desipramine, a similar, monomethylated, tricyclic antidepressant[15]. Similar techniques may be applicable to nortriptyline.

REFERENCES

1. N.F. XIII, 490-491 (1970).
2. Chem. Abstr., Subject Index.
3. The Merck Index, 750, Eighth Edition, Merck and Co., Inc., Rahway, N.J. (1968).
4. A. D. Kossoy and C. D. Underbrink, Eli Lilly and Company, Interpretation of spectra, TGA, and DTA.
5. Eli Lilly and Company, Belgium Patent Application, 628, 904 (1962).
6. L. R. Peters and G. F. Hennion, J. Med. Chem. 7, 390-392 (1964).
7. R. W. Shaffer, Eli Lilly and Company, personal communication.
8. Drugs of Today, 1, 25 (1965).
9. R. D. Hoffsommer, D. Taub, and N. L. Wendler, J. Org. Chem., 27, 4134-4137 (1962).
10. Analytical Development Dept. Eli Lilly and Company.
11. M. E. Amundson and J. A. Manthey, J. Pharm. Sci. 55, 277-280 (1966).
12. R. E. McMahon, F. J. Marshall, H. W. Culp, and W. M. Miller, Biochem. Pharm. 12, 1207-1217 (1963).
13. H. B. Hucker, Pharmacologist 4, 171 (1962).
14. G. L. Corona and R. M. Facino, Biochem. Pharm. 17, 2045-2050 (1968).
15. W. M. Hammer and B. B. Brodie, J. Pharmacol. and Exper. Therap. 157, 503-508 (1967).
16. F Sjoqvust, F. Berglund, O. Borga, W. Hammer, S. Andersson and C. Thorstrand, Clin. Pharm. and Therap.
17. B. Davidow, N. L. Petri, B. Quame, Amer. J. Of Clin. Path. 50, 714-719 (1968).
18. I. Zingales, J. Chrom. 34, 44-51 (1968).
19. Chem. Abstracts 65, 14283C (1966).
20. H. V. Street, J. Chromatog. 29, 68-79 (1967).
21. F. Sjoqvist, W. Hammer, C. M. Idestrom, M. Lind, D. Tuck, and M. Asberg, Proceedings of European Society Study of Drug Toxicity, Paris, 1967, Excerpt Med. Intermed. Congr., Ser. No. 145, pp. 246-257 (1968).
22. W. Hammer and F. Sjoqvist, Life Sci. 6, 1895-1903 (1967).
23. E. C. Munksgaard, Acta Pharmacol. et toxicol. 27, 129-134 (1969).

POTASSIUM PHENOXYMETHYL PENICILLIN

John M. Dunham

TABLE OF CONTENTS

1. Description
    1.1. Name, Formula, Appearance
    1.2. Definition of International Unit
2. Physical Properties
    2.1. Spectra
        2.11. Infrared Spectra
        2.12. Nuclear Magnetic Resonance Spectra
        2.13. Ultraviolet Spectra
        2.14. Mass Spectrometry
    2.2. Crystal Properties
        2.21. Change from Amorphous to Crystalline Form
        2.22. Differential Thermal Analysis
        2.23. Melting Range
    2.3. Solubility
    2.4. Ionization Constant, pKa
    2.5. Optical Rotation
3. Penicillin Degradation
    3.1. Modes of Penicillin Degradation
    3.2. Degradation of Phenoxymethyl Penicillin
        3.21. Degradation of crystalline Phenoxymethyl Penicillin
        3.22. Degradation in Solution
    3.3. Enzymatic Degradation
    3.4. Degradation of Frozen Systems
    3.5. Cupric Ion Hydrolysis
4. Synthesis
5. Purification and Analysis of Impurities
    5.1. Purification
    5.2. Countercurrent Distribution - Solvent Partitioning
    5.3. Assay Methods for Intermediates and Impurities
        5.31. Phenoxyacetic Acid
        5.32. Penicilloic Acid
        5.33. p-Hydroxyphenoxymethyl Penicillin
6. Methods of Analysis
    6.1. Identification Tests
    6.2. Ultraviolet Methods

6.3. Titrations
    6.31. Iodometric
    6.32. Nonaqueous
    6.33. p-Chloromercuribenzoate
6.4. Colorimetric Methods
    6.41. **Hydroxamic Acid**
    6.42. Dye Complex
    6.43. Nitration
    6.44. Folin Phenol Reagent
6.5. Chromatographic Analysis
    6.51. Paper
    6.52. Thin Layer
    6.53. Ion Exchange
    6.54. Gas Liquid
6.6. Electrophoretic Analysis
6.7. Polarography
6.8. Microbiological Methods
7. Serum Protein Binding
8. Drug Metabolism
9. References

1. Description

   1.1. Name, Formula, Appearance
   Potassium Phenoxymethyl Penicillin (Potassium Penicillin V)

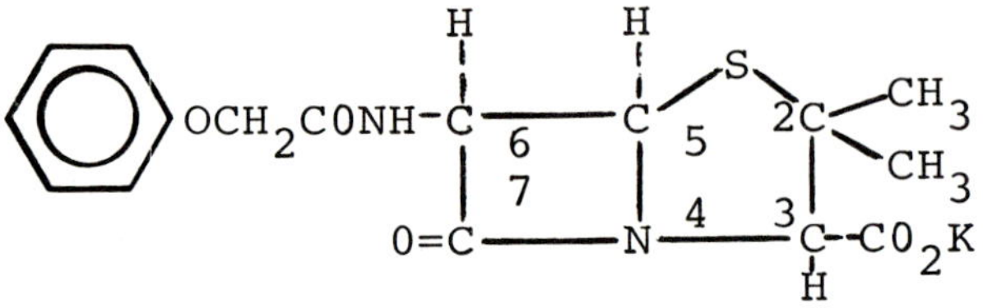

   $C_{16}H_{17}KN_2O_5S$ Molecular Weight = 388.49

Potassium phenoxymethyl penicillin is a white, practically odorless, crystalline powder.

   1.2. Definition of International Unit
   The International Standard for phenoxymethyl penicillin (free acid) is defined as having a potency of 1695 International Units per mg, or the International Unit is defined as the activity contained in 0.000590 mg of the International Standard[1].
   One mg. of potassium phenoxymethyl penicillin represents 1530 phenoxymethyl penicillin units[2].

2. Physical Properties

   2.1 Spectra

   2.11. Infrared Spectra
   The infrared spectrum of phenoxymethyl penicillin (free acid) is number 88 in Hayden's compilation of spectra measured on a Perkin-Elmer Model 21 sodium chloride prism spectrophotometer[3]. Infrared spectra of fourteen penicillins in the 1580 to 1880 cm^{-1} region are discussed by Ovechkin[4]. Figure 1 is the spectrum of the Squibb Primary Reference Substance of potassium phenoxymethyl penicillin recorded as a potassium bromide

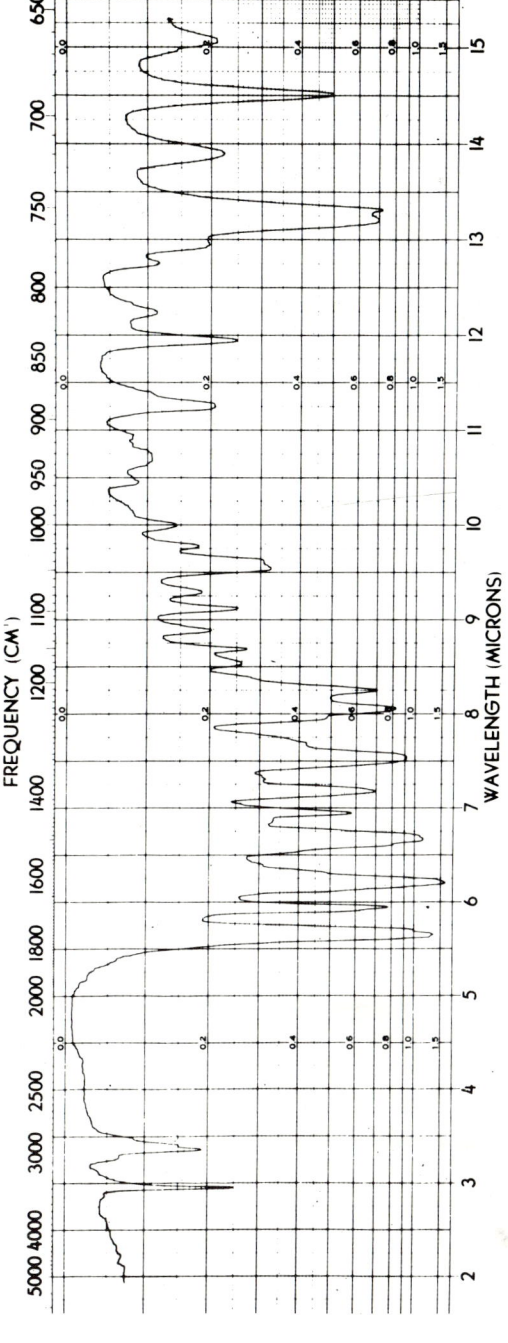

Figure 1. Infrared Spectrum. Potassium Phenoxymethyl Penicillin Squibb Primary Reference Substance 1.5 mg/300 mg KBr

pellet with a Perkin-Elmer Model 21 spectrophotometer.

2.12. Nuclear Magnetic Resonance Spectra

Structural assignments have been made for each peak in the proton magnetic resonance spectrum of sodium phenoxymethyl penicillin (fourth line of Table 5 in reference 5) by Green and co-workers. Proton resonance lines were measured in D_2O solution at 38°, with either t-butyl alcohol or sodium 3-(trimethylsilyl)propane-1-sulfonate (assumed to be 2 cycles per second below Me_4Si) as the internal standard, with the results referred to tetramethylsilane ($Me_4Si=10.0\tau$). Cohen and Puar[6] made structural assignments for the peaks of potassium phenoxymethyl penicillin measured in D_2O solution at 33° with the results referred to tetramethylsilane as the external standard ($Me_4Si=10.0\tau$). Figure 2 is the spectrum of the Squibb Primary Reference Substance from which proton assignments were made[6]. Spectra in both laboratories were measured on Varian A-60 spectrophotometers.

Assignment	Chemical Shift, τ (Coupling Constant, Hz)	
	Cohen and Puar[6]	Green et al[5]
2-Me_2	8.54	8.50
3-H	5.78	5.74
5-H	-	4.50d (J=4.0)
6-H	-	4.44d (J=4.0)
5-H, 6-H (AB quartet)	4.50 q (J=4.0)	-
Phenoxymethyl	-	2.60-3.72m, 5.48
Phenoxy	2.50-3.2m	-
-OCH_2-	5.48	-

d = doublet, m = multiplet, q = quartet

There is good correlation between the penicillin V assignments, taking into account differences in

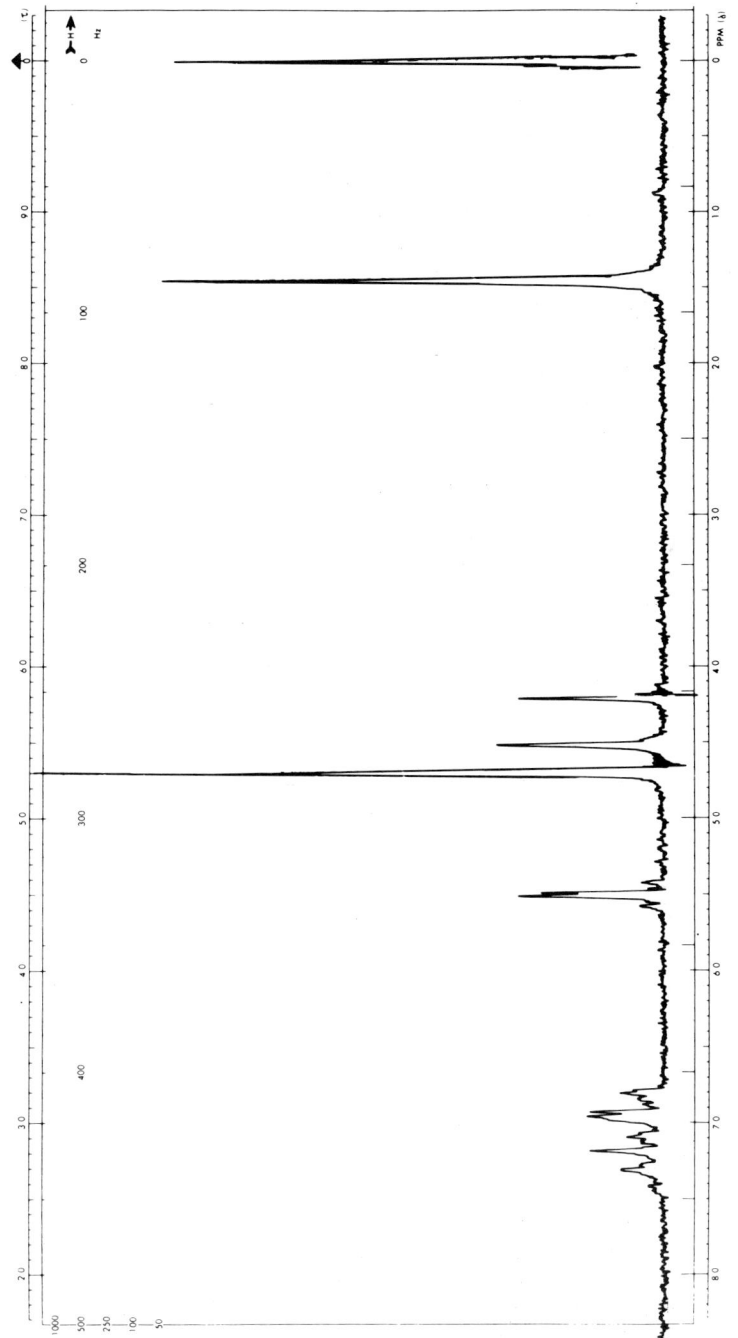

Figure 2. NMR Spectrum. Potassium Phenoxymethyl Penicillin Squibb Primary Reference Substance.

calibration, except for Green's upper limit of 3.72τ for the phenoxymethyl assignment. Cohen and Puar[6] find the coupling profile for the aromatic proton resonance of 2.5 to 3.2τ to be consistent with the phenoxy assignment and conclude the upper limit assigned by Green is in error.

Pek and co-workers[7] studied proton N.M.R. spectra for a large number of penicillins, including phenoxymethyl penicillin, at different temperatures. They conclude that intra- or intermolecular complexes are formed. Proton spectra have also been used[8] to assign the cis structure of the protons on C-5 and C-6 and to assign the proper conformation of the thiazolidine ring[9].

Recently, structural studies with ^{13}C nuclear magnetic resonance were reported[10] for phenoxymethyl and other penicillins and the related sulfoxides. ^{13}C chemical shift assignments were made for the different carbon atoms.

2.13. Ultraviolet Spectrum

Parker et. al.[11] in 1955 determined the ultraviolet properties of a highly purified sample of penicillin V (free acid). In water they reported:

λ nm	ε
268 max.	1330
272 min.	
274 max.	1100

Beer's law is obeyed at both maxima up to 0.04 per cent. However, they note the free acid begins to decompose in aqueous solution after only a few hours. Chloroform solutions of penicillin V have a similar spectrum to those in water with the maxima shifted slightly to 270 and 276 nm.

The ultraviolet spectrum of phenoxymethyl penicillin (free acid) in methanol is given as number 88 in Hayden's collection[3]. There are 3 maxima at 263, 268 and 275 nm.

Parker and his coworkers [11] prepared the readily soluble sodium salt of penicillin V in solution from the free acid and a solution of sodium bicarbonate. Its spectrum was very similar to that of the free acid with absorption maxima at 268 and 274 nm. Absorption at each maximum follows Beer's law up to 0.03%. In contrast to the free acid, solutions of the sodium salt showed no detectable changes in absorption intensities after 10 days at room temperature (pH not stated).

Rogers [12, 13] demonstrated that great care must be taken to measure the ultraviolet absorption of penicillin V at a fixed and narrow slit width. For phenoxymethyl penicillin and its salts, he determined the effect of slit width at the 268nm maximum:

	Maximum h (nm) for absorbance error less than		
Per Cent	0.2	1	2
h (nm)	0.7	0.9	1.1

The half width of the absorption band, h, is the range of wavelengths reaching the sample with at least one half the intensity of the nominal wavelength setting of the instrument.

Andersen [14] has also shown concern about the importance of slit width on the ultraviolet spectrum of potassium phenoxymethyl penicillin.

Residual precursor phenoxyacetic acid has been shown to give high absorbance values by Unterman and Mironeanu [15].

2.14. <u>Mass Spectrometry</u>

Biemann [16] in 1964 examined the methyl esters of both penicillin G and penicillin V by high-resolution mass spectrometry. This enabled him to determine the exact mass and thereby the elemental composition of all the fragment ions formed.

In 1967, Russian workers [17] studied the mass spectra of penicillin V derivatives with different substituents on C-3 by placing samples directly in the ion beam. Fragmentation initiated with the rupture of the β-lactam ring. Both rings of the molecular ion M^+ may also rupture.

2.2. <u>Crystal Properties</u>

2.21. <u>Change from Amorphous to Crystalline Form</u>

Mathews and his coworkers [18] described an unusual behavior for samples of freeze-dried potassium benzylpenicillin and potassium phenoxymethyl penicillin. These samples absorbed moisture from the air up to a maximum content of 7-14 per cent. Then the samples suddenly contracted in volume and began to lose weight until they had returned to their original weight.

They conclude the abrupt change at the point of maximum moisture content is a spontaneous crystallization of substantially amorphous material. At 120°, a freeze-dried sample of potassium benzylpenicillin decomposed within 16 hours. After exposure to air, it was stable for 96 hours at this temperature, as was a control sample crystallized from aqueous-butanol. The heat of solution of the freeze-dried sample decreased by a factor of five after air exposure.

This conclusion agrees with early observations that crystalline forms of penicillin were much more stable than amorphous forms [19c, 20]. Florey, et al. reported in 1949 that the alkali salts of the penicillins are hygroscopic in the impure state, but not when crystalline [19a].

2.22. Differential Thermal Analysis

Jacobson [21] recorded differential thermal analysis curves of potassium phenoxymethyl penicillin on a duPont Differential Thermal Analyzer, with a temperature rise of $15°$ per minute. A small endotherm occurs near $225°$. The temperature of the final exotherm is somewhat dependent on the sample examined. A highly purified sample had an exotherm at $265°$.

2.23. Melting Range

The melting range of phenoxymethyl penicillin (free acid) has been reported by a number of investigators:

Range	Reference	Comment
$127-128°$	22	Purified-countercurrent distribution
$119°$-dec.	23	Extensive purification
$124-127°$	24	Kofler Hot Stage
$120-123°$	24	Immersed in silicone oil
$128-132°$	24	Flowing N_2 atmosphere
$118-125°$	24	Sealed tube-N_2 atmosphere

Sheehan reported the melting range for the

potassium salt of penicillin V:

Range	Reference
263°dec.	25
264-265°dec.	26

2.3. Solubility

Solubilities in water of penicillin V and the potassium and calcium salts are respectively 0.06, >75 and 1.4% [27].

Weiss and his associates [28] determined the solubility of 18 antibiotics, including phenoxymethyl penicillin (free acid), in 24 solvents at room temperature (28 ± 4°). They indicate that a high order of accuracy is <u>not</u> likely because of the procedure used.

Solubility of Penicillin V Acid [28]

Solvent	mg/ml	Solvent (>20 mg./ml.)
isooctane	0.037	acetone
cyclohexane	0.08	benzyl alcohol
carbon tetrachloride	0.097	chloroform
		ethanol
petroleum ether	0.245	ethyl acetate
toluene	0.26	ethylene glycol
carbon disulfide	0.30	formamide
benzene	0.45	isoamyl acetate
water	0.90	isopropyl alcohol
diethyl ether	11.75	methanol
1,4-dioxane	12.60	methyl ethyl ketone
ethylene chloride	12.65	pyridine
isoamyl alcohol	16.95	

2.4. Ionization Constant, pKa

Rapson and Bird [29] determined the "apparent" ionization constant Ka for phenoxymethyl penicillin by titrating the salt with 0.4M hydrochloric acid.

$$K_a = \frac{(H^+)\,[A^-]}{[HA]}$$

(H^+) activity determined by pH meter
[A^-] and [HA] are concentrations

In water at 25°:

[A^-] initially	pKa
0.0098	2.73 ± 0.05
0.0054	2.74 ± 0.04

2.5. Optical Rotation

Phenoxymethyl penicillin has three asymmetric carbon atoms (formula section 1.1) and is strongly dextrorotatory. Investigators have determined the specific rotation under a number of conditions.

Potassium Phenoxymethyl Penicillin

[α]$_D$	Temp.	Concentration and Solvent	Comment	Ref.
+ 223°	25°	0.2% in water	synthetic sample	25
+ 223°	20°	1% in water	natural sample	25
+229.8°	27°	0.2% in water	Squibb primary ref. substance	30

Phenoxymethyl Penicillin (free acid)

+ 207°	25°	1% in methanol		31
+ 195°	25°	1% in ethanol		31
+ 188°	25°	1% in butanol		31
+ 193°	20°	1% in butanol	extensively purified	23

Sodium benzylpenicillin has a stronger rotation

for the mercury line, $[\alpha]_{5463} = +368°$, than for the sodium D line, $[\alpha]_D = 294°$ [19b]. This might also be true for potassium phenoxymethyl penicillin.

3. Penicillin Degradation
3.1. Modes of Penicillin Degradation

Penicillin chemistry is complex and has been widely studied during the last quarter century [19, 32, 33]. Before discussing penicillin V specifically, it will be profitable to examine penicillins in general. Their degradation may be summarized in Figure 3 [20, 34].

The β-lactam ring in the lactam-thiazolidine structure of penicillin (I) is much more sensitive to nucleophilic attack than simple β-lactams. The dibasic penicilloic acid (III) is the product formed under mild hydrolysis conditions in neutral and alkaline solutions. For penicillin G at constant temperature the reaction is first order with respect to penicillin and hydroxide ion concentrations. The mechanism in neutral solution is not completely understood, but the characteristic ultraviolet absorption [20] and polarography [33] indicate the degradation proceeds through penicillenic acid (II). Benzylpenicillenic acid (from penicillin G) has been isolated and shown to have an absorption maximum in its ultraviolet spectrum at 322 nm. During the hydrolysis of penicillin G at pH 7.5, the absorbance at 322 nm increases, indicating the presence of benzylpenicillenic acid. The latter compound is converted rapidly to the corresponding penicilloic acid (III), with a half-life of 6.5 minutes at pH 7.5 and 37°. Penicillin G is probably hydrolyzed directly to a salt of penicilloic acid in strongly alkaline solution [20]. In acid solution, both penicilloic acid

Figure 3. Penicillin Degradation

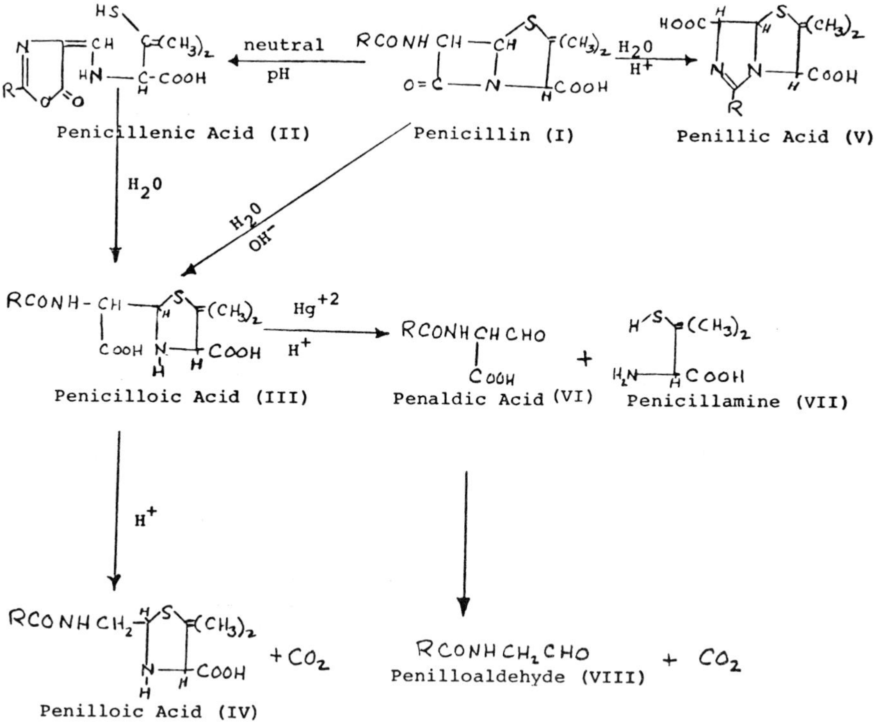

(III) and penillic acid (V) are formed concurrently [35].

Salts of penicilloic acid (III) are stable. However, on acidification, the resulting free penicilloic acids readily lose carbon dioxide to form the corresponding penilloic acid (IV).

Penicilloic acids (III) react instantly in aqueous solution with mercuric chloride [19d], giving penicillamine (VII) and penaldic acid (VI). Penaldic acids decarboxylate rapidly to form the corresponding penilloaldehyde (VIII). This degradation forms the basis for a sensitive analytical method for benzylpenicillin[36].

3.2. Degradation of Phenoxymethyl Penicillin

3.21. Degradation of Crystalline Phenoxymethyl Penicillin

The greater instability of amorphous penicillin, compared with the crystalline form, is discussed in section 2.21.

3.22. Degradation in Solution

The mechanism of the acid-catalyzed rearrangement of penicillin to penicillenic acid (II) and penillic acid (V) is represented and discussed by Doyle and Nayler[33]. Since the amide side chain is directly involved in the formation of these products, its nature has a great effect on the course of the hydrolysis. Substituting the electron-attracting phenoxy group for the benzyl group of penicillin G does give greatly increased acid stability [20, 33, 35, 37, 38, 39].

Acid Stability (pH 1.3, 35°, 50% aq. EtOH) [33]

Penicillin	Half-life, minutes
G	3.5
V	160

Hydrolysis of phenoxymethyl penicillin and 14 derivatives with benzene ring substituents was studied by Panarin and Solovskii [40] at pH 2.0 and 30, 35 and 40°. Enhanced acid stability was shown by electron-accepting substituents while the hydrolysis rate was increased by electron-donor substitution.

Under somewhat different conditions, phenoxymethyl penicillin in 0.1$\underline{N}$ hydrochloric acid at 37° was reported to have a half-life of 29 minutes [41]. At appropriate time intervals, 2 ml aliquots were removed and added to 8 ml of 0.1$\underline{M}$ phosphate buffer, pH 7.0, mixed and frozen for later bioassay. Because these samples were frozen before they were assayed, the validity of these data is subject to doubt (refer to section 3.4).

Kondrat'eva and Bruns [42] report phenoxymethyl penicillin shows maximum stability in aqueous solution in the pH range from 6.0 to 7.0. They prepared nomograms to show penicillin V inactivation as a function of pH and temperature.

Parker et al [11], in their study of the ultraviolet absorption spectrum of penicillin V (section 2.13), reported the effects of acidic and basic degradation. They reported complete decomposition of penicillin V at room temperature in 0.5$\underline{N}$ sodium hydroxide after 15 minutes.

Dennen and Davis [35] reported from the Lilly Research Laboratories in 1961 on their kinetic study of the formation of penillic acid (V) and penicilloic acid (III) from phenoxymethyl and five other penicillins in acid solution. They made no mention of penicillenic acid (II) as an intermediate in penicilloic acid (III) formation. Recently, Japanese [43] and Indian [44] workers have investigated the kinetics of the degradation of penicillin V and other penicillins in acid buffers. Joshi et al. measured absorbances at 322 nm and the iodine titer to follow the decompositions to the penicillenic acids (II). Recently, Blinova and Khokhlov have studied the isomerization of phenoxymethyl penicillin into the corresponding penicillenic acid (II) [45] and phenoxymethyl penicillic acid /penillic acid (V)/ [46]. However, an abstract to an earlier paper by Khokhlov and Blinova [47] reports phenoxymethyl penicillin does not isomerize to penillic acid.

The decomposition rate in an 8 to 10 pH range was followed by Rozenberg [48]. It is interesting to note that penicillin V degrades 2.2 times faster than penicillin G at pH 9.08 to 9.80 and that both show a 1.6-fold increase in degradation on increasing the sodium chloride concentration from 0.1 to 0.5$\underline{M}$. Mata and Gallego [31] reported at pH values more alkaline that 6.2 penicillin V is inactivated more rapidly than penicillin G while Doyle and Nayler[33] reported phenoxymethyl penicillin in neutral solution produces much less of the corresponding penicillenic acid (II) than does a neutral solution of benzylpenicillin.

A number of buffers are reported to catalyze the degradation of phenoxymethyl penicillin in aqueous solutions at 80° [49]. In other studies [50, 51] the presence of pharmaceutical excipients, surfactants, preservatives and suspending and thickening agents reduced the stability of penicillin V in aqueous solutions. Amino alkylcatechols have been studied as model compounds that simulate the action of penicillinase [136].

3.3. Enzymatic Degradation

Penicillin V is rapidly hydrolyzed by penicillinase, a term applied to any enzyme that specifically catalyzes the formation of the corresponding penicilloic acid. Early work on the occurrence, preparation and physical and chemical properties of penicillinases is summarized by Florey, et al.[19e]. Goodey, et al[52] reported the inactivation of phenoxymethyl penicillin by penicillinase, when examined microbiologically, was found to be complete in a short period. Oscillographic polarography was selected by Dusinsky[53] to measure the enzymatic inactivation of penicillins to penicilloic acids. His results were in good agreement with previously determined enzyme stabilities of phenoxymethyl and three other penicillins. Penicillinases were reviewed recently by Rauenbusch[54].

Two distinct types of enzymes (penicillin amidases) capable of removing the side chain attached to the amino position of 6-aminopenicillanic acid have been encountered in microorganisms[33,55]:

$$\text{Penicillin} + H_2O \longrightarrow \text{6-APA} + RCO_2H.$$

One type found among the actinomycetes and filamentous fungi, and present in certain yeasts, readily hydrolyzes pentyl-, heptyl- and phenoxymethyl penicillins, but splits benzylpenicillin only very slowly. The other type is of bacterial origin and hydrolyzes benzylpenicillin very rapidly, but penicillin V only slowly. Under suitable conditions, the reaction is reversible, permitting enzymatic synthesis of penicillins.

3.4. Degradation of Frozen Systems

Grant, Clark and Alburn[56] reported on

the imidazole and base-catalyzed hydrolysis of penicillin frozen systems. They found cleavage of the β-lactam ring occurred in frozen, but not in supercooled, systems.

β-Lactam Hydrolysis at Various Temperatures (a)

Catalyst	Substrate	% Hydrolysis					
		22°	0°(b)	-8°	-18°	-28°	-78°
Imidazole	Pen-G	23	4	71	64	20	4
OH⁻, pH 9.9	Pen-G	75	14	--	28	12	4
Imidazole	Pen-V	24	8	72	70	48	4

(a) Conditions 65 hours, pH 7.7, 0.01$\underline{M}$ catalyst and substrate. Buffer at pH 9.9 is 0.05$\underline{M}$ borate. (b) Unfrozen.

Solid and liquid phases in equilibrium at -2° were separated and assayed. Imidazole and penicillin (presumably Grant et al. used penicillin G) concentrations changed by less than 10% from their initial values. Subsequent storage of the solid at -2° led to as much as 77% hydrolysis while storage of the liquid at both -2° (unfrozen) and +22° gave no hydrolysis.

3.5 Cupric Ion Hydrolysis

Niebergall and his coworkers [58, 59, 60, 61] reported that cupric ion promoted the rapid hydrolysis of both penicillin G and V to a penicilloic acid, rather than simply forming a complex with penicillin. In the presence of excess penicillin, the reaction follows second-order kinetics and ceases when the available cupric ion has been consumed. Rearrangement of either penicillin in pH 5.5 acetate buffer to penicillenic acid (compound II in section 3.1) does not occur in the presence of an equimolar amount of cupric chloride. Penicilloic acid is formed directly. Thus, the mechanism as well as the rate of penicillin decomposition are influenced by the presence of cupric ions. A complex is formed with intact penicillin followed by rapid hydrolysis of the complex into the corresponding penicilloic acid-cupric ion complex. Association constants at room temperature for the complex with copper are:

	log K
Penicillin V	2.24 (in the absence of ionic strength control)
Penicillin V	2.09 (ionic strength 0.01$\underline{M}$)
Penicilloic V acid	4.50

No interaction was evident with calcium, cobalt, magnesium or nickel. Interaction with zinc was reported, but not studied.

4. Synthesis

Both benzylpenicillin and phenoxymethyl penicillin are produced commercially by fermentation. However, from the beginning of the vast British-American cooperative program during World

War II, attempts were made in many laboratories to synthesize penicillin [19, 32]. The first rational total synthesis of penicillin V was achieved by Sheehan and Henery-Logan [25, 26] in 1957, after a nine-year research effort [137]. Syntheses for both penicillins and cephalosporin C are summarized in a recent book by Manhas and Bose [138].

5. Purification and Analysis of Impurities

5.1. Purification

Purification procedures used for analysis are described under chromatographic analysis (section 6.5.) and electrophoretic analysis (section 6.6.). Parker, Cox and Richards [11] describe the purification and characterization of the phenoxymethyl penicillin (free acid) used later as the International Standard [1]. Samples were dissolved in water with sodium bicarbonate, precipitated by dilute hydrochloric acid and recrystallized from aqueous acetone. Samples appeared to be pure by paper chromatography and phase solubility analysis. Purification of the free acid was later studied by Unterman and Mironeanu [15].

For purification of the Squibb Primary Reference potassium phenoxymethyl penicillin, the starting material was dissolved in water and crystallized by the addition of a saturated potassium acetate solution. These crystals were then redissolved in water, diluted with butanol, and crystallized again by an azeotropic vacuum distillation to yield the final product [30]. Purification and isolation of the potassium salt were the subjects of two Czechoslovakian patents [62, 63].

Ege [135] purified sodium penicillin V by

precipitating the N,N'-dibenzyl-ethylenediamine salt.

5.2. Countercurrent Distribution - Solvent Partitioning

Thadani et al.[22] studied the purity of commercial penicillin V by countercurrent distribution in different solvent systems. They found 95.0% penicillin V and 2.9% penicillin J.

Partitioning between butyl acetate and aqueous acidic and basic solutions has been used as a separation procedure by many investigators, e.g., [64, 65]. Bethel and Bond [64] extract penicillin V acid into butanol, and the salt back into aqueous phosphate buffer. Such procedures are found in many of the papers describing analytical methods. In studies correlating serum protein binding with partition coefficients, the log of the n-octanol-water partition coefficient for phenoxymethyl penicillin (free acid) is given [66] as 2.09 and the distribution coefficient between silicone oil and water was determined [67].

5.3. Assay Methods for Intermediates and Impurities

5.31. Phenoxyacetic Acid

Phenoxyacetic acid was separated from fermentation fluids, crude phenoxymethyl penicillin and finished pharmaceutical products. Determinations of the separated acid were made by ultraviolet spectrophotometry or a bromometric titration using an amperometric end point [68]. A study of phenoxyacetic acid in penicillin V by ultraviolet spectroscopy has been described recently [15].

In four papers, Ida et al. described the determination of phenoxyacetic acid in penicillin production. After separation from penicillin V by an isopropyl ether extraction, it was determined colorimetrically with chromotropic acid[69]. In fermented broth, penicillin V was converted to penicilloic acid by alkali before chloroform extraction of the phenoxyacetic acid from the acidified solution. The latter compound was nitrated and determined polarographically[70]. For fermentation broth, a more elaborate solvent extraction clean-up was used before determining phenoxyacetic acid colorimetrically with chromotropic acid[65]. In the final paper[71], phenoxyacetic acid was separated from penicillin V, brominated with excess bromine and the excess determined iodometrically.

Additional colorimetric procedures have been described by other workers. Birner[72], after converting phenoxymethyl penicillin to penicilloic acid in base, extracted phenoxyacetic acid from the acidified solution into benzene. He nitrated the phenoxyacetic acid and measured the resulting yellow color in ammoniacal solution. Nogami and Kanazawa[73] reported that penicillin V and phenoxyacetic acid can be determined separately by a dye procedure using methyl green. Uberti[74] converted phenoxymethyl penicillin to penicilloic acid in base, extracted the phenoxyacetic acid from the acidified solution into toluene and transferred the latter material back into aqueous base. Here a red color is formed with 2,7-naphthalenediol.

Niedermayer developed a gas-chromatographic procedure for determining phenoxyacetic acid during penicillin production and in the finished drug[75,76]. The phenoxyacetic acid is extracted into benzene, esterfied with BF_3-methanol and the methyl phenoxyacetate reaction mixture chromatographed. A 6-foot column of Gas Chrom A treated with 2% by weight of phosphoric acid and coated with 10% by weight of diethylene glycol adipate was used at 210°. Kawai and Hashiba [77] hydrolyzed phenoxymethyl and six other penicillins by refluxing in aqueous base. The side-chain acid, e.g., phenoxyacetic acid, was converted to the methyl ester with diazomethane and chromatographed on a 1.5 meter column of 3.5% SE-30 at 120-150°.

5.32. Penicilloic Acid

Both paper chromatography [78] and thin layer chromatography [79] have been used to determine phenoxymethyl penicilloic acid in the presence of the intact penicillin.

Electrophoretic analysis of penicilloic acid in phenoxymethyl penicillin was described recently (refer to section 6.6.).

The oscillographic polarography of penicilloic acid in the presence of phenoxymethyl penicillin has been described [53].

5.33. p-Hydroxyphenoxymethyl Penicillin

p-Hydroxyphenoxymethyl penicillin has been determined by Birner [80] in penicillin V fermentation samples and by Vanderhaeghe et al. [141] as a metabolite of penicillin V in urine. Both papers describe paper chromatography procedures.

6. Methods of Analysis

A review article on the estimation of penicillins appeared in 1963 [84]. Discussions on the analysis of penicillin and explicit procedures are provided by Connors and Higuchi [85].

6.1. Identification Tests

Poet and Keeler [81] extracted phenoxymethyl penicillin from formulations and distinguished the drug from ampicillin, penicillin G and dicloxacillin by infrared. Bands at 6.7 and 9.45 μ are particularly useful for identifying penicillin V (acid).

Paper Chromatography (section 6.51.), thin layer chromatography (section 6.52.) and gas chromatography (section 6.54.) have all been used as identification procedures.

Parker, Cox and Richards [11], in their original paper characterizing penicillin V, described the color reaction with chromotropic and sulfuric acids. The decomposition of the phenoxyacetic acid moiety to formaldehyde is the basis of the distinctive color formed. Procedure:

Add a few crystals of chromotropic acid and 2 ml of concentrated sulfuric acid to a small amount of solid sample. Immerse the mixture in a 150° glycerol bath for 1 to 2 minutes. Remove and note the color. (Dilute with concentrated sulfuric acid if necessary).

Compound	Chromotropic Acid plus H_2SO_4	H_2SO_4 alone
Phenoxyacetic acid	Deep red	Brown
Penicillin V	Deep blue purple	Orange brown
Phenylacetic acid	Pale yellow	Colorless
Penicillin G	Brown	Light brown

Kawai and Hashiba [77] recommend the different penicillins be differentiated by gas chromatography. A sample is hydrolyzed under basic conditions, then the organic acid from the side chain at C-6 is converted to the methyl ester with diazomethane and chromatographed. The retention time serves to identify the original penicillin. Phenoxymethyl penicillin can be characterized by hydrolysis to the phenoxyacetic acid and conversion of the latter to the sparingly soluble p-bromophenoxyacetic acid [11]. After recrystallization from benzene, it melts at 158°.

Turback [82], after acidifying an aqueous solution of penicillin V, extracts the drug into butyl acetate. Addition of methanolic ammonia forms a white precipitate. This constitutes a positive test for phenoxymethyl penicillin.

Identification of various penicillins in pharmaceutical dosage forms was described recently by Weiss et al. [83]. Ultraviolet absorption, thin layer chromatography and the partition ratio between chloroform and an aqueous buffer were useful in identifying phenoxymethyl penicillin.

6.2. Ultraviolet Methods

The ultraviolet absorption properties of phenoxymethyl penicillin and its decomposition products are discussed in sections 2.13. and 3.22.

Phenoxymethyl penicillin in fermentation liquids may be determined selectively by measuring at 268 and 275 nm, after preliminary purification by solvent extraction and decomposition in base. Background ultraviolet absorbance is corrected by measurements made without basic decomposition [86]. Beer's law was valid for 0 to 280 units per ml.

For determining penicillins, including penicillin

V, in aqueous and protein solutions, the ultraviolet absorption of penicillenic acid (compound II in Figure 3) is measured at 322 nm [87, 88]. This rearranged product of penicillin is stabilized by forming the mercury salt of the sulfhydryl portion of the molecule.

Ultraviolet spectrophotometry has been used for quantitation of penicillin V following its separation by both paper and thin-layer chromatography (refer to sections 6.51. and 6.52.).

6.3. Titrations
6.31. Iodometric Titrations

Iodometric methods for the assay of penicillins were first described by Alicino [89] in 1946. Penicillin is inert to iodine in neutral aqueous solution. But after hydrolysis with alkali or penicillinase, the resulting penicilloic acid (III) consumes from 6 to 9 equivalents per mole, depending on the conditions used. The difference in consumption of iodine by penicillin preparations before and after alkaline hydrolysis was found proportional to the quantity of the drug. Alicino found penicillin G consumed 8.97 equivalents of iodine per mole under the conditions used in his assay procedure. He later demonstrated [90] other deactivated penicillins, including phenoxymethyl penicillin, consume nine equivalents of iodine per mole while 6-aminopenicillanic acid consumes only eight equivalents of iodine per mole.

Goodey et al. [52] in 1955 described the first application of Alicino's iodometric procedures to phenoxymethyl penicillin, using either alkali or penicillinase for the formation of penicilloic acid. Their purest sample after penicillinase deactivation had a consumption of only 8.61 equivalents of iodine per mole. Two other purified samples consumed an average of 8.47 equivalents per mole by this procedure, while after alkaline hydrolysis, they consumed 9.00 equivalents of iodine per mole as Alicino [90] demonstrated later.

Kleiner and Dendze-Pletman [23] after extensive purification determined phenoxymethyl penicillin iodometrically. They found the consumption of iodine by the products of alkaline hydrolysis is not stoichiometric but equal to 8.68 equivalents of iodine per mole.

Bethel and Bond[64] described two iodometric assay procedures for penicillin V in fermented broths. In the first, purification by extraction from acid into butanol and back into an aqueous phosphate buffer was followed by opening the lactam ring with base and an iodometric titration. Effects of impurities remaining after solvent extraction were rendered negligible by adjusting the pH to 3.6 before adding the iodine. To eliminate the steps involved in solvent extraction, a direct procedure was developed by opening the lactam ring with penicillinase. The difference in iodine consumption before and after penicillinase treatment was equated with penicillin content. After the alkaline hydrolysis procedure, we may calculate from their data that 9.29 equivalents of iodine were consumed per mole of potassium penicillin V.

This somewhat erratic iodine consumption has led both the Federal Register[91] and the British Pharmacopeia[92] to require the use of a working standard for this titration.

The reaction mechanism in the iodometric determination of phenoxymethyl penicillin has been studied recently by Glombitza and Pallenbach[93, 94, 95].

In the iodometric procedures described up to this point, an excess of iodine is added to penicilloic acid. After an appropriate time interval (15 to 30 minutes), the excess is back-titrated with thiosulfate. In 1967, Kalinowski and Czlonkowski[96] described a direct coulometric titration of unhydrolyzed potassium benzylpenicillin with chlorine. Samples after alkaline hydrolysis were also titrated coulometrically with chlorine, bromine or iodine. It is interesting to note in these methods, eight equivalents of halogen reacted with one mole of penicillin G. The back titrations described earlier usually required nine equivalents per mole. No information was reported on the direct coulometric titration of penicillin V or phenoxymethyl penicilloic acid. These same investigators[97] later described a titration in which an excess of iodine was generated coulometrically after this drug was hydrolyzed by alkali. After standing for 15 minutes, a known excess of sodium thiosulfate solution was added and the generation of iodine was resumed. The method was suitable for routine use.

6.32. Nonaqueous Titrations

Since phenoxymethyl penicillin is a strong acid (pKa in water is 2.74), Mohoric found a nonaqueous titration to be more accurate and faster than the iodometric method. A 100-mg sample dissolved in 20 ml of dimethylformamide is titrated with 0.1$\underline{N}$ sodium methoxide to a thymol blue end point.

Floris and Simonyi[99] prefer using tetramethylammonium hydroxide as the titrant with the penicillin V dissolved in dimethylformamide. An alternate procedure used anhydrous methanol as the solvent. Phenoxymethyl penicillin has been titrated to a potentiometric end point with sodium methoxide[100].

6.33. p-Chloromercuribenzoate

Siegmund and Korber [101] described the spectrophotometric titration of penicillin with p-chloromercuribenzoate after its cleavage by penicillinase. The thiazolidine moiety reacts with excess mercuribenzoate. After 30 minutes the excess is titrated with a solution of cysteine or glutathione. The endpoint is determined by measuring the absorbance at 250 nm. As little as 0.1 μM of penicillin can be determined. Preparation of a stable p-chloromercuribenzoate solution has been described recently by Muftic [102].

6.4. Colorimetric Methods
6.41. Hydroxamic Acid

Reaction of penicillin with hydroxylamine leads to the formation of a hydroxamic acid. The colored chelate formed with ferric ion and the hydroxamic acid of benzylpenicillin was developed into an assay procedure in 1949 [103]. The color was stable if solutions were read within 5 minutes. Extraction of the colored complex into n-butanol [104] gave a stable color. Details of the hydroxamic acid procedure for a number of penicillins, including phenoxymethyl penicillin, were published in the Federal Register[91].

In 1960, the method was adopted [105] as an automated procedure using the Technicon Autoanalyzer. This procedure is used routinely for the determination of penicillin V [106]. Details of the automated procedure applied to phenoxymethyl penicillin have now been published [107, 108].

Pligin and Portnov [109] added cobalt nitrate to the hydroxamic acid and extracted the chelate into butanol. The resulting solutions obeyed Beer's law and were stable for 2 hours.

6.42. Dye Complex

The phenoxymethyl penicillin dye complex with methyl green can be extracted from a pH 3 buffer into benzene and determined colorimetrically [73]. Penicillin V and phenoxyacetic acid can be determined separately. The procedure is claimed to be simple, rapid and accurate, and the values obtained agree with potencies obtained by biological tests.

6.43. Nitration

Penicillin V may be nitrated by 10% potassium nitrate in concentrated sulfuric acid [72, 110]. The yellow derivative in ammonia solution can be determined colorimetrically. Birner's [72] method estimates both penicillin V and phenoxyacetic acid in fermentation broths. Selectivity is achieved by solvent extraction techniques.

6.44. Folin Phenol Reagent

Silverman [111] reported that several penicillins, including phenoxymethyl penicillin, interfere with the Folin colorimetric protein determination. Since penicillin V gives a pronounced color formation in the absence of protein, it might be possible to use this reaction for its determination.

6.5. Chromatographic Analysis

6.51. Paper Chromatography

The following systems have been described for the paper chromatography of phenoxymethyl penicillin:

Solvent System	Paper	R_f	References
1. H$_2$O-saturated ether	pH 6.6 phosphate buffer on Whatman #4 paper (a,e)	0.31	57, 114
2. Butanol-2:HCOOH:H$_2$O 75:15:10	Schleicher and Schule 2043 b paper (b)	0.92	78
3. Propanol-2:H$_2$O 60:40 70:30	Same	0.81 0.70	78 78
4. Propanol-1:EtOH:H$_2$O 30:40:30 50:20:30 50:30:20	Same	0.80 0.71 0.65	78 78 78

	Solvent System	Paper	R_f	References
5.	Butanol-1:Propanol-1:H_2O	Same		
	25:40:35		0.60	78
	20:50:30		0.60	78
	25:50:25		0.55	78
6.	H_2O	Whatman No. 1 paper (c)	0.93	112
7.	H_2O saturated n-Butanol	Same	0.33	112
8.	H_2O saturated EtOAc	Same	0.80	112
9.	H_2O saturated Benzene	Same	0.00	112
10.	3% NH_4Cl in H_2O	Same	0.84	112
11.	Isoamyl acetate: MeOH:HCOOH:H_2O 65:20:5:10 (upper phase)	Same	0.89	112

Solvent System	Paper	R_f	References
12. n-Butyl acetate:methyl ethyl ketone: 0.15M phosphate buffer, pH 7.4, 50:25:5 (upper phase)	Same	0.43	112
13. EtOAc:n-hexane:0.15M phosphate buffer, pH 6.0, 65:15:20 (upper phase)	Same	0.00	112
14. Isopropyl ether:isopropanol: H_2O 70:30:100 (two phases)	Paper impregnated with pH 5.0 phthalate buffer(d)	0.33	113
15. Ether saturated with 28% $(NH_4)_2SO_4$	pH 5.5 citrate buffer on Whatman No.1 (f)	0.60	80

Detection systems:

(a) Bioautographic plates, Bacillus subtilis 288.
(b) Alkaline hydrolysis + $AgNO_3$.
(c) Bioautographic plates, Bacillus subtilis.
(d) A number of chemical systems suggested.

(e) Bioautographic plates, Staphylococcus aureus 209P
(f) Spray with starch-iodine solution after basic hydrolysis

Rohr 78 described the separation of penicillin V and its enzymatic hydrolysis products. Separation of eight penicillins by paper and thin-layer chromatography was reported by Hellberg 113 while the paper chromatographic separation of five penicillins was achieved by Watanabe, Endo and Iida 139.

6.52. Thin-Layer Chromatography

Systems for thin-layer chromatography are summarized here:

Solvent System	Plate	R_f	Reference
1. 0.1M NaCl	Cellulose (a)	0.82	115
2. 0.3M Citric acid saturated with n-BuOH	Same	0.76(0.56)	115
3. Organic phase of isoamyl acetate:MeOH:HCO$_2$H:H$_2$O 65:20:5:10	Silica Gel G (a) activated 110°	0.66	115
4. Acetone:HOAc 95:5	Same very temperature-dependent	0.75	115

Solvent System	Plate	R_f	Reference
5. pH 6.0 phosphate buffer	Sephadex G-15 (b) buffered at pH 6.0	0.9 (b)	119
6. CCl_4: Isopropanol:H_2O 65:35:4	Kieselgahn buffered at pH 5.3 (c)	0.5	113

Detection Systems

a. Spray reagent is freshly prepared $FeCl_3-K_3Fe(CN)_6$
b. Bioautographic plates, Bacillus subtilis ATCC 6633; R_f relative to penicillin G.
c. Various spray reagents

Separation of eight penicillins by paper and thin layer chromatography was described by Hellberg 113. Thin-layer methods are described for the intact penicillins and the methyl esters of the penicilloic acids.

Mc Gilveray and Strickland 115 reported four thin-layer systems and spray reagents for differentiating ten penicillins. The comparative behavior of 18 other antibiotics was presented also. Nussbaumer 116, 117, 118 described thin layer chromatography of phenoxymethyl penicillin in tablets, followed by removal of the spot and quantitation by spectrophotometry. Sephadex thin-layer chromatography was applied to 17 antibiotics.

119

An extremely sensitive procedure[120] can detect 0.76 ng of phenoxymethyl penicillin. After hydrolysis, the secondary amine of penicilloic acid is coupled with 9-isothiocyanatoacridine to form a fluorescent compound. Following thin-layer chromatography, fluorescence of the spot is measured, with linearity in the 3 to 30 ng range.

Reversed-phase thin-layer chromatography has been used[121, 122] to study structure-antibacterial activity relationships of cephalosporins and penicillins[123]. The stationary phase was Silica Gel G impregnated with Silicone DC 200. Mobile phases with buffers and 0 to 50% acetone gave R_f values for penicillin V from 0.1 to 0.9.

6.53. Ion-Exchange Chromatography

Russian investigators[124, 125] have described the absorption capacity and selectivity of two strong anion exchange resins for phenoxymethyl penicillin and three other penicillins. A phosphate buffer is the eluant.

6.54. Gas-Liquid Chromatography

The methyl esters of benzyl-and phenoxymethyl penicillins can be separated on columns of either 0.75% QF-1 or 0.4% SE-52[126,127]. Lightly loaded SE-30, SE-52 and QF-1 columns were studied, with SE-52 being the most suitable liquid phase. No evidence of thermal decomposition was found, as shown by the single reproducible peaks. Products collected from the exit of the QF-1 column had the same thin-layer and gas chromatographic properties as the starting materials. Samples of the methyl esters in acetone were injected directly onto 130-cm by 4-mm glass columns containing the stationary phase coated on 100-200 mesh acid-washed silanized 'Gas Chrom P.'

Stationary Phase	QF-1	SE-30	SE-52	SE-52
Loading	0.75%	0.4%	0.4%	0.4%
Column Temperature	225°	200°	180°	240°
Relative Retention Time, Me ester:				
Penicillin G	1.00	1.00	1.00	1.00
Penicillin V	1.17	1.24	1.35	1.18
Retention Time Me ester Pen.G	11 min.	6 min.	28 min.	1.5 min.

Preparation and properties of some phenoxymethyl penicillin esters, including the methyl ester, was the subject of a recent study [131].

These studies were concerned with separation and gave no details on quantitation. Claims of thermal stability and reproducible peaks [127] may make the method applicable for quantitative analysis. Quantitative conversion to the methyl ester or other suitable volatile derivative would be required.

6.6. Electrophoretic Analysis

Thomas and Broadbridge [128] have developed a low-voltage electrophoresis method for the rapid separation of a penicillin from its penicilloic acid. Phenoxymethyl penicillin was one of 12 penicillins studied. The method was found to be particularly useful for rabbit sera and urine and the products of bacterial hydrolysis of penicillins in broth cultures.

6.7. **Polarography**

Dusinsky[53] reported 1 cathodic incision on the oscillographic polarogram in pH 7 phosphate buffer for phenoxymethyl and three other penicillins. This incision decreases and disappears as the penicillin is inactivated by penicillinase. Three more positive incisions that appear are the result of the penicilloic acid formed.

6.8. **Microbiological Methods**

Goodey et al.[52] reported in 1955 on the comparative microbiological assays of penicillin V and penicillin G. Dose-response curves for three species, Staphylococcus aureus 209P, Bacillus subtilis 288 and Sarcina lutea, were determined.

The Federal Register[91] specifies the use of Staphylococcus aureus 6538P for the determination of potency of potassium phenoxymethyl penicillin. The S. aureus 6538P strain is the same as S. aureus 209P.

George[129] modified the bio-assay procedure developed by Platt, Weisblatt and Guevrekian[130] for use in assaying penicillin V in bulk powders for solutions and tablets. This automated, turbidometric procedure relates the weight of antibiotic to the time required to dilute a sample to a concentration yielding a specified growth inhibition. The method uses Streptococcus faecalis ATCC 10,541 as the test organism. For the assay of penicillin V in oral suspensions, Levin[132] describes a procedure using Staphylococcus aureus 209P.

7. Serum Protein Binding
Binding of drugs, including penicillin V, with plasma proteins has been reviewed recently [133, 134]. Studies correlating serum protein binding with partition coefficients were summarized in section 5.2.

8. Drug Metabolism
Many penicillins, including penicillin V, are metabolized in the human body to other active compounds. Rolinson and Batchelor [140] administered the drugs orally or by intramuscular injection and evaluated both blood and urine samples by paper chromatography. Active metabolites were detected by bioautographic plates. Vanderhaeghe, Parmentier and Evrard [141] identified the major metabolite of penicillin V as p-hydroxyphenoxymethyl penicillin. It represented about 10% of the microbiological activity in the urine. A small amount of o-hydroxyphenoxymethyl penicillin was also observed while in the urine of some patients another metabolite was detected. They speculated this might be the dihydroxy derivative.

References Cited

1. Humphrey, J., J. Lightbrown and M. Mussett, Bull. W.H.O. 20, 1221 (1959).

2. United States Pharmacopeia XVIII, 526 (1970).

3. Hayden, A., O. Sammul, G. Selzer and J. Carol, J. Assoc. Offic. Agr. Chem. 45, 797 (1962).

4. Ovechkin, G., Zh. Obshch. Khim. 33, 1923 (1963); C. A. 59, 11190 b (1963).

5. Green, G., J. Page and S. Staniforth, J. Chem.

Soc. 1595 (1965).

6. Cohen, A. and M. Puar, Squibb Institute, private communication.

7. Pek, G., V. Bystrov, E. Kleiner, I. Blinovo and A. Khokhlov, Izv. Akad. Nauk SSSR, Ser. Khim. 1968, 2213; C.A. 70, 28229 (1969).

8. Manhas, M. and A. Bose, "Synthesis of Penicillin, Cephalosporin C and Analogs", Marcel Dekker, New York, 1969.

9. Cooper, R., P. De Marco, J. Cheng and N. Jones, J. Amer.Chem Soc. 91, 1408 (1969).

10. Archer, R., R. Cooper, P. De Marco and L. Johnson, Chem. Commun. 1291 (1970).

11. Parker, G., R. Cox and D. Richards, J. Pharm. Pharmacol. 7, 683 (1955).

12. Rogers, A., ibid 11, 291 (1959).

13. Rogers, A., ibid 16, 433 (1964).

14. Andersen, H. Dan. Tidsskr. Farm. 38, 1(1964); C. A. 60, 7871b (1964).

15. Unterman, H. and T. Mironeanu, Rev. Chim. (Bucharest) 18, 530 (1967); C.A. 68, 43130(1968)

16. Richter, W. and K. Biemann, Monatsh. Chem. 95, 766 (1964); C.A. 61, 14015 g (1964).

17. Bochkarev, V., N. Ovchinnikova, N. Vul'fson, E. Kleiner and A. Khokhlov, Dokl. Akad. Nauk SSSR 172, 1079 (1967); C.A. 67, 116843(1967).

18. Mathews, A., C. Schram and D. Minty, Nature 211, 959 (1966).

19. Florey, H., E. Chain, N. Heatley, M. Jennings, A. Saunders, E. Abraham and M. Florey, "Antibiotics," Vol. 2, Oxford University Press, London, 1949. (a) p. 787, (b) p. 789, (c) p. 795 (d) p. 804 and (e) p. 1090.

20. Schwartz, M. and F. Buckwalter, J. Pharm. Sci. 51, 1119 (1962).

21. Jacobson, H., Squibb Institute, private communication.

22. Thadani, S., G. Sen and D. Ghosh, Hindustan Antibiot. Bull. 3, 69 (1960); C.A. 55, 9794b (1961).

23. Kleiner, G. and B. Dendze-Pletnam, Med. Prom. SSSR 13, 42 (1959); C.A. 53, 18389 d (1959).

24. Kuhnert-Brandstatter, M. and L. Muller, Microchem. J. 13, 20 (1968).

25. Sheehan, J. and K. Henery-Logan, J. Amer. Chem. Soc. 81, 3089 (1959).

26. Sheehan, J. and K. Henery-Logan, ibid 84, 2983 (1962).

27. Ganapathi, K. Hindustan Antibiot. Bull. 1, 29 (1958); C.A. 54, 21637 b (1960).

28. Weiss, P., M. Andrew and W. Wright, Antibiot. Chemother. 7, 374 (1957).

29. Rapson, H. and A. Bird, J. Pharm. Pharmacol. 15, 222T (1963).

30. Mc Credie, R., Squibb Institute, private communication.

31. Mata, J. and A. Gallego, *An. Inst. Farmacol. Espan.* **4**, 87 (1955); *C.A.* **52**, 1480b (1958).

32. Clarke, H., J. Johnson and R. Robinson, Eds. "The Chemistry of Penicillin," Princeton University Press, Princeton, 1949.

33. Doyle, F. and J. Nayler, *Advan. Drug Res.* **1**, 1 (1964).

34. White, A. in "Textbook of Organic Medicinal and Pharmaceutical Chemistry", 5th ed; C. Wilson, O. Giswold and R. Doerge, Eds.; J.B. Lippincott, Philadelphia, 1966, p. 326.

35. Dennen, D. and W. Davis, *Antibact. Agents Chemother.* **1961**, 531.

36. Katz, S. and A. Winnett, *J. Agr. Food Chem.* **10**, 284 (1962).

37. Elias, W. and H. Merrion, *Antibiot. Annu.*, 510 (1955-1956).

38. Narasimhachari, N. and G. Rao, *Hindustan Antibiot. Bull.* **6**, 114 (1964); *C.A.* **63**, 16134 c (1965).

39. Narasimhachari, N., and G. Rao and J. Tatke, *ibid* **7**, 127 (1965); *C.A.* **63**, 16134 d (1965).

40. Panarin, E. and M. Solovskii *Antibiotiki (Moscow)* **15**, 426 (1970); *C.A.* **73**, 44514 (1970)

41. Forist, A., L. Brown and M. Royer, J. Pharm. Sci. 54, 476 (1965).

42. Kondrat'eva, A. and B. Bruns, Khim.-Farm.Zh. 1, (12), 30 (1967); C.A. 68, 72247 (1968).

43. Umemura, K. and M. Nagasawa, Meiji Seika Kenkyu Nempo 1965 (7), 6; C.A. 64,8973 d (1966).

44. Joshi, V., S. Krishnan and N. Narasimhachari, Hindustan Antibiot. Bull. 9, 16 (1966); C.A. 66, 68885 (1967).

45. Blinova, I. and A. Khokhlov, Antibiotiki 11, 99 (1966); C.A. 64, 17568 g (1966).

46. Blinova, I. and A. Khokhlov, ibid 12, 275 (1967); C.A. 67, 64289 (1967).

47. Khokhlov, A. and I. Blinova, ibid 8, 35(1963) C.A. 61, 2908 a (1964).

48. Rozenberg, A., Med. Prom. SSSR 17, 43 (1963); C.A. 59, 11193 a (1963).

49. Finhold, P., R. Erickson and R. Pedersen, Medd. Norsk Farm. Selsk. 30, 69 (1968); C.A. 69, 109781 (1968).

50. Thoma, K., E. Ullmann and G. Zelfel, Acta Pharm. Hung. 35, 1 (1965); C.A. 63, 1664 g (1965).

51. El-Nakeeb, M. and R. Yousef, Acta Pharm. Suec. 5, 1 (1968); C.A. 68, 89857 (1968).

52. Goodey, R., K. Reed and J. Stephens, J.Pharm. Pharmacol. 7, 692 (1955).

53. Dusinsky, G., *Sci. Pharm.*, *Proc. 25th* **2**, 241 (1965); *C.A.* **70**, 18568 (1969).

54. Rauenbusch, E., *Antibiot. Chemother.* **14**, 95 (1968).

55. Singh, K., S. Sehgal and C. Vezina, *Appl. Microbiol.* **17**, 643 (1969).

56. Grant, N., D. Clark and H. Alburn, *J. Amer. Chem. Soc.* **83**, 4476 (1961).

57. Stephens, J. and A. Grainger, *J. Pharm. Pharmacol.* **7**, 702 (1955).

58. Niebergall, P., D. Hussar, W. Cressman, E. Sugita and J. Doluisio, *ibid* **18**, 729 (1966).

59. Cressman, W. and P. Niebergall, *ibid* **19**, 774 (1967).

60. Cressman, W., E. Sugita, J. Doluisio and P. Niebergall, *ibid* **18**, 801 (1966).

61. Cressman, W., E. Sugita, J. Doluisio and P. Niebergall, *J. Pharm. Sci.* **58**, 1471 (1969).

62. Sokol, M., D. Zemla and B. Okanik, *Czech.* 124,845, October 15, 1967; *C.A.* **69**, 34690 (1968).

63. Chorvat, I., *Czech.* 127,853, June 15, 1968; *C.A.* **70**, 90739 (1969).

64. Bethel, M. and C. Bond, *Analyst* **86**, 448 (1961).

65. Ida, K., *J. Antibiot.* (*Tokyo*) Ser. B. **11**, 189 (1958); *C.A.* **53**, 17224 f (1959).

66. Bird, A. and A. Marshall, Biochem. Pharmacol. 16, 2275 (1967).

67. Biagi, G., Antibiotica (Rome) 5, 198 (1967).

68. Doskocilova, D., H. Parizkova and J. Doskocil, Pharmazie 12, 608 (1957); C.A. 52, 3260 f (1958).

69. Ida, K., T. Onga, K. Kinoshita and S. Kawaji, J. Antibiot. (Tokyo) Ser. B. 10, 37 (1957); C.A. 53, 17224 c (1959).

70. Ida, K., S. Zushi and S. Kawaji, ibid, 40; C.A. 53, 17224 e (1959).

71. Ida, K., ibid 11, 192 (1958); C.A. 53, 17224 g (1959).

72. Birner, J., Anal. Chem. 31, 271 (1959).

73. Nogami, H. and S. Kanazawa, Yakugaku Zasshi 80, 1101 (1960); C.A. 54, 25581 d (1960).

74. Uberti, L., Squibb International, private communication.

75. Niedermayer, A., Squibb Institute, private communication.

76. Niedermayer, A., Squibb Institute, private communication.

77. Kawai, S. and S. Hashiba, Bunseki Kagaku 13, 1223 (1964); C.A. 62, 6339 f (1965).

78. Rohr, M., Mikrochim. Acta, 705 (1965).

79. Birner, J., J. Pharm. Sci. 59, 757 (1970).

80. Birner, J., *ibid* 57, 1606 (1968).

81. Poet, R. and B. Keeler, Squibb Institute, private communication.

82. Turback, C., Squibb International, private communication.

83. Weiss, P., B. Taliaferro, R. Huckins and R. Chastonay, J. Ass. Offic. Anal. Chem. 50, 1294 (1967).

84. Hamilton-Miller, J., J. Smith and R. Knox, J. Pharm. Pharmacol. 15, 81 (1963).

85. Connors, K. and T. Higuchi in "Pharmaceutical Analysis", T. Higuchi and E. Brochmann-Hanssen, Eds., New York, Interscience, 1961, p. 593.

86. Unterman, H., Rev. Chim. 15, 283 (1964); C.A. 61, 7654 d (1964).

87. Brandriss, M., E. Denny, M. Huber and H. Steinman, Antimicrob. Agents Chemother. 626 (1962).

88. Saccani, F. and G. Pitrolo, Boll. Chim. Farm 108, 89 (1969); C.A. 71, 33454 (1969).

89. Alicino, J., Ind. Eng. Chem., Anal. Ed. 18, 619 (1946).

90. Alicino, J., Anal. Chem. 33, 648 (1961).

91. Fed. Regist. 33, 4099 (March 2, 1968), 141 a. 81, 141 a. 82, 141 a. 1, 141 a.10f.

92. British Pharmacopeia, 1968, 756.

93. Glombitza, K. W. and D. Pallenbach, Pharmazie 23, 157 (1968).

94. Glombitza, K. W. and D. Pallenbach, Arch. Pharm. (Weinheim) 302, 695 (1969); C.A. 71, 116568 (1969).

95. Glombitza, K. W. and D. Pallenbach, ibid, 985; C.A. 72, 90359 (1970).

96. Kalinowski, K. and F. Czlonkowski, Acta Pol. Pharm. 24, 31 (1967); C.A. 67, 67632 (1967).

97. Kalinowski, K. and F. Czlonkowski, ibid 25, 29 (1968); C.A. 68, 117162 (1968).

98. Mohoric, J., Farm. Vestn. (Ljubljana) 14, 74 (1963); C.A. 60, 5280 f (1964).

99. Floris, G. and I. Simonyi, Acta Pharm. Hung. 39, 74 (1969); C.A. 71, 42386 (1969).

100. Regosz, A., Farm. Pol. 24, 797 (1968); C.A. 71, 42391 (1969).

101. Siegmund, P. and F. Korber, Clin. Chem. 14, 808 (1968).

102. Muftic, M., Anal. Biochem. 36, 539 (1970).

103. Boxer, G. and P. Everett, Anal. Chem. 21, 670 (1949).

104. Henstock, H., Nature 164, 139 (1949).

105. Niedermayer, A., F. Russo-Alesi, C. Lendzian and J. Kelly, Anal. Chem. 32, 664 (1960).

106. Russo-Alesi, F., Squibb Institute, private communication.

107. Stevenson, C., L. Bechtel and L. Coursen, *Advan. Automat. Anal., Technicon Int. Cong.* 2, 251 (1969); *C.A.* 73, 80528 (1970).

108. Oedegaard, E. and K. Oeydvin, *Medd. Nor. Farm. Selsk.* 31, 57 (1969); *C.A.* 72, 15791 (1970).

109. Pligin, S. and A. Portnov, *Issled. Obl. Farm.* 1959, 40; *C.A.* 54, 25570 c (1960).

110. Fursov, A., *Aptech. Delo* 11, 63 (1962); *C.A.* 60, 10480 b (1964).

111. Silverman, D., *Anal. Biochem.* 27, 189 (1969).

112. Betina, V., *J. Chromatogr.* 15, 379 (1964).

113. Hellberg, H., *J. Ass. Offic. Anal. Chem.* 51, 552 (1968).

114. Roberts, H., Squibb Institute, private communication.

115. Mc Gilveray, I. and R. Strickland, *J. Pharm. Sci.* 56, 77 (1967).

116. Nussbaumer, P., *Pharm. Acta Helv.* 37, 65 (1962); *C.A.* 57, 961b (1962).

117. Nussbaumer, P., *ibid* 38, 245 (1963); *C.A.* 59, 4977 h (1963).

118. Nussbaumer, P., *ibid*, 758; *C.A.* 60, 1541e (1964).

119. Zuidweg, M., J. Oostendorp and C. Bos, J. Chromatogr. *42*, 552 (1969).

120. Sinsheimer, J., D. Hong and J. Burckhalter, J. Pharm. Sci. *58*, 1041 (1969).

121. Biagi, G., A. Barbaro, M. Gamba and M. Guerra, J. Chromatogr. *41*, 371 (1969).

122. Biagi, G., A. Barbaro and M. Guerra, ibid *51* 548 (1970).

123. Biagi, G., M. Guerra, A. Barbaro and M. Gamba, J. Med. Chem. *13*, 511 (1970).

124. Vedeneeva, V., T. Vikhoreva and G. Samsonov, Tr. Leningrad. Khim.-Farm. Inst. *1968*, 85; C.A. *71*, 53511 (1969).

125. Samsonov, G., V. Vedeneeva, T. Vikhoreva, A. Pashkov, A. Selezneva and E. Trostyanskaya, U.S.S.R. *213*, *261*, March 12, 1968; C.A. *69*, 18033 (1968).

126. Vanderhaeghe, H., E. Evrard and M. Claesen, Inst. Kongr. Pharm. Wiss; Vortr. Originalmitt. 23rd., *1963*, 405; C.A. *62*, 5140 d (1965).

127. Evrard, E., M. Claesen and H. Vanderhaeghe, Nature *201*, 1124 (1964).

128. Thomas, A. and R. Broadbridge, Analyst *95*, 459 (1970).

129. George, M., Squibb Institute, private communication.

130. Platt, T., H. Weisblatt and L. Guevrekian, Ann. N. Y. Acad. Sci. 153, 571 (1968).

131. Gomis, P., M. Izquierdo and A. Jurado, Bull. Soc. Chim. Fr. 1968, 420; C.A. 69, 2898 (1968).

132. Levin, J., Squibb Institute, private communication.

133. Meyer, M. and D. Guttman, J. Pharm. Sci. 57, 895 (1968).

134. Scholtan, W., Antibiot. Chemother. 14, 53 (1968).

135. Ege, H., Dan. 116,361 May 25, 1970; C.A. 73, 98934 (1970)

136. Kinget, R. and M. Schwartz, J. Pharm. Sci. 58, 1102 (1969).

137. Sheehan, J., Ann. N.Y. Acad. Sci. 145, 216 (1967).

138. Manhas, M. and A. Bose, "Synthesis of Penicillin, Cephalosporin C and Analogs", Marcel Dekker, New York, 1969.

139. Watanabe, T., S. Endo and Y. Iida, J. Antibiot. (Tokyo) Ser. A, 15, 112 (1962); C.A. 57, 1213 g (1962).

140. Rolinson, G. and F. Batchelor, Antimicrob. Ag. Chemother. 1962, 654.

141. Vanderhaeghe, H., G. Parmentier and E. Evrard, Nature 200, 891 (1963).

PROPOXYPHENE HYDROCHLORIDE

B. McEwan

CONTENTS

1. Description
    1.1 Name, Formula, Molecular Weight
    1.2 Appearance, Color, Odor, Other Physical Properties
2. Synthesis
3. Identification Techniques With Spectral Data
    3.1 Crystal Properties
    3.2 Infrared Spectra
    3.3 Nuclear Magnetic Resonance Spectra
    3.4 Ultraviolet Spectra
    3.5 Mass Spectra
    3.6 Differential Thermal Analysis
    3.7 Thermogravimetric Analysis
    3.8 Microchemical Tests
4. Methods of Analysis
    4.1 Non-aqueous Titration
    4.2 Infrared Absorption
    4.3 Ultraviolet Absorption
    4.4 Gas Chromatography
    4.5 Visible Absorption (With an Autoanalyzer)
    4.6 Thin Layer Chromatography
5. Stability
6. Metabolism
7. Pharmacokinetics
8. References

1. Description

   1.1 Name, Formula, Molecular Weight
   Propoxyphene Hydrochloride is (+)-α-4-(Dimethylamino)-3-methyl-1,2-diphenyl-2-butanol Propionate Hydrochloride. Since nomenclature can become very burdensome, references will be made to the common name, propoxyphene hydrochloride, whenever possible.

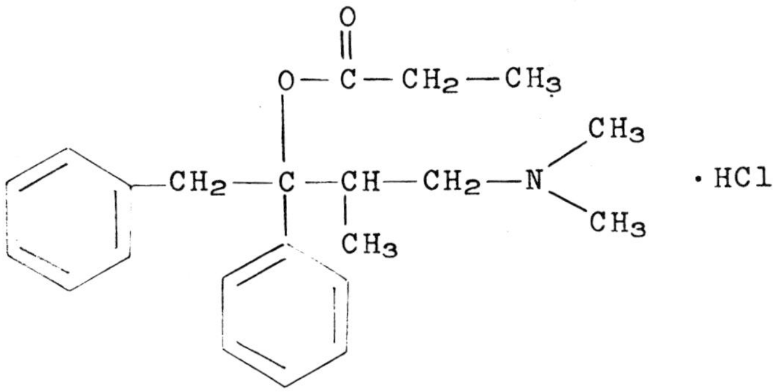

$$\text{structure shown}$$

$C_{22}H_{29}NO_2 \cdot HCl$ M.W. = 375.94

This molecule has two asymmetric centers, thus four diastereoisomers are possible. Here we are concerned only with the "α-d" isomer.

   1.2 Appearance, Color, Odor, Other Physical Properties
   Propoxyphene hydrochloride is a white, crystalline powder with no noticeable odor and a bitter taste. The melting range is 162.5°C - 168.5°C. The specific rotation for a one percent solution should be between +52° and +57°[1]. Propoxyphene hydrochloride has a unique characteristic, in that the specific rotation value has an apparent dependency upon the concentration of the solution being measured. An $[\alpha]_D^{25}$ value of +59.8° is given in literature[2] (for a 0.6 percent solution).

2. Synthesis

The original synthesis was of the "α-dl" mixture[3]. In brief, this synthesis is as follows: A solution of benzylmagnesium chloride (in ether) was added to a solution of α-methyl-β-dimethylaminopropiophenone (in ether), this solution was refluxed one hour. The reaction mixture was decomposed with saturated aqueous ammonium chloride. The ether solution containing the 1,2-diphenyl-2-hydroxy-3-methyl-4-dimethylaminobutane was decanted and then treated to form the hydrochloride. This product was then refluxed with propionic anhydride (in pyridine) for five hours. The reaction product was then precipitated and purified. The final product was α-dl-1,2-diphehyl-2-propionoxy-3-methyl-4-dimethylaminobutane hydrochloride.

Later investigations showed that the "α-d" form was the analgesic, while the "α-l" form exhibited no analgesic properties. A description of the resolution of the "α-dl" mixture is found in literature[2]. Essentially, the resolution is achieved by fractional crystallization of the d-camphorsulfonic acid salt. This technique allows production of the pure "α-d" isomer now known as propoxyphene hydrochloride.

3. Identification Techniques With Spectral Data

3.1 Crystal Properties

A comprehensive study of the crystallography of propoxyphene hydrochloride is documented[4]. All the following crystallographic data are from this article.

Crystallization from ethyl acetate resulted in orthorhombic prisms elongated parallel to the b axis. These crystals showed the prism {110} and the orthodome {101}.

PROPOXYPHENE HYDROCHLORIDE

The interfacial angles are 110>$\bar{1}$10 (polar), 94° 18' (optical), 93° 58' (X-ray); 101>$\bar{1}$01 (polar), 86° 03' (optical), 86° 10' (X-ray).

Optical crystallographic data (5893Å, 25°C) α = 1.560, β = 1.582, γ = 1.638

Optical Axial Angle (+)2V = 66° (Calculated from α, β, and γ).

Optic Axial Plane 100

Acute Bisectrix γ = C

X-ray crystallographic data
Unit cell dimensions a_o = 12.83Å, b_o = 13.75Å, c_o = 12.00Å
Formula weight per cell is 4.
Formula weight is 375.94
Density 1.173 gm/cc (flotation); 1.181 gm/cc (X-ray).
Axial Ratio a:b:c = 0.9331:1:0.8727
Space Group $P2_12_12_1$
Tabular data on X-ray powder diffraction:

d(Å)	I/I$_1$	d(Å)	I/I$_1$
9.50	3	3.01	7
8.79	33	2.92	13
7.40	20	2.86	7
6.39	3	2.78	3
6.02	100	2.70	3
5.60	3	2.62	7
5.38	3	2.52	13
5.06	27	2.46	7
4.55	20	2.40	3
4.38	20	2.34	7b
4.08	67b	2.27	3
3.86	7	2.23	3
3.76	20	2.14	3b
3.65	7	2.09	3b
3.50	20	2.02	3

3.35	7	2.00	3
3.21	13	1.95	3
3.10	7	1.89	3

3.2 Infrared Spectra

An IR spectrum of propoxyphene hydrochloride is shown in figure No. 1. This is a solid state spectrum (KBr pellet) run on a Perkin-Elmer 221. The broad band at 2360 cm^{-1} is typical or tertiary amine hydrochlorides. The bands at 1725 cm^{-1} and 1170 cm^{-1} are indicative of a saturated ester linkage. The bands at 762 cm^{-1} and 703 cm^{-1} show the presence of a monosubstituted benzene ring. Solution spectra (in $CHCl_3$) are very similar in detail to that shown by the KBr pellet spectrum. Lilly Lot No. P-88455 is the sample used for all of the spectral data. The operating parameters are as follows: Sample weight 1.26 mg in 200 mg of KBr, prism-NaCl, resolution - 927, response - 1100, gain - 6.0, speed - 30, suppression - 5.5, and scale - 1X.

3.3 Nuclear Magnetic Resonance Spectra

A NMR spectrum of propoxyphene hydrochloride is shown in figure No. 2. This is a solution ($CDCl_3$) spectrum run on a Varian A-60 spectrometer. A triplet at ~1 δ indicates $-CH_2-CH_3$. A doublet at ~1.3 δ is typical of $-CH-CH_3$. The unresolved quartet at ~2.3 δ is due to $CH_3-CH_2-\overset{O}{\overset{\|}{C}}-$. The large peak at ~2.7 δ is actually due to two overlapping doublets which arise from the $-N\overset{HCl}{\underset{CH_3}{\diagup\kern-0.5em CH_3}}$. In theory these protons should merge into one single peak, but a small excess or deficiency of ·HCl leads to the overlapping doublets or apparent singlet. The singlet at ~3.8 δ indicates $\equiv C-CH_2-\overset{|}{C}-$. The unresolved multiplet at ~3.5 δ is due to $CH_3-\overset{|}{C}H-CH_2-$. The multiplet at ~7.2 δ identifies aromatic hydrogens. The broad peak at ~11.5 δ

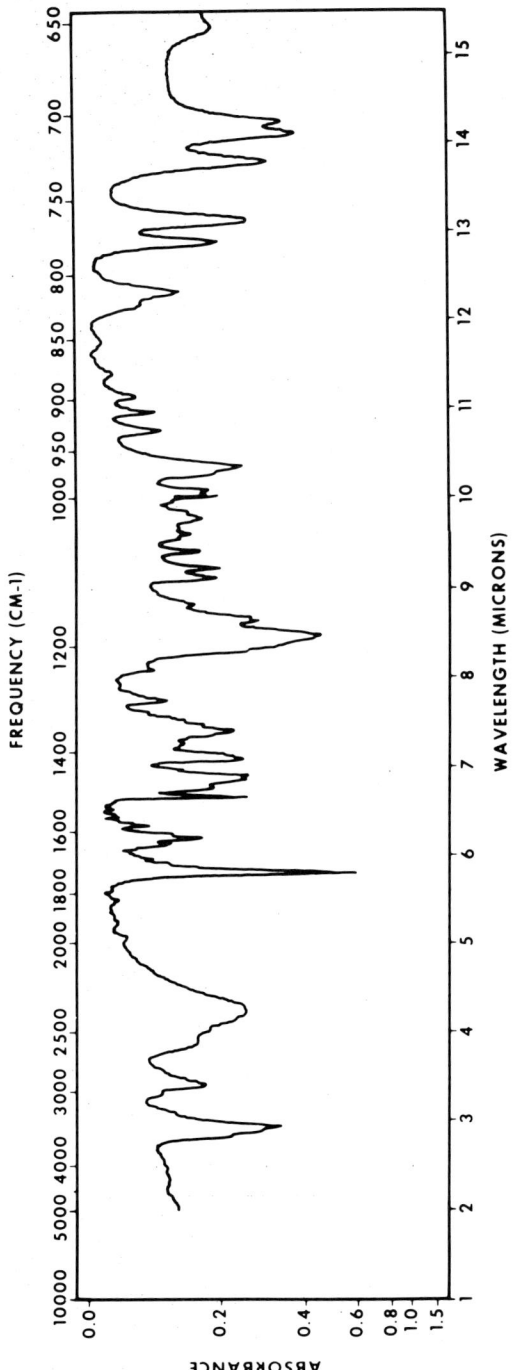

Fig. 1.

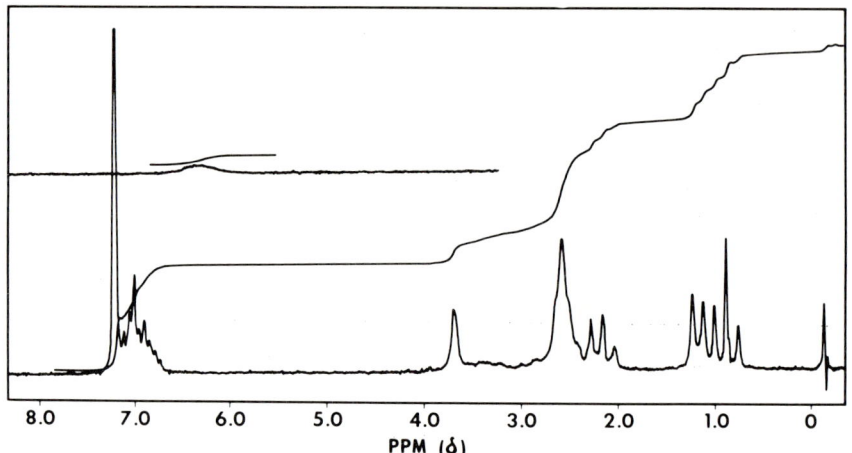

Fig. 2.

is due to ·HCl. All the integrated areas (shown by upper line in figure) correlate with the above assignments. The operating parameters of the A-60 are: temperature-ambient, filter band width - 4, R. F. field - 0.14, sweep time - 250, sweep width - 500, sweep offset - 300, spectrum amplitude - 5.0, and integral amplitude 20.0.

3.4 Ultraviolet Spectra

An ultraviolet spectrum of propoxyphene hydrochloride is shown in figure No. 3. This is a solution spectrum (41.04 mg 25 ml; 10 25 dilution using 95% ethanol) run on a Cary 14. The λ max of 242, 247, 252, 258, 264, and 267 nm are typical of fine structure due to isolated benzene chromophore. Using the spectrum and the molar concentration of propoxyphene hydrochloride (1.94 x 10^{-3} moles/liter), the following table can be constructed.

λ(nm)	Absorbance (corrected)	Σ
242	0.266	137
247	0.389	200
252	0.541	279
258	0.698	360
264	0.559	288
267	0.303	156

The operating parameters for this spectrum are: slit program-50, dynode-2, and speed 5A/sec, with use of 1-cm silica cells.

3.5 Mass Spectra

The mass spectrum of propoxyphene hydrochloride was run on a Hitachi Perkin-Elmer RMU-6 mass spectrometer (low resolution). Since the salt (·HCl) ionizes so rapidly, no m/e 376 is seen. Propionoxy is also lost so rapidly that no m/e 339 (for the free base) is found. Three likely cleavages are: (1) The loss of $CH_3-CH_2-\overset{O}{\overset{\|}{C}}-O-$ (m/e 266), (2) loss of $CH_3-CH_2-\overset{O}{\overset{\|}{C}}-O-$ and $-CH_2-N-(CH_3)_2$ (m/e 208) and (3) loss of $CH_3-CH_2-\overset{O}{\overset{\|}{C}}-O-$, $-CH_2-N-(CH_3)_2$, and $-CH_3$ (m/e 193).

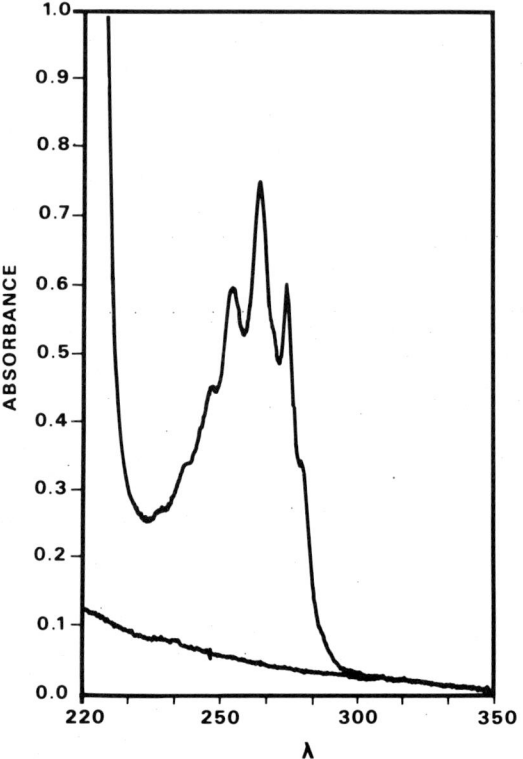

Fig. 3.

Using the m/e 208 as 100%, the intensity (relative to m/e 208) of m/e 193 is 67% and of m/e 266 is 1.3%. The sampling was done at 130°C.

3.6 Differential Thermal Analysis

Figure No. 4 is a DTA curve of propoxyphene hydrochloride. A DuPont 900 D.T. Analyzer using 2mm microtubes and N_2 atmosphere is used to obtain this curve. Two sharp phase transitions are shown, the first at 169°C and the second at 233°C. Melting is associated with the first transition, and decomposition with second. The operating parameters are: glass beads for reference, rate of heating 20°C/min, scale 20°C/in and ΔT of 0.5°C/in.

3.7 Thermogravimetric Analysis

Figure No. 5 shows a TGA curve of propoxyphene hydrochloride. Instrumentation is a DuPont 950 TGA unit. This curve shows a 1% weight loss at 169°C, 5% at 191°C, and 20% at 205°C. A sample size of 10.25 mg, temperature scale of 20°C/in, heating rate of 5°C/min, and a N_2 atmosphere are the operating parameters.

3.8 Microchemical Tests

In addition to instrumental methods of identification, an analyst may make use of some method of microchemical testing. An excellent discussion of the latter is available[5]. This work is a systematic study of precipitate tests, color tests, and crystal formation for modern analgesics. Through manipulation of the various testing parameters, Clarke identified a large number of analgesics. Even with the sophisticated instrumentation now available, "wet" methods of identification are still very useful.

While all types of spectral data can be compiled, the data of most interest to the analytical chemist are those which lend themselves to a quantitative assay. Infrared, ultraviolet, and visible absorption spectroscopy plus

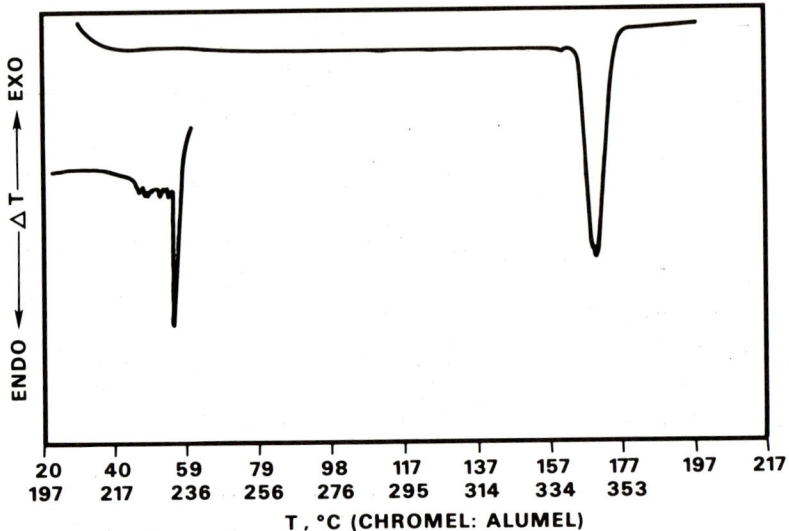

Fig. 4.

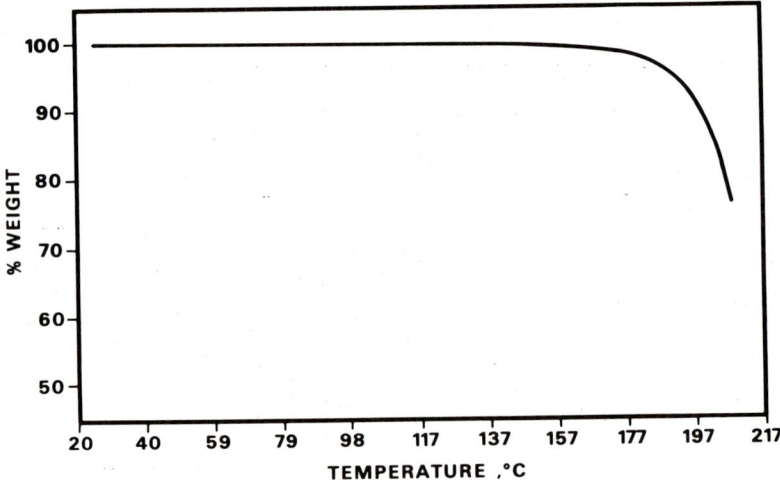

Fig. 5.

gas chromatography are excellent methods for the assay of propoxyphene hydrochloride.

4. Methods of Analysis

4.1 Non-aqueous Titration

The non-aqueous titration (NAT) of propoxyphene hydrochloride is a fast, simple procedure[1]. Accurately weigh about 600 mg. of sample, dissolve in 40 ml. of glacial acetic acid and add 10 ml. of mercuric acetate T.S. Add crystal violet T.S. and titrate with 0.1N perchloric acid (taking care to use the same end point as in the standardization of the perchloric acid). Perform a blank titration and correct the sample titration when necessary. Each ml. of 0.1N perchloric acid is equivalent to 37.59 mg of propoxyphene hydrochloride.

4.2 Infrared Absorption

In most forms of quantitative spectroscopy, the sample and standard should be handled identically. In this IR assay[6], one accurately weighs about 130 mg. of sample and of USP Propoxyphene Hydrochloride reference standard; transfer both (quantitatively) to 125 ml. separatory funnels, containing 25 ml. of water. Add 0.4 ml. of sodium hydroxide solution (1 in 2) and 50 ml. of chloroform. Extract for 3 minutes and then allow the layers to separate. Drain the organic phase through anhydrous sodium sulfate into a 250 ml. beaker. Repeat the extraction with 3 - 50 ml. portions of chloroform and pool them in the 250 ml. beaker. Evaporate (on steam bath with air) to a small volume; then transfer (quantitatively) to a 50 ml volumetric flask, dilute to volume with chloroform and mix. Using a suitable infrared spectrophotometer and 1 mm. cells, read the sample and standard at the maximum (about 5.80 μ) using chloroform as the blank. The calculation is then: $\frac{samp.\ Abs.}{Std.\ Abs.} \times \frac{Std.\ wt.}{samp.\ wt.} \times 100 = \%$ propoxyphene hydrochloride.

As an alternate IR assay, accurately weigh about 130 mg. of sample and <u>USP Propoxyphene Hydrochloride</u> reference standard. Quantitatively transfer both to 50 ml. volumetric flasks and dilute with chloroform. Read both at the maximum (about 5.75 μ) with chloroform as the blank, using 1 mm. cells and a suitable spectrophotometer. The calculation is the same as for the "extracted" IR assay. For a true measure of propoxyphene hydrochloride content in a sample, the extraction technique is better. The extraction technique removes many (if not all) of the acidic contaminants which absorb in the carbonyl region.

4.3 <u>Ultraviolet Absorption</u>
Accurately weigh about 25 mg of sample and of <u>USP Propoxyphene Hydrochloride</u> reference standard; transfer both (quantitatively) to 100 ml. volumetric flasks, dilute to volume with purified water and mix. Determine the absorbance of both solutions at the maximum (about 257 mμ)[7] using 1 cm. silica cells with purified water as the blank, on a suitable UV spectrophotometer. The calculation is:

$$\frac{\text{samp. Abs.}}{\text{Std. Abs.}} \times \frac{\text{Std. wt.}}{\text{samp. wt.}} \times 100 = \%\ \text{propoxyphene hydrochloride.}$$

4.4 <u>Gas Chromatography</u>
Most spectroscopic techniques quantitate results by comparison of sample values with standard values. In gas chromatography, the preferred quantitative technique makes use of an internal standard. For the assay of propoxyphene hydrochloride, pyrroliphene hydrochloride, is an excellent choice[8]. The multi-extraction of sample and standard is described, as well as the operating parameters of the gas chromatograph. Individual analysts usually have preferences on operating conditions. For this assay a flame ionization detector is desirable, because of its sensitivity. Detection of microgram quantities is practicable under the conditions described by

Wolen and Gruber[8]. Other investigators have used much the same technique[9].

4.5 Visible Absorption (With an Autoanalyzer)

Propoxyphene hydrochloride can be assayed by visible absorption spectroscopy. An automated assay of this type is found in literature[10]. Since a variety of dyes will complex with tertiary amines, studies were undertaken to determine which dye is most specific and has the least retention on the analytical train. Bromocresol purple seems best suited for this type assay. Essentially the assay is as follows: an aqueous sample (or standard) is mixed with reagent, chloroform is used to extract the dye complex, the extract is then read on a colorimeter (at 420 mµ) and recorded on a linearized recorder. Comparison of sample and standard peak heights enable the analyst to quantitate the propoxyphene hydrochloride. The same assay can be performed by manual extraction, but the automated assay increases assay output nearly twenty fold. Twenty samples per hour can be assayed with a relative standard deviation of 0.5 - 1.2% (for 5 observations). Thus good precision is demonstrated. Accuracy equal to that of a manual IR assay is shown by tabular data.

4.6 Thin Layer Chromatography

TLC of propoxyphene hydrochloride (and other analgesics) is well documented[11]. Emmerson and Anderson give Rf values for thirteen solvent systems. The use of slightly alkaline absorption layers and ammonium saturated developing chambers is discussed in detail. The slight alkalinity facilitates spot movement on the plates. An iodoplatinate spray was used to locate the spots. For true quantitative results the spots should be removed and the propoxyphene hydrochloride determined by UV or other instrumental methods[7].

Another source[12] gives Rf values for five solvent systems. These systems are unique because they are "salted" (with either ammonium chloride or sodium chloride), which results in greatly reduced tailing and zone diffusion.

5. Stability

An interesting use was made of solubility analysis in a stability study of propoxyphene hydrochloride[13]. In this study, propoxyphene hydrochloride was stored at 80°C, 105°C, and 130°C. Both the solubility analysis and the alternate IR assay showed the thermal stability of propoxyphene hydrochloride. Gas chromatography can be used to measure the slow hydrolysis of propoxyphene hydrochloride in aqueous solutions. The hydrolytic products are propionic acid and (+)-α-4-(Dimethylamino)-3-methyl-1,2-diphenyl-2-butanol. With suitable gas chromatographic conditions either product can be quantitated. A chloroform extraction of an alkaline solution of propoxyphene hydrochloride (where degradation has occurred) can be assayed using the previously described IR method.

6. Metabolism

At present only one metabolite of propoxyphene hydrochloride has been discovered. This metabolite, des-N-methyl propoxyphene, is produced by enzymatic N-demethylation in the liver. To study this metabolic process, propoxyphene hydrochloride labelled with C^{14} in the N-methyl position was utilized[14]. Laboratory rats and human beings were dosed with the labelled analgesic[15]. $C^{14}O_2$ was detected in incubates of rat liver and in the expired air of rats. The dinitrophenyl derivative of des-N-methyl propoxyphene was isolated from human urine.

7. Pharmacokinetics

Emmerson, Welles, and Anderson[16] used C^{14} labelled propoxyphene hydrochloride to study tissue distributions. The patterns of tissue distribution differ according to the route of administration. Thus, such parameters as elimination rates and distribution rates depend on the route of administration as well as the specific tissue being examined. A pharmacokinetic model would have to be formulated for each route of administration in conjunction with the tissue in question. These models and the associated mathematics are beyond the scope of this paper.

References

1. U.S. Pharmacopeia XVIII, p. 556.
2. A. Pohland and H. R. Sullivan, J. Am. Chem. Soc. $\underline{77}$, 3400-1 (1955).
3. U.S. Patent 2,728,779 (Patented Dec. 27, 1955).
4. Harry A. Rose, J. Am. Pharm. Assoc. $\underline{47}$, 228 (1958).
5. E. G. C Clarke, Bull. Narcotics $\underline{11}$, No. 1, 27-44 (1959).
6. National Formulary XIII, p. 606-8.
7. S. J. Mule, Anal. Chem. $\underline{36}$, 1907-14 (1964).
8. R. L. Wolen and C. M. Gruber, Anal. Chem. $\underline{40}$, 1243 (1968).
9. L. B. Foster and C. S. Frings, Clin. Chem. $\underline{16}$, 177-179 (March 1970).
10. N. Kuzel, J. Pharm. Sci., $\underline{57}$ (5), 852-5 (1968).
11. J. L. Emmerson and R. C. Anderson, J. Chromatog. $\underline{17}$ (3), 495-500 (1965).
12. J. A. Manthey and M. E. Amundson, J. Chromatog. $\underline{19}$, 522-526 (1965).
13. J. P. Comer and L. D. Howell, J. Pharm. Sci., $\underline{53}$ (3), 335-7 (1964).
14. A. Pohland and H. R. Sullivan, J. Am. Chem. Soc. $\underline{79}$, 1442-4 (1957).
15. H. Lee, E. Scott, and A. Pohland, J. Pharmacol. Exptl. Therap. $\underline{125}$, 14-18 (1959).
16. J. L. Emmerson, J. S. Welles, and R. C. Anderson, Toxicol. Appl. Pharmacol. $\underline{11}$ (3), 482-8 (1967).

Special thanks go to Mr. C. D. Underbrink and Dr. A. D. Kossoy of the Analytical Development Department at Eli Lilly and Company for their help in gathering and interpreting the spectral data in sections 3.2 through 3.7.

SODIUM CEPHALOTHIN

R. J. Simmons

CONTENTS

1. Description
    1.1 Name, Formula, Molecular Weight
    1.2 Appearance, Color, Odor
2. Physical Properties
    2.1 Infrared Spectra
    2.2 Nuclear Magnetic Resonance Spectra
    2.3 Mass Spectra
    2.4 Ultraviolet Spectra
    2.5 Optical Rotation
    2.6 Solubility
    2.7 Differential Thermal Analysis
    2.8 Thermogravimetric Analysis
    2.9 Crystal Properties
3. Synthesis
    3.1 Biosynthesis
    3.2 Chemical Synthesis
4. Stability - Degradation
5. Drug Metabolic Products - Pharmacokinetics
6. Methods of Analysis
    6.1 Elemental Analysis
    6.2 Microbiological Assay
    6.3 Iodometric Titration
    6.4 Colorimetric Analysis
    6.5 Ultraviolet Absorption
    6.6 Non-Aqueous Titration
    6.7 Polarographic Analysis
    6.8 Chromatographic Analysis
        6.81 Paper
        6.82 Thin Layer
7. Determination in Body Fluids and Tissues
8. References

SODIUM CEPHALOTHIN

1. **Description**

   1.1 **Name, Formula, Molecular Weight**
   Cephalothin is 3(Hydroxymethyl)-8-oxo-7-[2-(2-thienyl)-acetamido]-5-thia-1-azabicyclo[4.2.0]oct-2-ene-2-carboxylic acid acetate, and is also known as 7-(2-thienylacetamido) cephalosporanic acid. The antibiotic is supplied as the sodium salt.

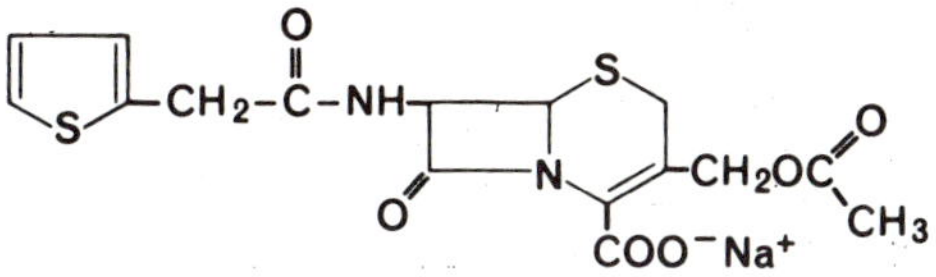

   $C_{16}H_{15}N_2NaO_6S_2$ Mol. Wt.: 418.43

   1.2 **Appearance, Color, Odor**
   White to off-white crystalline powder having essentially no odor.

2. **Physical Properties**

   2.1 **Infrared Spectra**
   The infrared spectrum of sodium cephalothin presented in Fig. 1 was taken in a KBr pellet.[1] A spectrum of the same standard taken in a Nujol Mull was essentially identical to the one presented. Characteristic stretching frequencies (cm^{-1}) of sodium cephalothin were as follows:
   a. N-H stretching band: 3300
   b. β-lactam carbonyl: 1760
   c. ester carbonyl: 1735
   d. secondary amide carbonyl: 1660 and 1535

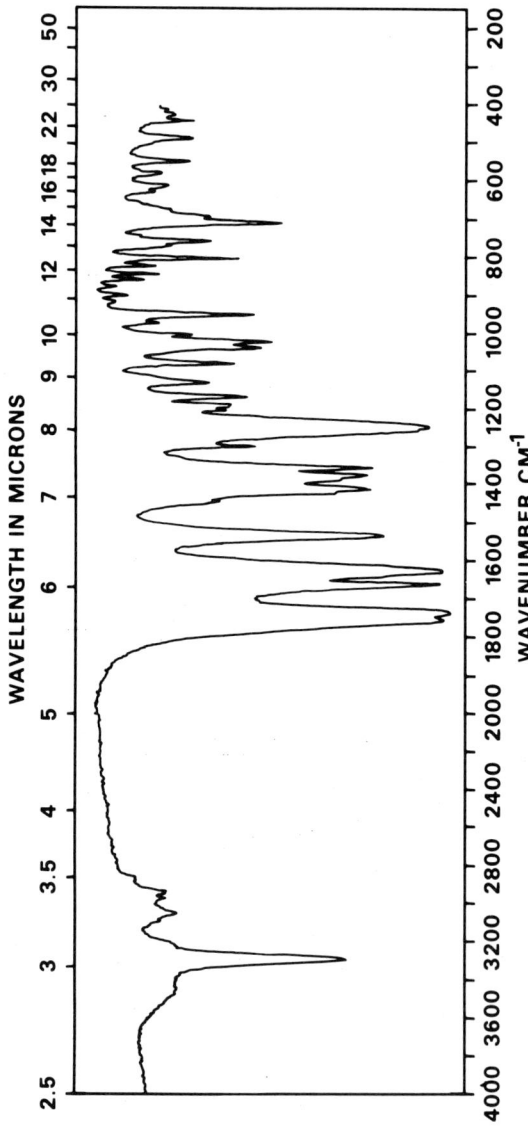

Fig. 1. Infrared spectrum of sodium cephalothin.

e. carboxyl carbonyl: 1630

f. C-O stretching band ($CH_3\text{-}\overset{\overset{O}{\|}}{C}\text{-}O$): 1250

The carbonyl stretching region, which exists between 1500 and 1800 cm^{-1}, is the most characteristic region in the infrared spectra of cephalothin. Ester cleavage, lactone formation, and opening of the β-lactam can be indicated by changes in this part of the spectrum. Of the three carbonyl stretching frequencies (β-lactam, ester, amide), the one of most diagnostic value is that originating from the β-lactam carbonyl. The importance of this stretching frequency has been discussed in a recent review[2], and characteristic stretching frequencies for the different carbonyl groups in many derivatives of cephalosporins has been reported[3].

2.2 Nuclear Magnetic Resonance Spectra

The NMR spectrum Fig. 2 was obtained by preparing a solution of sodium cephalothin in deuterated water. The spectral assignments shown in Fig. 2 have been discussed in detail by DeMarco and Nagarajan.[2] Interpretation and assignment of the absorption and resonance frequencies to the different atomic features of many cephalosporin derivatives has been discussed.[2,3]

NMR spectroscopy has been the most useful tool in cephalosporin C chemistry. In cephalosporins the carbons are unsaturated or highly substituted with heteroatoms, and the protons are usually widely separated in chemical shift and have simple coupling patterns. Recently, solvent induced chemical shifts, nuclear Overhauser effects, and the anisotropy of the sulfoxide bond have been utilized in chemical studies of cephalosporin C derivatives.[2] Analytical information may be derived from NMR spectra of cephalothin by observing the contribution of the β-lactam protons, thiophene protons, methylene groups, and methyl protons (from acetate).

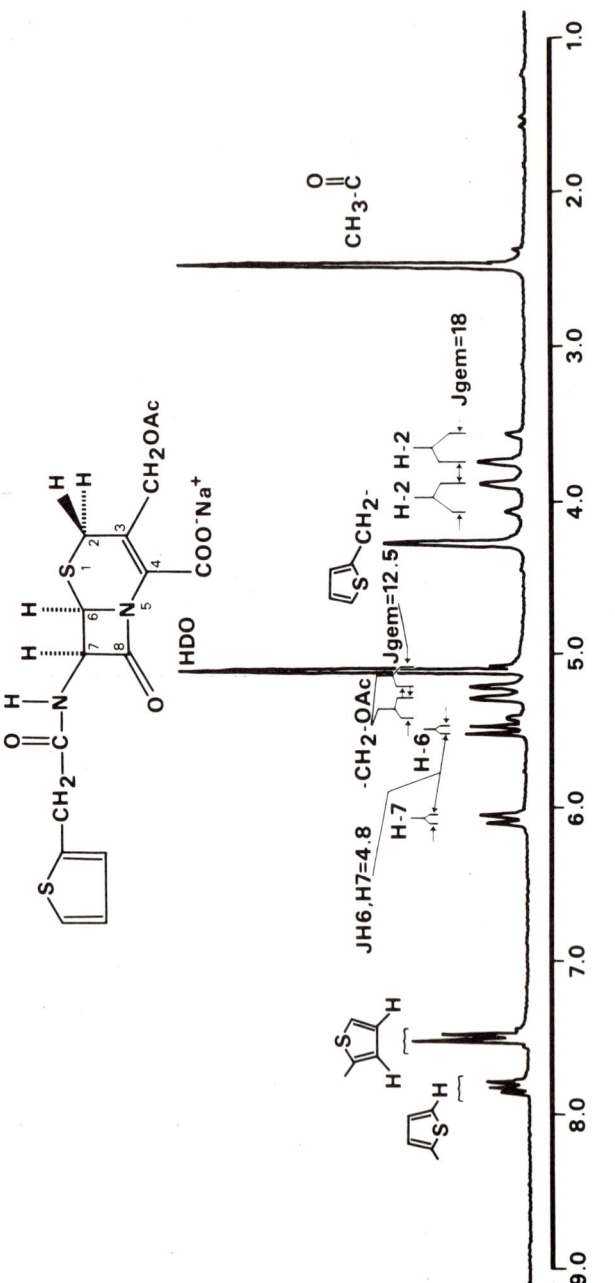

Figure 2. NMR Spectrum of sodium cephalothin.

2.3 Mass Spectra

On electron impact in the mass spectrometer, the sodium salt of cephalothin did not yield a satisfactory spectrum. However, the free acid of cephalothin gave a mass spectrum when the sample was flashed into the ion source using an ion source temperature of about 300 C and a separately heated direct probe into the ion source.[4]

The spectrum of the free acid showed no molecular ion. Ions of highest mass, m/e 304 and 292, probably arose from the free acid by the hydrolytic elimination of acetic acid and sulfur, and acetic acid and carbon dioxide respectively. An intense peak at m/e 44 (CO_2) supported this conclusion. Peaks at m/e 216 and 215, expected from the fission across the β-lactam ring, were not observed. Three peaks which may be assigned to the side chain and part of the β-lactam ring were m/e 97 ($[S]$-CH_2^+), m/e 124 ($[S]$-CH-C≡O$^+$), and m/e 181 ($[S]$-CH_2-CO-NH-CH=C=O$^+$). Occolowitz concluded that the spectrum arose by ionization of hydrolysis products of the free acid.

For detailed information relating to the mass spectral analysis of cephalosporin C derivatives, the reader is referred to the paper by Richter and Biemann[5] and to the review by DeMarco and Nagarajan[2].

2.4 Ultraviolet Spectra

When scanned between 220 and 310 nm, an aqueous solution (25 μg/ml) of sodium cephalothin exhibited maxima at 237 nm ($E_{1\,cm}^{1\%}$ = 336) and at 265 nm ($E_{1\,cm}^{1\%}$ = 204).[6] The absorption at 237 nm is mainly thienyl contribution whereas the absorption at 265 nm is due to the fused ring system, 7-aminocephalosporanic acid.

2.5 Optical Rotation

The specific rotation in a 5% (w/v) aqueous solution was $[\alpha]_D^{25}$ + 129° ± 5° calculated with reference to the anhydrous substance.[6]

2.6 Solubility

Marsh and Weiss[7] determined the solubilities in 26 solvents at 21 ± 1 C. In general, sodium cephalothin is very soluble in water, formamide, dimethyl sulfoxide, and ethylene and propylene glycol, slightly soluble in lower alcohols, and insoluble in non-polar organic solvents.

2.7 Differential Thermal Analysis

A differential thermal analysis was performed on sodium cephalothin (Eli Lilly reference standard #P-89448). At a heating rate of 20 C/min. an exotherm peaked at 220 C indicating decomposition.[8]

2.8 Thermogravimetric Analysis

A thermogravimetric analysis performed on sodium cephalothin (Eli Lilly reference standard #P-89448) showed no weight loss until 154 C; at 154 C weight loss begins resulting in decomposition.[8] The measurement was performed under nitrogen sweep at a heating rate of 5 C/min.

2.9 Crystal Properties

The optical crystallographic properties of cephalothin was determined by Rose.[9] The data reported were as follows:
Optical Crystallographic Data:
Refractive indices: $\alpha=1.568$, $\beta=1.592$, $\gamma=1.684$
Optical axial angle: $\oplus 2V=56° 34'$
Orientation: OAP=010, $\gamma=a$
X-Ray Diffraction Data:
Unit cell dimensions: $a_o=11.00$ Å., $b_o=34.20$ Å., $c_o=5.05$ Å.
Formula weights per cell: 4
Formula weight: 418.4
Density: 1.477 Gm/ml by flotation, 1.463 Gm/ml by X-ray
Axial Ratio: 0.3216:1:0.1477
Space Group: $D_2^4-P2,2,2,$

The X-ray powder diffraction pattern of cephalothin is presented in Table I. Data were obtained using copper radiation and nickel filter with a camera 114.6 mm in diameter. Indexing of the powder pattern was done on the basis of single crystal rotation patterns around both the a and c axes, and a wavelength of 1.5418 Å was used in the calculations.

3. Synthesis

3.1 Biosynthesis

Cephalothin is a semisynthetic β-lactam antibiotic prepared by the reaction of thiophene-2-acetyl chloride with 7-aminocephalosporanic acid (7-ACA).[10] The cephalosporanic nucleus is obtained from cephalosporin C which is produced by a fungus identified as Cephalosporium acremonium No. 49137 (Imperial Mycological Institute, England), or strains thereof.

Although cephalosporin C is divisable into α-aminoadipic acid, cysteine, and valine, the actual mechanism whereby Cephalosporium sp. incorporates the three amino acids into cephalosporin C has not been established. Arnstein and Morris[11] isolated δ (α-aminoadipyl) cysteinyl valine from mycelia of Penicillium chrysogenum and suggested that the tripeptide is a precursor in all penicillin biosynthesis. This same tripeptide also appears to be found in the intracellular pool of Cephalosporium sp.[12] The final postulated step in the biosynthesis of penicillin is an acyl transfer reaction, or the production of 6-aminopenicillanic acid if precursor is not added. Cephalosporium sp. apparently do not produce sidechain amidases or acyl transferases, and no 7-ACA has been reported found in the fermentation. Thus, to obtain clinically useful antibiotics, chemical manipulation of cephalosporin C is necessary. Synthesis of many 7-acyl derivatives was possible once a practical cleavage reaction made available large amounts of 7-ACA from cephalosporin C.[13] Of these derivatives, sodium cephalothin was the first

TABLE I

X-Ray Powder Diffraction

d	I/I_1	hKl	d (calcd.)
16.52	0.27	020	17.10
10.53	0.50	110	10.47
9.21	0.50	120	9.25
6.71	0.20	140	6.75
5.71	0.07	060	5.70
5.44	0.07	210	5.43
5.20	0.27	220	5.24
5.05	0.27	160	5.06
4.82	0.07	021	4.84
4.49	1.00b	114	4.55
		170	4.47
		121	4.43
4.25	0.67	080	4.28
		131	4.26
4.05	1.00	051	4.06
		141	4.04
3.78	0.03	061	3.78
3.67	0.07	211	3.70
3.59	0.03	190	3.59
		320	3.59
3.49	0.50	071	3.51
		330	3.49
3.36	0.13	340	3.37
3.24	0.20	081	3.26
3.12	0.20	181	3.13
3.01	0.13	1110	2.99
2.91	0.20	2100	2.90
2.86	0.03	0120	2.85
2.78	0.03	341	2.80
2.72	0.13	420	2.72
2.65	0.07	0111	2.65
2.61	0.07	440	2.62
2.53	0.13	002	2.53
2.43	0.03		
2.39	0.03		
2.36	0.03		
2.29	0.07		
2.24	0.03		

one marketed (Patent: Eli Lilly and Co., by
E.H. Flynn, Belg. 618, 663, Dec. 7, 1962; U.S.
Appl. June 8, 1961; Chem. abstracts 59, 5176,
1963).

3.2 Chemical Synthesis

After the structural details of cephalosporin C were reported [14,15], many chemists attempted its synthesis. Efforts were directed toward fusion of a β-lactam to a dihydrothiazine ring[16], conversion of a penicillin nucleus into a cephalosporin nucleus[17], and construction of the β-lactam ring from L(+) cysteine as the initial monocycle from which to synthesize the cephalosporin nucleus[18]. The latter effort resulted in the stereospecific total synthesis of cephalosporin C and cephalothin. A review[19] conveniently supplies details of the biosynthesis and chemical synthesis of cephalosporins.

4. Stability - Degradation

Solid dry sodium cephalothin, stored in tightly closed glass containers and protected from moisture, is stable for at least three years at 25 C. Aqueous solutions held at 25 C for 24 hours lost approximately 8% activity, and rate of loss of activity was about the same in buffers between pH values of 3.0 and 7.0. Ampoules of cephalothin reconstituted in saline, U.S.P. water for injection, or 5% dextrose maintained label potency after 3 days storage at 4 C. Cephalothin in water solution only slowly hydrolyzed to produce deacetylcephalothin, and under mild acid conditions the deacetyl compound and cephalothin was converted into cephalothin lactone.

Deacetylcephalothin

Cephalothin lactone

Alkaline solutions above pH 8.0 rapidly lose biological activity on standing at room temperature. The β-lactam is more stable to strong acids than to strong bases, but under vigorous conditions the entire nucleus was disrupted to biologically inactive thienylacetamidoacetaldehyde.[20]

The cephalosporin nucleus was reported[21] to be labile to ultraviolet light (260 nm). The decomposition of an aqueous solution of cephalosporin C, measured by loss of biological activity, was 90% within half an hour.

Activity of cephalothin in human serum was fully maintained for a period of 14 days at -20 C. At 5 C, 12% inactivation occurred in 2 days and 50% in 14 days. At room temperature inactivation was rapid and only 15% of the original activity was detected after 2 days. Wick[22] reported that cephalothin and its deacetyl metabolite were stable in human sera incubated at 37 C for 2.5 hours. Lee et al.[23] reported finding very little, if any, deacetylcephalothin in serum and heparinized whole blood after incubation with cephalothin for 1 hour at 37 C. This suggested that neither serum nor whole blood contained a significant amount of esterase capable of hydrolyzing cephalothin.

Chemical degradations of cephalosporin C and its analogues did not lead to structural equivalents of penicillin degradation products.[24] Cephalosporins did not give the expected analogue of penicilloates or penicillenates, nor did they form penicillamine. Reactions with alcohols, in either the presence or absence of metal salts, failed to cleave the lactam ring, whereas reaction with alkoxide resulted in expulsion of the acetate grouping.[25]

Opening of the β-lactam ring of cephalothin with β-lactamases was found to be accompanied by expulsion of acetate and striking changes in ultraviolet absorption spectra.[26] The reaction yielded labile compounds with maximum at 230 nm that disappeared in several hours. The tentative structures of the compounds formed on hydrolytic degradation with β-lactamase or ammonia have been reported[27], and support for the proposed structures has been obtained from studies of proton-magnetic-resonance spectra.[28]

Acyl esterases may attack cephalothin at the acetyl function attached to the C-3 methyl to yield deacetylcephalothin. Although quite stable under physiological conditions[22], the deacetyl compound was unstable when its β-lactam was opened.[26] However, cephalothin lactone was exceptional. The presence of the five-membered lactone caused the reaction with β-lactamases to proceed as with the penicillins and a stable reaction product formed.[29]

Amidases were capable of splitting off relatively non-polar sidechains from cephalosporanates.[30,31] However, the naturally occurring α-aminoadipic amide of cephalosporin C was resistant to amidases.

5. Drug Metabolic Products - Pharmacokinetics

Cephalothin was partially converted to deacetylcephalothin after parenteral administration to experimental animals and to man.[23] In the dog, initial excretion was distributed equally between cephalothin and its deacetyl metabolite; later excretion showed a preponderance of the metabolite over the parent compound. Cephalothin persisted over a longer period of time when administered by the intramuscular route than when given intravenously. In man, the total amount excreted in the urine was measured after intramuscular administration of 1 gram of cephalothin. Recovery of cephalothin and its deacetyl metabolite averaged 460 and 240 mg, respectively, a ratio of 2:1 in favor of cephalothin.

Deacetylcephalothin is biologically active. In vitro studies[22] demonstrated that the metabolite has an antimicrobial spectrum similar to cephalothin, but from 2 to 16 times more metabolite is needed for inhibition of representative strains of test organisms. Although data on therapeutic effectiveness of deacetylcephalothin in man are not available, treatment of experimental infections in mice has been studied.[22] The curative effect of the metabolite was less than that of cephalothin, a finding in agreement with results of the in vitro evaluation.

Urinary excretion was the major route of elimination for cephalothin both in men and in dogs.[23,32] From 60 to 90% of the dose appeared in the urine during the 6 hours following injection. Renal clearance studies in dogs indicated that the antibiotic activity was secreted through the renal tubules, and this tubular secretion could be completely blocked by probenecid.

Serum reached peak levels within ½ hour after administration.[23,32] The microbiological half-life in serum of dogs averaged 42.3 ± 2.5 minutes for cephalothin[23] as compared with 25.9 ± 0.8 minutes for deacetylcephalothin[22]. The short half-life in serum was attributable to rapid urinary excretion.[32]

Lee and co-workers[23,33] reported the distribution of cephalothin in blood and tissues and its passage to body fluids. Since the average relative volume distribution was greater than one (1.413 ± 0.244 ml/Gm), the authors suggested that cephalothin distributed beyond extracellular space and concentrated in the tissues.

6. Methods of Analysis

6.1 Elemental Analysis

Element	% Theory	Found[34]
C	45.93	46.05
H	3.61	3.82
N	6.70	6.41
O	22.94	23.26
S	15.33	15.52
Na	5.49	5.22

6.2 Microbiological Assay

Potency of cephalothin is routinely determined by agar-diffusion (plate) assay using Staphylococcus aureus or Bacillus subtilis, and by turbidimetric (tube) assay using S. aureus.[35] Of the two plate methods, the B. subtilis assay produces better defined zones of inhibition. The sharp, single-edged zones obtained offers greater ease and accuracy of measurement. The turbidimetric assay has the advantages of speed, reproducibility, and accuracy. Although good precision is obtainable by manual performance of the turbidimetric assay, even better precision (±2-3%) is possible by use of semi-automated methods such as the Autoturb® System[36], or its equivalent[37].

Deacetylcephalothin and cephalothin lactone are microbiologically active hydrolysis products of cephalothin. Since these hydrolysis products give dose-response curves with the same slope as cephalothin, their presence can interfere with potency determinations of the parent compound. In the presence of high levels of hydrolysis products, the plate method using B. subtilis is recommended[35] for assaying cephalothin because the method is quite insensitive to these substances. The presence of 15%, or less, of deacetylcephalothin does not interfere with the assay for cephalothin, when measured relative to a cephalothin standard curve, and the low activity of cephalothin lactone precludes its interference.

6.3 Iodometric Titration

In lieu of microbiological assays described above, the iodometric assay is used as an alternate method to determine the potency of cephalothin. Alicino[38] originally developed a manual titration procedure for quantitative assay of penicillins then, later, applied the assay to the determination of cephalosporin C. Based on the original method, Stevenson and Bechtel[39] developed an automated iodometric assay for cephalosporins. The iodometric technique is rapid, reliable, reproducible to 1-2%, and compares favorably with the microbiological cylinder-plate assay. The standard used in assaying cephalothin should correspond to the sample both in concentration and in composition. Cephalothin degradation products having an intact β-lactam ring and intermediates used in synthetic processes such as 7-aminocephalosporanic acid titrate as well as the parent compound.[6]

6.4 Colorimetric Analysis

Cephalothin can be determined by means of the colored complex formed on the addition of a ferric reagent to the corresponding hydroxamic acid produced by treatment with hydroxylamine. The method used is essentially the same as the procedure described for penicillins.[40] The ferric hydroxamate procedure is not specific for cephalothin or penicillins. For example, many amides, esters, and anhydrides form hydroxamic acids when reacted with hydroxylamine.[41] This type of interference is eliminated by the blank determination wherein cephalothin is rendered incapable of forming hydroxamic acid by use of basic hydrolysis or enzymatic hydrolysis with cephalosporinase.[6] Since cephalothin degradation products having an intact β-lactam ring react as well as the parent compound, the method measures total β-lactam content.

Redstone[42] developed a specific colorimetric assay to determine cephalosporin derivatives containing the acetyl moiety. In this assay the acetoxyl portion of cephalothin is displaced by nicotinamide and the resulting derivative is reacted with 1,3-dihydroxyacetone. The final product has a chromophore which exhibits a maximum absorbance at 360 nm. Thus, deacetylcephalothin and cephalothin lactone, if present, do not interfere in the determination of the parent compound.

6.5 Ultraviolet Absorption

Ultraviolet absorption in the range 220-310 nm of a 0.0025% (w/v) solution is useful for identification. The spectrum should compare qualitatively to that of a 0.0025% solution of cephalothin sodium standard. An increase in the ratio of the extinction at the maximum at 237 nm to that at 265 nm provides an indication of the onset of deterioration of aqueous solutions.[6]

6.6 Non-Aqueous Titration

Cephalothin content can be determined by non-aqueous titration.[6] Dissolve an accurately weighed sample (approximately 200 mg) in 40 ml of glacial acetic acid, add 1 drop of crystal violet indicator (2 percent in acetic acid), and titrate the solution with standardized 0.1N perchloric acid. Calculate percent cephalothin as follows:

$$\text{Percent cephalothin} = \frac{\text{ml of } HClO_4 \times N \text{ of } HClO_4 \times 418.4 \times 100}{\text{sample weight in mg}}$$

6.7 Polarographic Analysis

Hall[43] determined the half-wave potential of sodium cephalothin. The $E\frac{1}{2}$ value was dependent upon both pH and concentration. The equation for the pH dependence at a concentration of 0.45 mM in McIlvaine buffers was $E\frac{1}{2}$ = -0.992-0.078 pH. All data obtained were versus the saturated calomel electrode at 25 C. The method was not considered sufficiently accurate for use as a quantitative assay since the relative standard deviation was about 5%.

6.8 Chromatographic Analysis

Solvent systems used for paper chromatographic analysis of cephalosporin derivatives and their degradation products are supplied in a review by Betina.[44] Paper and thin layer chromatographic procedures having particular application for cephalothin are given below.

6.81 **Paper**

For identification, modification of a system reported by Loder et al.[45] is commonly used:
  Solvent: water-saturated ethyl acetate for both the mobile and stationary phase.
  Paper: Whatman No. 1 buffered with 1.25M phosphate buffer, pH 5.5.
  Developing Time: 5 hours descending in chambers equilibrated at least 24 hours.
  Load: 1 µl of a 1 mg/ml aqueous solution.
  R_f: 0.50
For detection of metabolites:[46]
  Solvent: butanol/ethanol/water 4:1:5.
  Paper: Whatman No. 1 buffered with 0.05M sodium phosphate, pH 6.0.
  Developing Time: 18 hours descending.
  Comparative R_f: deacetylcephalothin, 0.52; sodium cephalothin, 0.67; cephalothin lactone 0.83.
For quantitation of deacetyl and lactone derivatives in the presence of cephalothin:[47]
  Solvent: water-saturated methylethylketone; both phases of solvent in bottom of chamber.
  Paper: Whatman No. 1 strips 0.25 x 20 inches.
  Developing Time: 3 hours descending at 23 C without prior equilibration.
Hoehn et al.[48] used a modification of this system.

Paper electrophoresis has had limited usefulness for cephalothin. The problem is that although cephalothin can be made to migrate as an anion or to remain immobile as the free acid, many of the chemical differences between degradation products and impurities do not result in a significant difference in electrophoretic mobility at a given pH. Reference to electrophoresis of cephalothin is found in an article by Martin and Shaw.[49]

6.82 *Thin Layer*
Experience with thin layer chromatography of cephalothin on silica gel is presented in Table II.[50] Hussey[50] used solvent system 3 to follow the stability of cephalothin. Vanillin dissolved in a mixture of phosphoric acid/methanol, 1:1, was used as a spray for detection. When deacetylcephalothin mobility needed to be increased, system 4 was used.

TABLE II

R_f Values of Cephalothin on Silica Gel

Solvent System*

1	2	3	4	5	6	7
0.65	0.45	0.30	0.45	0.60	0.70	0.60

*Solvent system:
1. Acetonitrile/H_2O, 4:1
2. Ethyl acetate/acetone/H_2O, 2:4:2
3. Acetone/chloroform/acetic acid, 50:50:7
4. Ethanol/chloroform/acetic acid, 50:100:7.5
5. Acetone/acetic acid, 20:1
6. Ethyl acetate/acetic acid/H_2O, 3:1:1
7. Methanol/ethyl acetate/acetic acid, 50:100:5

Methods useful to detect cephalosporins on chromatograms are short wave (254 nm) ultraviolet light, ninhydrin spray, sodium hydroxide-iodine-starch spray[51], and bioautographs using *Bacillus subtilis* ATCC 6633,

<u>Staphylococcus</u> <u>aureus</u> 6538P, or, for maximum sensitivity, <u>Sarcina</u> <u>lutea</u> ATCC 9341.

7. <u>Determination in Body Fluids and Tissues</u>

The conventional plate assay using <u>Sarcina</u> <u>lutea</u> ATCC 9341 has application for assay of urine, serum, and tissue extracts because of its sensitivity.[35] The method can measure activities as low as 0.2 µg/ml of cephalothin and 0.4 µg/ml of deacetylcephalothin. In this assay, deacetylcephalothin is one-half as active as cephalothin, and since the dose-response curve of the deacetyl compound parallels that of the parent compound the method measures total activity of mixtures of these substances in terms of one of the pure standards.

Paper chromatographic methods have been devised to determine cephalothin and its microbiologically active metabolite deacetylcephalothin in body fluids. Miller[47] developed a method that is satisfactory for analysis of urine samples. Hoehn <u>et al.</u>[48] described a method that affords quantitative disassociation of cephalothin from plasma proteins, and developed a chromatographic technique to measure low levels of cephalothin and its metabolite in urine, plasma, synovial fluid, and cerebrospinal fluid.

Benner[52] used a polarographic method to estimate a number of β-lactam antibiotics, including cephalothin, in serum. The entire operation of venipuncture, separation of serum, ultrafiltration of serum, and polarographic analysis was accomplished in two hours.

8. References
   1. C.D. Underbrink, Lilly Research Laboratories, personal communication.
   2. P.V. DeMarco and R. Nagarajan, In "The Cephalosporin and Penicillin Compounds: Their Chemistry and Biology", In Press.
   3. C.F.H. Green, J.E. Page, and S.E. Staniforth, J. Chem. Soc., 1595 (1965).
   4. J.L. Occolowitz, Lilly Research Laboratories, unpublished research.
   5. W. Richter and K. Biemann, Monatsh. Chem. 96, 484 (1965).
   6. L.P. Marrelli, Lilly Research Laboratories, personal communication.
   7. J.R. Marsh and P.J. Weiss, J.A.O.A.C. 50 No. 2, 457 (1967).
   8. T.E. Cole, Lilly Research Laboratories, personal communication.
   9. H.A. Rose, J. Pharm. Sci. 52, 1008 (1963).
   10. R.R. Chauvette, E.H. Flynn, B.G. Jackson, E.R. Lavagnino, R.B. Morin, R.A. Mueller, R.P. Pioch, R.W. Roeske, C.W. Ryan, J.L. Spencer, and E. Van Heyningen, J. Am. Chem. Soc. 84, 3401 (1962).
   11. H.R.V. Arnstein and D. Morris, Biochem. J. 76, 357 (1960).
   12. E.P. Abraham and G.G.F. Newton, In "Advances in Chemotherapy", Vol. 2, p. 44. Academic Press, New York and London (1965).
   13. R.B. Morin, B.G. Jackson, E.H. Flynn, and R.W. Roeske, J. Am. Chem. Soc. 84, 3400 (1962).
   14. E.P. Abraham and G.G.F. Newton, Biochem. J. 79, 377 (1961).
   15. D.C. Hodgkin and E.N. Maslen, Biochem. J. 79, 393 (1961).
   16. R. Heymès, G. Amiard, and G. Nominé, Compt. Rend. Acad. Sc. Paris Série C 263, 170 (1966).

17. R.B. Morin, B.G. Jackson, R.A. Mueller, E.R. Lavagnino, W.B. Scanlon, and S.L. Andrews, J. Am. Chem. Soc. $\underline{85}$, 1897 (1963); J. Am. Chem. Soc. $\underline{91}$, 1401 (1969).
18. R.B. Woodward, K. Heusler, J. Gostelli, P. Naegeli, W. Oppolzer, R. Ramage, S. Ranganathan, and H. Vorbrüggen, J. Am. Chem. Soc. $\underline{88}$, 852 (1966).
19. E. Van Heyningen, Advances in Drug Research Vol. $\underline{4}$, 1 (1967).
20. H.R. Sullivan and R.E. McMahon, Biochem. J. $\underline{102}$, 976 (1967).
21. A.L. Demain, Nature $\underline{210}$, 426 (1966).
22. W.E. Wick, Antimicrobial Agents and Chemotherapy $\underline{1965}$, 870 (1966).
23. C.C. Lee, E.B. Herr, Jr., and R.C. Anderson, Clin. Med. $\underline{70}$, 1123 (1963).
24. E.P. Abraham, Am. J. Med. $\underline{39}$, 692 (1965).
25. S.H. Eggers, T.R. Emerson, V.V. Kane, and G. Lowe, Proc. Chem. Soc. (London), p. 248 (1963).
26. L.D. Sabath, M. Jago, and E.P. Abraham, Biochem. J. $\underline{96}$, 739 (1965).
27. J.M.T. Hamilton-Miller, G.G.F. Newton, and E.P. Abraham, Biochem. J. $\underline{116}$, 371 (1970).
28. J.M.T. Hamilton-Miller, E. Richards, and E.P. Abraham, Biochem. J. $\underline{116}$, 385 (1970).
29. G.G.F. Newton, E.P. Abraham, and S. Kuwabara, Antimicrobial Agents and Chemotherapy $\underline{1967}$, 449 (1968).
30. M. Cole, Nature 203, 519 (1964).
31. W. Kaufmann and L. Bauer, Nature $\underline{203}$, 520 (1964).
32. R.S. Griffith and H.R. Black, J. Am. Med. Assoc. $\underline{189}$, 823 (1964).
33. C.C. Lee and R.C. Anderson, Antimicrobial Agents and Chemotherapy $\underline{1962}$, 695 (1963).
34. G. Maciak, Lilly Research Laboratories, personal communication.

35. R.J. Simmons, In "Analytical Microbiology", Vol. II, (ed. by F. Kavanagh) Academic Press, In Press.
36. N.R. Kuzel and F. Kavanagh, J. Pharm. Sci., In press.
37. F. Kavanagh, In "Analytical Microbiology", Vol. II, (ed. F. Kavanagh) Academic Press, In Press.
38. J.F. Alicino, Ind. Eng. Chem. Anal. Ed. 18, 619 (1946); Anal. Chem. 33, 648, (1961).
39. C.E. Stevenson and L.D. Bechtel, In Press.
40. G.E. Boxer and P.M. Everett, Anal. Chem. 21, 670 (1949).
41. F. Lipmann and L.L. Tuttle, J. Biol. Chem. 159, 21 (1945).
42. M.O. Redstone, In Press.
43. D.A. Hall, Lilly Research Laboratories, unpublished research.
44. V. Betina, In "Chromatographic Reviews". (M. Lederer ed.), Elsevier Publishing Co., New York, Vol. 7, p. 119 (1965).
45. B. Loder, G.G.F. Newton, and E.P. Abraham, Biochem. J. 79, 408 (1961).
46. C.H. O'Callaghan and P.W. Muggleton, Biochem. J. 89, 304 (1963).
47. R.P. Miller, Antib. Chemotherapy 12, 689 (1962).
48. M.H. Hoehn, H.W. Murphy, C.T. Pugh, and N.E. Davis, Appl. Microbiol. 20, 734 (1970).
49. J.L. Martin and W.H.C. Shaw, In "Proceedings of the S.A.C. Conference, Nottingham, 1965", W. Heffer and Sons, Cambridge, p. 7, (1965).
50. R.L. Hussey, Lilly Research Laboratories, unpublished results.
51. R. Thomas, Nature 191, 1161 (1961).
52. E.J. Benner, Abstract 124, Tenth Interscience Conference On Antimicrobial Agents And Chemotherapy, (1970).

SODIUM SECOBARBITAL

I. Comer

CONTENTS

1. Description
   1.1 Name, Formula, Molecular Weight
   1.2 Appearance, Color, Odor
2. Physical Properties
   2.1 Infrared Spectrum
   2.2 Nuclear Magnetic Resonance Spectrum
   2.3 Ultraviolet Spectrum
   2.4 Mass Spectroscopy
   2.5 Melting Range
   2.6 Differential Thermal Analysis
   2.7 Thermogravimetric Analysis
   2.8 Solubility
   2.9 Crystal Properties
   2.10 pH Range
   2.11 pK
   2.12 Heat of Solution
3. Synthesis
4. Stability - Degradation
5. Drug Metabolic Products
6. Methods of Analysis
   6.1 Elemental Analysis
   6.2 Gravimetric Analysis
   6.3 Direct Spectrophotometric Analysis
   6.4 Nonaqueous Titration
   6.5 Chromatographic Analysis
      6.51 Paper
      6.52 Thin Layer
      6.53 Gas
      6.54 Ion Exchange and Column
7. References

SODIUM SECOBARBITAL

1. Description

   1.1 Name, Formula, Molecular Weight
   Sodium secobarbital is 5-allyl-5-(1-methylbutyl) barbituric acid sodium salt[1]. It is also known as sodium propyl-methyl-carbinyl allyl barbiturate[2]; sodium allyl 1-methyl-butyl barbiturate[2,3]; sodium allyl (methyl propyl carbonyl) barbiturate[3]; 5-allyl-5-(1-methylbutyl) malonylurea sodium salt[4]; sodium 5-allyl-5-(1-methylbutyl)-barbiturate[4].

   $C_{12}H_{17}N_2O_3Na$ Mol. wt.: 260.27

   1.2 Appearance, Color, Odor
   White, hygroscopic, odorless powder[1,4].

2. Physical Properties

   2.1 Infrared Spectrum
   The infrared spectrum, Fig. 1, was obtained from a KBr pellet of the sample[5]. Several papers have been published using infrared spectroscopy in reference to barbiturates[6,7], their derivatives[8,9,10] and degradation products[11].
   Manning and O'Brien[6] attribute the band at 3.1 microns in barbiturates to the hydrogen bonded NH and the strong band at 6.4 microns is present when the di-alkyl substituted carbon atom carries the propyl or butyl group.

   2.2 Nuclear Magnetic Resonance Spectrum
   The NMR spectrum of sodium secobarbital in deuterated water is shown in Fig. 2. Underbrink[12] has given the spectral assignments,

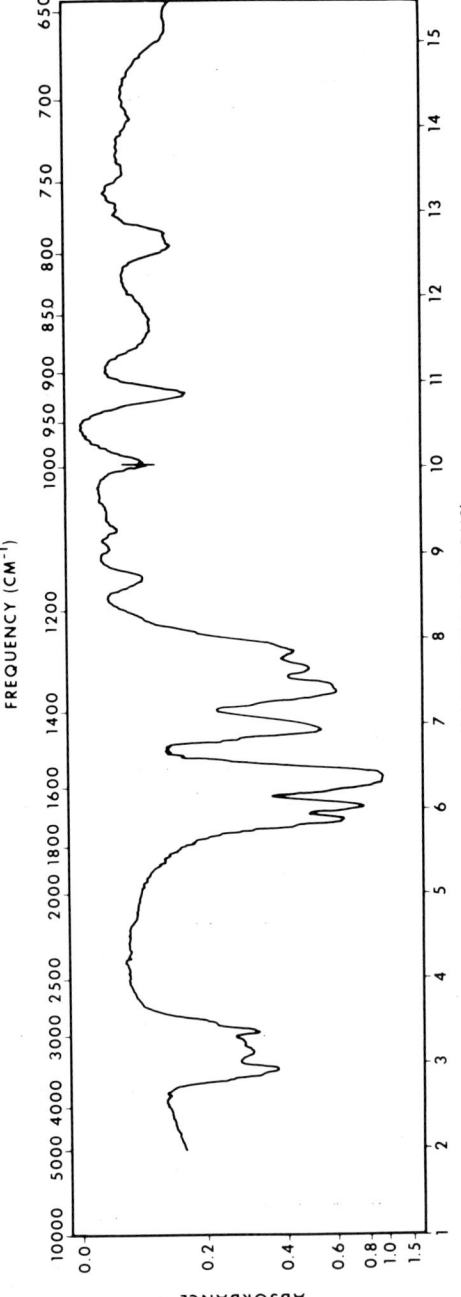

Fig. 1. Infrared spectrum of sodium secobarbital KBr pellet; instrument: Perkin-Elmer 221

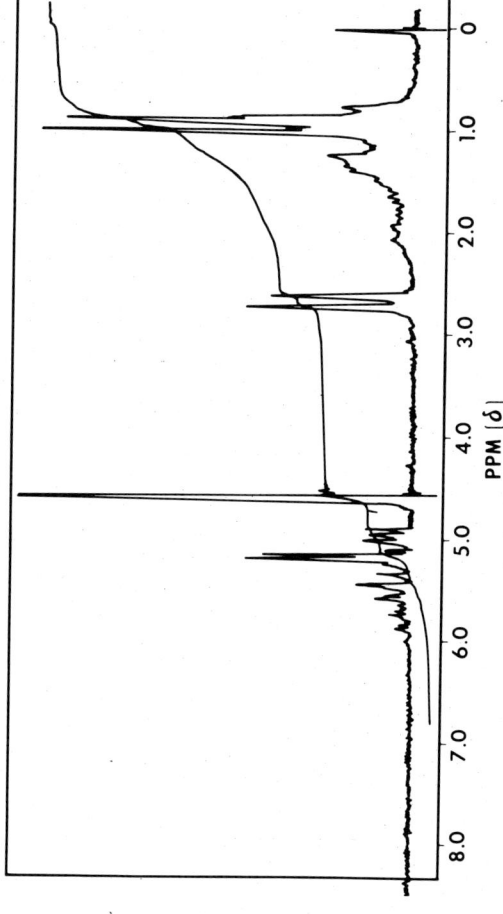

Fig. 2. Nuclear magnetic resonance spectrum of sodium secobarbital; instrument: Varian A-60

Table I, in delta units measured in ppm downfield from the 3-(trimethylsilyl)propanesulfonic acid sodium salt reference peak. The approximate coupling constants (J) are given in Hz where appropriate.

Avdovich and Neville[13] gave chemical shift data of barbiturates from proton magnetic resonance spectroscopy in different solvents.

TABLE I

NMR Spectral Assignments of Sodium Secobarbital

Group	Shape	Chemical Shift	J(Hz)
$CH_3CH_2CH_2\underline{CH}-$ $\quad\quad\quad\quad\quad CH_3$	Triplet	0.90 ppm	6.5
$CH_3CH_2CH_2CH-$ $\quad\quad\quad\quad\quad \underline{CH_3}$	Doublet	0.96 ppm	6.5
$CH_3\underline{CH_2CH_2}CH-$ $\quad\quad\quad\quad\quad CH_3$	Broad Multiplet	Approx. 1.30 ppm	
$CH_2=CH\underline{CH_2}-$	Doublet	2.68	6
$\underline{CH_2=CH}-$	Multiplet	4.9 - 5.8	

Neville[14] used NMR for structure studies of the methylated derivatives of 5,5-disubstituted barbituric acids by various methylation techniques.

2.3 Ultraviolet Spectrum

The maximum absorbance of a 0.0013% solution in 0.003% sodium hydroxide occurs at (about) 241 nm; $A_{1cm}^{1\%}$ = 362[15].

(See also Section 6.3.)

2.4 Mass Spectroscopy

Coutts and Locock[16,17] in a study of the mass spectra of barbiturates made the following observations concerning secobarbital. The parent peaks m/e 238 and 224 were absent, and the M+1 ions m/e 239 and m/e 225 were present in less than 1% abundance. The ion m/e 168 is formed by loss of $CH_3CH_2CH_2CH=CH_2$ from the molecular ion. The formation of the ion m/e 167 results from the loss of the radical- $CH_3CH_2CH_2CHCH_3$. Since the mass spectra of secobarbital and 5-allyl-5(2-methyl propyl) and 5-allyl-5(1-methyl propyl) barbiturate are virtually identical,[16,17] the spectra are not reproduced here. Minor differences are observed above m/e 168[16]. The loss of a propyl radical in the alkanyl side chain from the molecular ion gives an ion of significant abundance (26%) of m/e 195. (m/e calculated for $C_9H_{11}N_2O_3$: 195.0775; measured: 195.0764.) The loss of an ethyl radical gives an ion of low abundance (3%) of m/e 209.

2.5 Melting Range

The m.p. of the acid is 98-100°C.[2,4,18] and of the p-nitrobenzyl derivative is 163°C.[19]

2.6 Differential Thermal Analysis

The differential thermal analysis of sodium secobarbital at the rate of 20°C./min. showed an endothermic phase transition at 305°C.[5]

2.7 Thermogravimetric Analysis

The thermogravimetric analysis performed on sodium secobarbital at the rate of 5°C./min. under nitrogen sweep showed a 1% weight loss at 59°C., a 5% weight loss at 301°C., and a 20% weight loss at 320°C.[5]

2.8 Solubility

Sodium secobarbital is very soluble in water, soluble in alcohol, and practically insoluble in ether[1]. Burlage[20] reported the solubility in isopropyl alcohol as 60.7620 g./100 ml. of solution.

2.9 Crystal Properties

Castle[21] reported the optical crystallographic properties of the free acid secobarbital:

Lath-shaped crystals; monoclinic; Z ∥ b; 2V = 31° calculated; horizontal dispersion r>v; optic sign is negative; α = 1.487, β = 1.557 and γ = 1.563.

The indices of refraction of sodium secobarbital were published by Eisenberg[22]:

α = 1.490, β = 1.500 (intermediate index) and γ = 1.525.

The X-ray diffraction pattern for sodium secobarbital was published in the "X-Ray Powder Data File"[23] and is shown as follows:

dA	I/I_1	dA	I/I_1
14.1	67	6.11	3
12.2	53	5.72	3
11.1	20	5.49	3
10.3	100	5.21	3
8.87	3	4.66	7
8.45	3	4.45	3
6.74	3	3.84	3

2.10 pH Range

The pH of a 5% solution is between 9.8 and 10.1[15].

2.11 pK

For secobarbital the pK_1 was 7.92 at 20°C.[24,25] and the pK_2 12.60 at 38°C.[26]

human urine. They reported detecting other metabolic products and identified one as 5-(methylbutyl) barbituric acid. Waddell[31] in a later study reported three major metabolites of secobarbital: 5-(2,3 dihydroxypropyl)-5-(1 methylbutyl) barbituric acid and two diastereoisomers of hydroxysecobarbital or 5-allyl-5-(3-hydroxy-1-methylbutyl) barbituric acid. A fourth metabolite 5-(1-methylbutyl) barbituric acid was found in the urine of some human patients. Horning[32] used gas chromatography and mass spectroscopy to isolate hydroxysecobarbital from the urine of newborn infants.

6. Methods of Analysis

6.1 Elemental Analysis

The elements and % composition[4] are as follows: C_{12} - 55.37%, H_{17} - 6.58%, N_2 - 10.76%, O_3 - 18.44%, Na - 8.84%.

The sodium and nitrogen assay methods are described in New and Nonofficial Remedies[3].

6.2 Gravimetric Analysis

Historically the gravimetric procedure has undergone changes in procedure, from extraction solvents to methods of drying the residue[3,33,34,35,36,37,38]. The present gravimetric procedure is given in U.S.P. XVIII[1].

6.3 Direct Spectrophotometric Analysis

(See also Section 2.3.)

Goldbaum[39] extracted the barbiturate from biological fluids into chloroform and then extracted the chloroform with 0.5 N sodium hydroxide, and read this solution at 255 nm for quantitation. To differentiate from other barbiturates, the optical density at 235, 230 and 225 nm are recorded and ratios to the optical density at 255 nm are calculated.

For quantitation of barbiturates in blood, Walker et al.[40] used dilutions of 12.5

mcg. per ml. in pH 10 solutions read at 239 nm. To correct for interference in blood extracts the optical density in acid solution pH 2 was subtracted.

Goldbaum[41] introduced the optical density difference method for differentiating barbiturates. The absorption spectra in strong alkali 0.45 N sodium hydroxide and in pH 10.5 were recorded. The absorption spectra in strong alkali show characteristic maxima at 255 nm and minima at about 235 nm. In a buffer solution at pH 9.8 to 10.5 higher maxima occur at 240 nm with no minima. The optical density differences at specific wavelengths were recorded. The differences are greatest at 260 nm decreasing through zero at about 250 nm (isobestic point) to a maximum negative at about 235 nm and increasing through zero at about 227 nm (isobestic point) to a positive difference at the lower wavelengths. The optical density differences at specific wavelengths are divided by the optical density difference at 260 nm to provide ratios to be used for differentiation of barbiturates. For quantitation the optical density difference at 260 nm in alkaline solution containing 2-30 mcg. per ml. is used.

Rotondaro[42] described an extraction procedure for pharmaceuticals and listed various methods of assay, one of which is the ultraviolet spectrophotometric procedure. Dilutions were made to 1-2 mg. per 100 ml. in the final solution having an alkalinity of 0.01-0.02 N sodium hydroxide or 0.1 to 1.0% NH_3 and a pH 10.5-11.5 to give a maximum at 240 nm ±1. These readings were made in reference to a barbiturate of known purity prepared under the same conditions.

6.4 Nonaqueous Titration

Various methods of extraction of secobarbital prior to titration with nonaqueous titrants have been reported[43,44,45,46]. Endpoints were determined visually or potentiometrically. Johnson and Byers[47] titrated the sodium salts in glacial acetic acid using 0.1 N perchloric acid in glacial acetic acid and methyl violet as the indicator. A nonaqueous titration method is given for the assay of secobarbital in U.S.P. XVIII[48].

6.5 Chromatographic Analysis

6.51 Paper Chromatography

Some data from various authors on paper chromatography are shown in Table II and refer to identification of barbiturates in biological fluids[49,50], in pharmaceuticals[51], and organic acidic compounds[52].

TABLE II

Paper Chromatography of Secobarbital

Ref.	Solvent[a]	Method	Detection[b]	Rf x 100
49	A	Ascending	2,3,4	76
50	A	Descending	2,3,4	76
51	B	Ascending	5	56
52	C	Descending	1	80

[a] A, n-Butyl Alcohol saturated with 5 N Ammonia; B, Ethylene chloride; C, Chloroform.

[b] 1, Examine under ultraviolet light; 2, 1% silver acetate solution; 3, Lemaire Reagent; 4, $KMnO_4$; 5, 0.05 N silver nitrate in alcohol solution.

For the identification of barbiturates, Algeri and Walker[49] applied 50-100 mcg. samples in the acid form to the paper. These samples could be obtained by further extraction of samples used in the ultraviolet spectrophotometric procedure.

Hilf et al.[50] used the paper chromatography method by Algeri and Walker[49] using descending chromatography. After development and detection of the barbiturate spot, the silver-barbiturate salt spot was cut out of the paper, shredded and extracted in order to obtain confirmation of identity by ultraviolet spectral ratios.

Sabatino[51] applied the barbiturate in acid form in a suitable solvent about 1 inch from the edge of an 8" x 8" Whatman No. 1 sheet of paper. A standard barbiturate solution (0.001 ml. of a 20 mg./10 ml. alcohol) was spotted along with the sample. The paper was mist sprayed with 0.5 M sodium carbonate and placed in the chromatograph tank containing redistilled ethylene chloride as the mobile solvent. After development of 4-6" above the spotting line, the paper was removed and dried at 100°C. in a forced draft oven. It was then sprayed with 0.05 N silver nitrate in alcohol solution and placed in the oven until tan spots on a brown background appeared.

6.52 <u>Thin Layer Chromatography</u>

The rapidity and economy of thin layer chromatography provide a convenient method for the analysis of secobarbital, as evidenced by the numerous publications. The data shown in Table III from the paper of Comer[53], has been expanded to include some of the references since 1967.

The following method was outlined from the paper of Heaton and Blumberg[57]. They used the popular solvent system, $CHCl_3$-acetone (9:1). (See Table III for 7 references using this system.) They have successfully employed the method for detection of barbiturates, narcotics, and amphetamines in urine of patients receiving psychotropic drugs. The procedure can be used for pharmaceutical dosage forms.

Shake 5 ml. urine with 2.5 ml. 1M potassium phosphate buffer pH 6 and 30 ml.

TABLE III

TLC of Secobarbital

Ref.	Solv.[a,b]	Rf x100	Ref.	Solv.[a,b]	Rf x100
30	A	64	60	D	75
54	A	55	61	E	29
55	A	41	55	F	39
56	A	55	55	G	59
57	A	x	58	H	78
58	A	51	62	I	63
59	A	64	62	J	43
30	B	46	63	K	83
58	B	36	63	L	63
30	C	54	64	M	85
58	C	44	59	N	73

[a] A, $CHCl_3$-Acetone (9:1); B, Benzene-Acetic Acid (9:1); C, Dioxane-Benzene-aq.NH_3 (20:75:5); D, Diisopropyl Ether- $CHCl_3$ (1:1); E, Acetone-n-Butyl Alcohol-NH_4OH (9:9:2) prel. treat samp. with H_2SO_4; F, Benzin-Dioxane (5:2) DMF stationary phase; G, Benzol-Ether (1:1); H, Isopropanol-NH_4OH — $CHCl_3$ (45:10:45); I, Ethyl Acetate-n-Hexane-NH_4OH (20:9:10); J, $CHCl_3$-CCl_4 (1:1); K, Isopropanol-$CHCl_3$-25% NH_4OH (60:30:10); L, isopropyl ether aq. sat.; M, n-amyl methyl ketone; N, Ethyl Acetate-CH_3OH-NH_4OH (85:10:5).
[b] Adsorbent used was silica gel except for solvent J-Kiesilguhr with formamide and solvent M-cellulose.

---

ether for 2 min. Separate, filter, and evaporate the ether layer. Dissolve the residue in 0.2 ml. $CHCl_3$ and spot on a 20 x 20 cm. Silica Gel G glass plate with appropriate standards. Develop for 10 cm. in $CHCl_3$-Acetone (9:1). Dry and spray with a suitable detection reagent.

Secobarbital was detectable for 4-6 days after a single 3 grain dose by this method[57].

Detection reagents may be selected for specificity based on other drugs or metabolites expected in the sample. A sensitivity of 1-10 mcg. is normally obtainable. The following means of visualization have been used: 0.5-1% silver nitrate or acetate-white[58,62] mercuric nitrate-white[30,63], $HgSO_4$ then diphenylcarbazone-blue[54,56] 0.02 M and 0.2% $KMnO_4$-yellow[58,30], 5% cobalt nitrate in ethanol, then NH_3 vapor-violet[60], iodine vapor-brown[65].

Morrison and Chatten[63] developed a quantitative method by measuring secobarbital coupled mercury with dithizone at 475 nm.

6.53 Gas Chromatography

The analysis of barbiturates including secobarbital has been the subject of many papers. A good review with 77 references was presented by Brochmann-Hanssen[66].

Pharmaceutical Dosage Forms: The method outlined is from the paper of Sibert and Fricke[67].

Sample Preparation: Stir the powdered dosage form (tablets) containing the secobarbital equivalent to 50 mg. of barbiturate in 25 ml. of hot methanol for 5 min. Cool and decant or filter through S and S No. 589 black filter paper into a 100-ml. volumetric flask. Wash with methanol, add 10 ml. of internal standard, and dilute to volume with methanol.

Use a similar concentration of a barbiturate (amobarbital), not contained in the sample and that is completely resolved from secobarbital, as the internal standard. Inject 5 μl of the sample plus internal standard solution into the gas chromatograph, and then a standard solution of secobarbital containing the internal standard. For dosages containing sodium secobarbital, extract the secobarbital from 10% HCl with $CHCl_3$, evaporate and continue as above with methanol.

Column: 6' x 4 mm. packed with 4.5% cyclohexanedimethanol succinate from $CHCl_3$ solution on 80/100 mesh Gas Chrom Q, conditioned 24 hr. at 240°C. with a flow of 60 ml./min. nitrogen.

Operating Conditions: Column 240°C., flame ionization detector 260°C., injection port 250°C., nitrogen 85 ml./min., detector hydrogen 60 and air 550 ml./min. (A Packard Series 7800 instrument was used[67].)

Calculation: Mg. secobarbital per dose = ratio area of sample to internal std. in sample soln. x conc. std. x diln. factor stds. x ave. wt. dosage form divided by ratio area std. to area of internal std. in std. soln. x wt. sample.

Rader and Aranda[68] used a similar column for a general quantitative procedure for various drugs including secobarbital. Allen[69] used 10% SE-30 on Gas Chrom Q column at 180°C. for mixtures of barbiturates in tablets.

Biological Materials: Kazyak and Knoblock[70], in their work in analytical toxicology, recommended that the specimen be adjusted to pH 4.0-7.5, extracted with chloroform, and filtered. The $CHCl_3$ is washed with phosphate buffer pH 7.4, filtered and evaporated. Parker and Kirk[71], and Anders[72] used $CHCl_3$ and Jain and Kirk[73] used acetone-ether as the initial solvent with modifications of the procedure for purification.

The $CHCl_3$ solvent residue may be redissolved in the appropriate solvent for injection into the gas chromatograph.

Columns: The column described under dosage form analysis may be used. Jain and Kirk[73] used a 1% cyclohexanedimethanol succinate on 100/120 mesh silanized Gas Chrom P, 3' x 1/8" O.D. glass column, with a flow rate of carrier gas at 220°C. of 15.8 ml./min.

The ultimate in details for column preparation for maximum sensitivity was described by McMartin and Street[74]. Using a 6 ft. SE-30 on Chromosorb W column they were able to detect 0.01 mcg. of secobarbital. Trimethylsilyl[75] and methyl[14,76,77,78] derivatives may be used to change retention time for identification purposes.

6.54 Ion Exchange and Column Chromatography

Ion exchange and column chromatography have been used to separate secobarbital from interfering substances in pharmaceutical preparations[46,79,80] and biological samples[81], Table IV, prior to quantitative determination.

TABLE IV

Ref.	Resin of adsorbant	Solvent[a]	Eluant[a]
46	Cation Resin Amberlite IRC-50	A	A
79, 80	Anion Resin Dowex 2-X8	B	D
81	Adsorbant Florisil	C	E

[a] A, Dimethylformamide; B, 50% Ethanol; C, Chloroform; D, 50% Acetic Acid in 95% Ethanol; E, 10% Anhydrous Methanol in Chloroform.

References

1. "The United States Pharmacopeia," 18th revision, Mack Publishing Co., Easton, Pa. 18042 (1970) p. 662.
2. United States Patent 1,954,429; Patented Apr. 10, 1934.
3. "New and Nonofficial Remedies," American Medical Association, Chicago, Ill. 60610 (1942), p. 472.
4. "The Merck Index," 8th ed., Merck and Co., Inc., Rahway, N.J. 07065 (1968), p. 939.
5. Underbrink, C. D. and Kossoy, A.D., personal communication, Eli Lilly and Company, Indianapolis, Indiana 46206.
6. Manning, J. J. and O'Brien, K. P., Bull. Narcotics, $\underline{10}$, 25 (1958).
7. Cleverley, $\underline{B}$., Analyst, $\underline{85}$, 582 (1960).
8. Chatten, L. G. and Levi, $\underline{L}$., Applied Spectroscopy, $\underline{11}$, 177 (1957).
9. Levi, L. and Hubley, C. E., Anal. Chem., $\underline{28}$, 1591 (1956).
10. Umberger, C. J. and Adams, G., Anal. Chem., $\underline{24}$, 1309 (1952).
11. Watson, J. R. and Pernarowski, M., J. Assoc. Offic. Anal. Chem., $\underline{45}$, 609 (1962).
12. Underbrink, C. D., personal communication, Eli Lilly and Co., Indianapolis, Indiana 46206.
13. Avdovich, H. W. and Neville, G. A., Can. J. Pharm. Sci., $\underline{4}$, 51 (1969).
14. Neville, G. A., Anal. Chem., $\underline{42}$, 347 (1970).
15. "New and Nonofficial Remedies," American Medical Association, Chicago, Ill. 60610 (1952), p. 688.
16. Coutts, R. T. and Locock, R. A., J. Pharm. Sci., $\underline{57}$, 2096 (1968).
17. Coutts, R. T. and Locock, R. A., J. Pharm. Sci., $\underline{58}$, 775 (1969).
18. Shonle, H. A., J. Am. Chem. Soc., $\underline{56}$, 2490 (1934).

19. Castle, R. N. and Poe, C. F., J. Am. Chem. Soc., 66, 1440 (1944).
20. Burlage, H. M., J. Am. Pharm. Assoc., Sci. Ed., 37, 345 (1948).
21. Castle, R. N., J. Am. Pharm. Assoc., Sci. Ed., 38, 47 (1949).
22. Eisenberg, W. V., J. Assoc. Offic. Anal. Chem., 36, 730 (1953).
23. "X-ray Powder Data File," Am. Soc. for Testing Materials, (5-0085), Phila., Pa. (1960).
24. Clowes, G. H. A., Keltch, A. K. and Krahl, M. E., J. Pharmacol. and Exptl. Therap., 68, 312 (1940).
25. Krahl, M. E., J. Phy. Chem., 44, 449 (1940).
26. Butler, T. C., Ruth, J. M. and Tucker, Jr., G. F., J. Am. Chem. Soc., 77, 1486 (1955).
27. Udani, J. H. and Autian, J., J. Am. Pharm. Assoc., Sci. Ed., 49, 376 (1960).
28. White, E. C., Briggs, J. R. and Sunshine, I., Amer. J. Clin. Path., 29, 506 (1958).
29. Kapadia, A. J. and Autian, J., J. Am. Pharm. Assoc., Sci. Ed., 49, 380 (1960).
30. Cochin, J. and Daly, J. W., J. Pharmacol. and Exptl. Therap., 139, 154 (1963).
31. Waddell, W. J., J. Pharmacol. and Exptl. Therap., 149, 23 (1965).
32. Horning, M. G., Abstracts 9th Nat'l. Meeting A.Ph.A. Academy Pharmaceutical Sciences, Washington, D.C., Nov. 1970.
33. "The United States Pharmacopeia," 15th revision, Mack Publishing Co., Easton, Pa. 18042 (1955) p. 626.
34. "The United States Pharmacopeia," 16th revision, Mack Publishing Co., Easton, Pa. 18042 (1960) p. 639.
35. "The United States Pharmacopeia," 17th revision, Mack Publishing Co., Easton, Pa. 18042 (1965) p. 642.
36. Warren, L. E., J. Assoc. Offic. Anal. Chem., 25, 799 (1942).
37. Warren, L. E., J. Assoc. Offic. Anal. Chem., 26, 101 (1943).

38. Warren, L. E., J. Assoc. Offic. Anal. Chem., 27, 352 (1944).
39. Goldbaum, L. R., J. Pharmacol. and Exptl. Therap., 94, 68 (1948).
40. Walker, J. T., Fisher, R. S. and McHugh, J. J., Am. J. Clin. Path., 18, 451 (1948).
41. Goldbaum, L. R., Anal. Chem., 24, 1604 (1952).
42. Rotondaro, F. A., J. Assoc. Offic. Anal. Chem., 38, 809 (1955).
43. Ryan, J. C., Yanowski, L. K. and Pifer, C. W., J. Am. Pharm. Assoc., 43, 656 (1954).
44. Swartz, C. J. and Foss, N. E., J. Am. Pharm. Assoc., 44, 217 (1955).
45. Chatten, L. G., J. Pharm. Pharmacol., 8, 504 (1956).
46. Vincent, M. C. and Blake, M. I., J. Am. Pharm. Assoc., 48, 359 (1959).
47. Johnson, L. Y. and Byers, T. E., J. Assoc. Offic. Anal. Chem., 43, 255 (1960).
48. "The United States Pharmacopeia," 18th revision, Mack Publishing Co., Easton, Pa. 18042 (1970) p. 600.
49. Algeri, E. J. and Walker, J. T., Am. J. Clin. Path., 22, 37 (1952).
50. Hilf, R., Lightburn, G. A. and Castano, F. F., J. Lab. Clin. Med., 54, 320 (1959).
51. Sabatino, F. J., J. Assoc. Offic. Anal. Chem., 37, 1001 (1954).
52. Schmall, M., Wollish, E. G., Colarusso, R., Keller, C. W., and Shafer, E. G. E., Anal. Chem., 29, 791 (1957).
53. Comer, J. P. and Comer, I., J. Pharm. Sci., 56, 413 (1967).
54. Stolman, A., "Progress in Chemical Toxicology," Academic Press, Inc., New York, N.Y. (1965) vol. 2, p. 321.
55. Sahli, M. and Oesch, M., J. Chromatog., 14, 526 (1964).
56. Sunshine, I., Am. J. Clin. Path., 40, 576 (1963).
57. Heaton, A. M. and Blumberg, A. G., J. Chromatog., 41, 367 (1969).
58. Mule, S. J., J. Chromatog., 39, 302 (1969).

59. Dole, V. P., Kim, W. K. and Eglitis, I., J. Amer. Med. Assoc., 198, 349 (1966).
60. Shellard, E. J. and Osisiogu, I. U., Lab. Pract., 13, 516 (1964).
61. Petzold, J. A., Camp, W. J. R. and Kirch, E. R., J. Pharm. Sci., 52, 1106 (1963).
62. Ahmed, Z. F., El-Darawy, Z. I., Aboul-Enein, M. N., El-Naga, M. A. A., and El-Leithy, S. A., J. Pharm. Sci., 55, 433 (1966).
63. Morrison, J. C. and Chatten, L. G., J. Pharm. Pharmacol. 17, 655 (1965).
64. Curry, A. S. and Fox. R. H., Analyst, 93, 834 (1968).
65. Huang, J. T. and Wang, K. T., J. Chromatog., 31, 587 (1967).
66. Brochmann-Hanssen, E. "Theory and Appl. of Gas Chromatog. in Ind. and Med.," Kroman, H. S. and Bender, S. R., Eds., Grune and Stratton, New York, N.Y. (1968) p. 182.
67. Sibert, J. L. and Fricke, F. L., J. Assoc. Offic. Anal. Chem., 51, 1326 (1968).
68. Rader, B. R. and Aranda, E. S., J. Pharm. Sci., 57, 847 (1968).
69. Allen, J. L., J. Assoc. Offic. Anal. Chem., 51, 619 (1968).
70. Kazyak, L. and Knoblock, E. C., Anal. Chem., 35, 1448 (1963).
71. Parker, K. D. and Kirk, P. L., Anal. Chem., 33, 1378 (1961).
72. Anders, M. W., Anal. Chem., 38, 1945 (1966).
73. Jain, N. C. and Kirk, P. L., Microchem. J., 12, 249 (1967).
74. McMartin, C. and Street, H. V., J. Chromatog., 22, 274 (1966).
75. Street, H. V., J. Chromatog., 41, 358 (1969).
76. Stevenson, G. W., Anal. Chem., 38, 1948 (1966).
77. Martin, H. F. and Driscoll, J. L., Anal. Chem., 38, 345 (1966).
78. Brockmann-Hanssen, E. and Oke, T. O., J. Pharm. Sci., 58, 370 (1969).
79. Blake, M. I. and Siegel, F. P., J. Pharm. Sci., 51, 944 (1962).

80. Blake, M. I. and Nona, D. A., J. Pharm. Sci., 53, 570 (1964).
81. Stokes, D. M., Camp, W. J. R. and Kirch, E. R., J. Pharm. Sci., 51, 379 (1962).

TRIAMCINOLONE

K. Florey

Reviewed by N. E. Rigler

CONTENTS

1. Description
   1.1 Name, Formula, Molecular Weight
   1.2 Appearance, Color, Odor
2. Physical Properties
   2.1 Infrared Spectra
   2.2 Nuclear Magnetic Resonance Spectra
   2.3 Ultraviolet Spectra
   2.4 Mass Spectra
   2.5 Optical Rotation
   2.6 Melting Range
   2.7 Differential Thermal Analysis
   2.8 Thermogravimetric Analysis
   2.9 Solubility
   2.10 Crystal Properties
3. Synthesis
4. Stability, Isomerization, Degradation
5. Drug Metabolic Products
6. Methods of Analysis
   6.1 Elemental Analysis
   6.2 Direct Spectrophotometric Analysis
   6.3 Colorimetric Analysis
   6.4 Polarographic Analysis
   6.5 Chromatographic Analysis
      6.51 Paper
      6.52 Thin Layer
      6.53 Column
7. Determination in Body Fluids and Tissues
8. References

TRIAMCINOLONE

1. Description

1.1 Name, Formula, Molecular Weight

Triamcinolone is 9α-Fluoro-11β,16α,17,21-tetrahydroxypregna-1,4-diene-3,20-dione. It is also known as Δ^1-9α-fluoro-16α-hydroxyhydrocortisone; 9α-fluoro-16α-hydroxyprednisolone; Δ^1-16α-hydroxy-9α-fluorohydrocortisone and 16α-hydroxy-9α-fluoroprednisolone.

$C_{21}H_{27}FO_6$ Mol. Wt.: 434.49

1.2 Appearance, Color, Odor

White to off white, odorless crystalline powder.

2. Physical Properties

2.1 Infrared Spectra

Depending on solvent of crystallization triamcinolone exists in two polymorphic forms (see also 2.7; 2.9; 2.10 for further discussion of polymorphs) with different infrared spectra. The infrared spectra "A" and "B"[1] presented in figures 1 and 2 correspond to the spectra "A" and "B" of Mesley[2] and "I" and "II" of Smith et.al.[3] respectively.

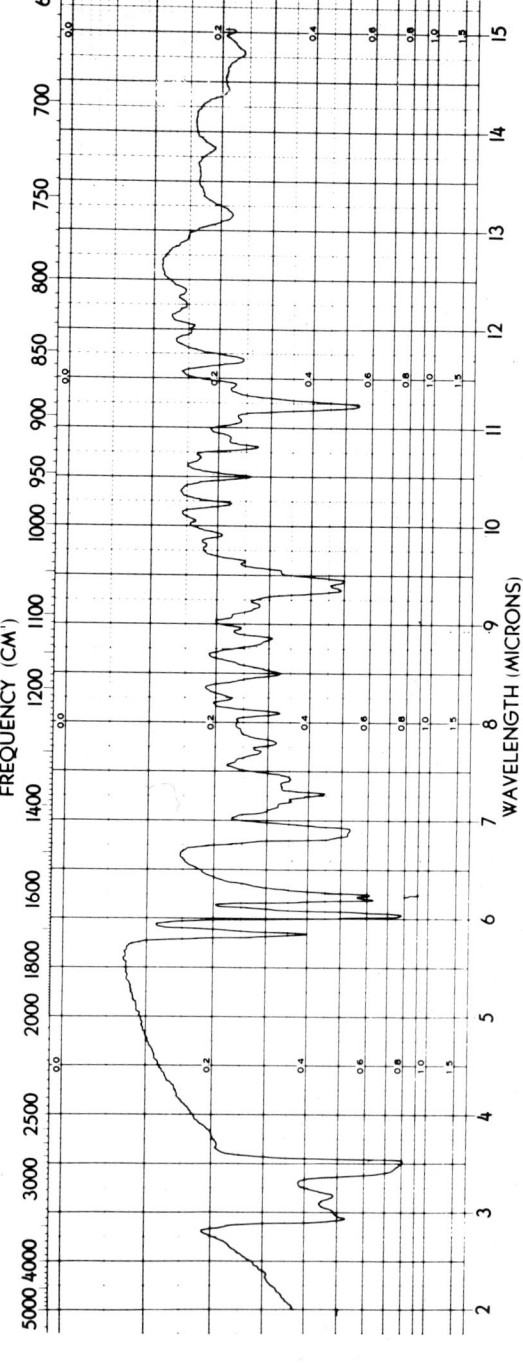

Fig. 1 Triamcinolone polymorph "A". Sample #4789/35-1 (Dr. Michel) recryst. from 60% isopropanol/H$_2$O. I.R.curve #28523. Taken in mineral oil; Instrument: Perkin-Elmer 621.

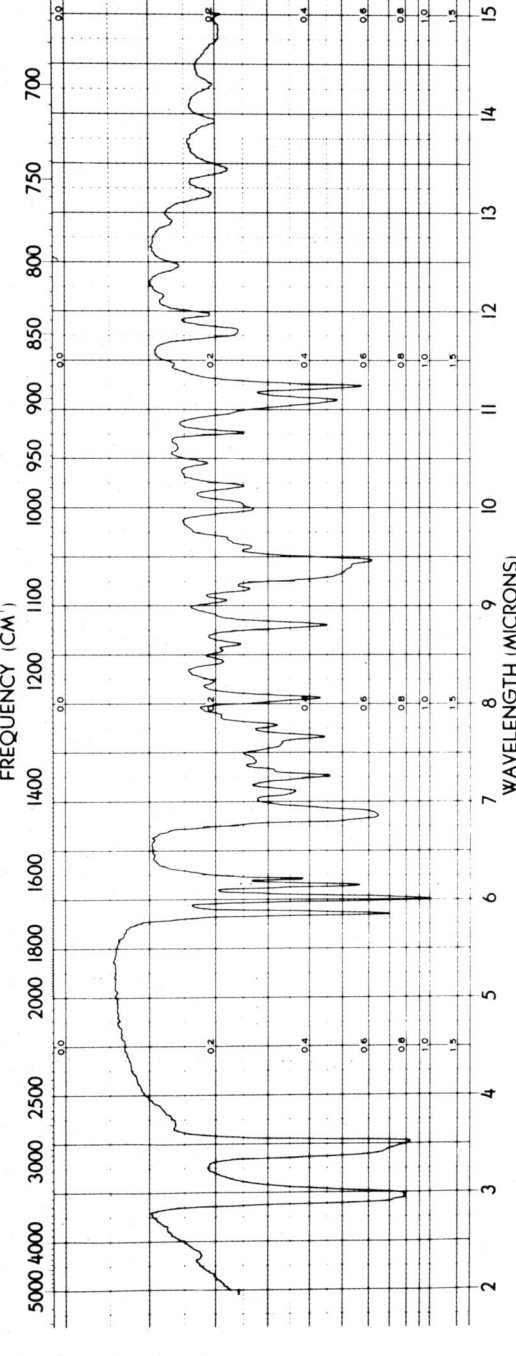

Fig. 2 Triamcinolone polymorph "B". Sample #4783-33-1 (Dr. Michel) recryst. from dimethylacetamide/water. I.R. curve #28522. Taken in mineral oil. Instrument: Perkin-Elmer 621

R. J. Mesley[2] in a discussion of "The infrared spectra of steroids in the solid state" assigns the following bands (cm^{-1}) to triamcinolone:

	Type A	Type B
a. characteristic for 11β-OH:	1043	1039
b. " for 17α-OH:	1132,1119	1136,1121
c. " for 21-OH:	1105,1090, 1059	1104,1090, 1056
d. " for 1,4-diene-3-ketones	1402,1300, 1244,954, 937,926, 890,887, 854,828, 710	1406,1304, 1242,955, 943,928, 889,848, 833,699

These assignments as well as those previously made[3,4,5] essentially agree with the peaks or shoulders presented in the spectra of figures 1 and 2.

2.2 Nuclear Magnetic Resonance Spectra

The NMR spectrum[6] of triamcinolone, dissolved in deuterodimethylsulfoxide containing tetramethylsilane as an internal reference, shows the presence of the cross conjugated dienone, C-1 proton resonance at 2.74 τ (doublet, $J_{1,2}$ = 10 Hz), C-2 proton resonance at 3.80 τ (quartet, $J_{2,4}$ = 1.5 Hz, $J_{1,2}$ = 10 Hz) and C-4 proton resonance at 3.99 τ (multiplet) (cf Figure 3). The C-18 protons are assigned to the 3-proton singlet at 9.15 τ while the C-19 protons are assigned to the 3-proton singlet at 8.52 τ. By adding deuterium oxide to the solution, exchange of the hydroxyl protons for deuterium is achieved and with the coupling of the hydroxyl protons to the protons on carbons bearing the hydroxyl groups eliminated, a simplified spectrum of triamcinolone is achieved (cf Figure 4). The AB quarter of the C-21

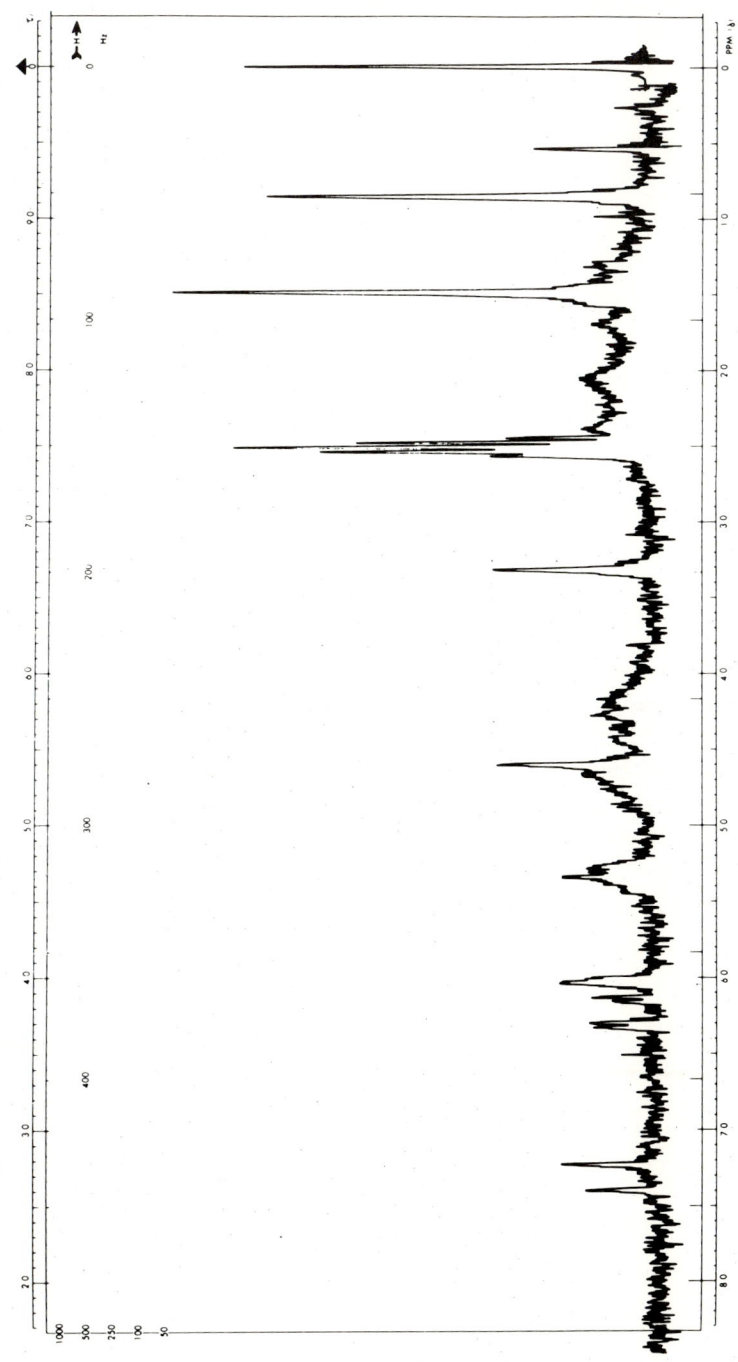

Fig. 3 NMR spectrum of triamcinolone, Squibb House Standard #30197-503 in deuterodimethylsulfoxide containing tetramethylsilane as internal reference. Instrument: Varian A-60

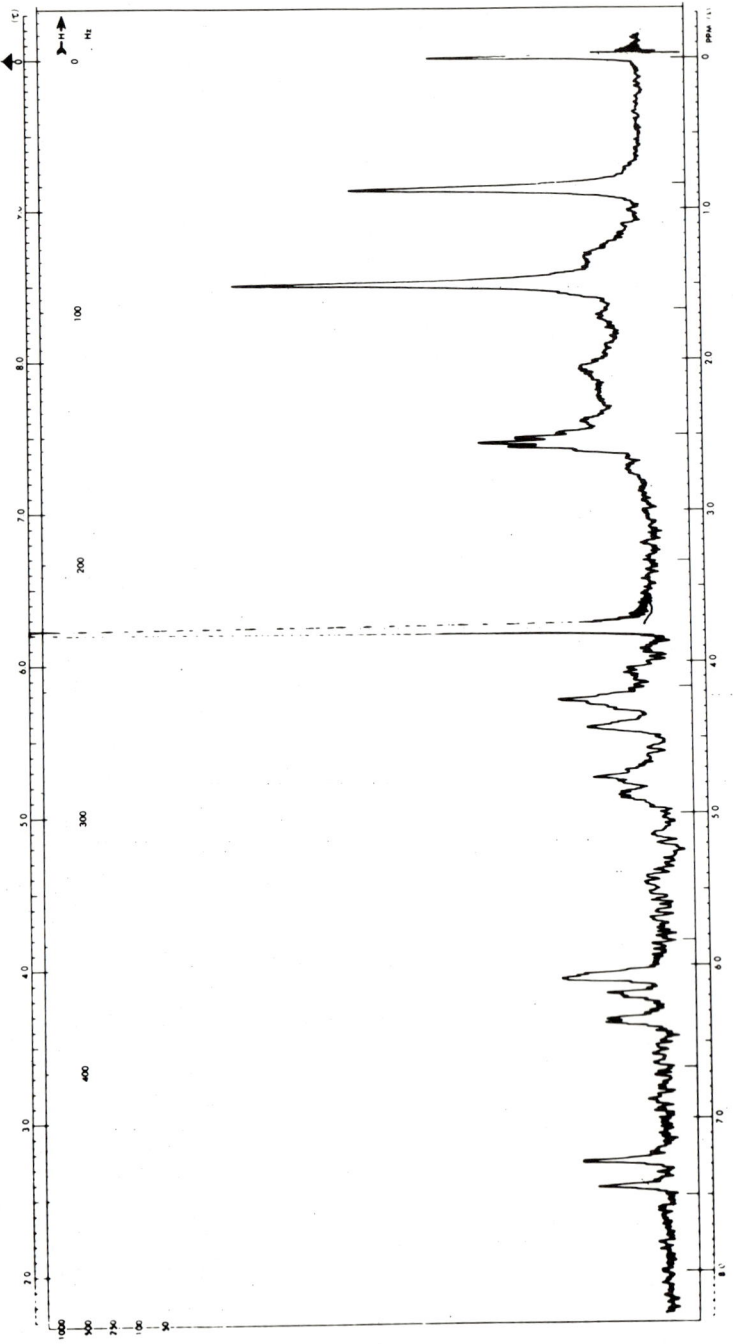

Fig.4 NMR spectrum of triamcinolone, Squibb House Standard #30197-503 in deuterodimethylsulfoxide containing tetramethylsilane as internal standard and deuterium oxide. Instrument: Varian A-60

geminal proton centered at 5.69 τ is observed and the doublet at 5.2 τ is assigned to the resonance of the C-16β proton. (see table I). These data agree with published data[57].

TABLE I

NMR Spectral Assignments of Triamcinolone

Protons at	Chemical Shift τ	
C-1	2.74	d; $J_{1,2} = 10$
C-2	3.80	q; $J_{2,4} = 1.5$
		$J_{1,2} = 10$
C-4	3.99	m
C-16	5.2	d
C-18	9.15	s
C-19	8.52	s
C-21	5.69	AB_q

s = singlet; d = doublet; m = multiplet; AB_q = AB quartet; J = coupling constant in Hz.

2.3 **Ultraviolet Spectra**
   The following ultraviolet spectral data have been reported:

	Ref. 3	Ref. 4,7	Ref. 8
λmax. in ethanol	239 mµ	238 mµ	239 mµ
ε	15,224	15,800	15,900

Squibb House Standard 30194-503 (0.0016% in absolute ethanol) when scanned between 340 and 210 mµ

(Instrument: Cary 15) exhibited a single band (E_1^1 381) peaking at 239 mμ[8]. This band is due to the 1,4-diene-3-keto system.

2.4 Mass Spectra

The high resolution mass spectrum[6] of triamcinolone was taken on a MS-9 instrument and is summarized in Table II. Unlike its acetonide derivative, this more polar compound degrades thermally extensively. The peak at the molecular ion (m/e 394) is almost imperceptible in the low resolution spectrum and was not detected at all in the high resolution spectrum. A peak occurs at m/e 374 which corresponds to the loss of HF from the M$^+$. The most significant peak in the high mass range is the m/e 326 ion, corresponding to the formula $C_{20}H_{22}O_4$ which appears to result from the loss of the CH_2OH, H_2O and HF groups. The base peak of triamcinolone acetonide of m/e 375 ($C_{22}H_{28}O_4F$) arises from the cleavage of the C-20 and C-21 carbons while the base peak of triamcinolone occurs at m/e 122 ($C_8H_{10}O$), which together with the ion at m/e 121 demonstrates the presence of the $\Delta^{1,4}$-e-keto group. Significant peaks of the low resolution spectrum are shown in Figure 5.

2.5 Optical Rotation

The following specific rotations have been reported:

$[\alpha]_D^{25}$ + 75° (C = 0.20 in acetone)[4,7]

$[\alpha]_D^{23}$ + 71° (C = 0.35 in acetone)[5]

$[\alpha]_D^{22}$ + 67.1° ± 1.9° (C = 0.5 in MeOH)[3]

$[\alpha]_D^{21}$ + 70.5° (C = 1 in dimethylformamide)[9]

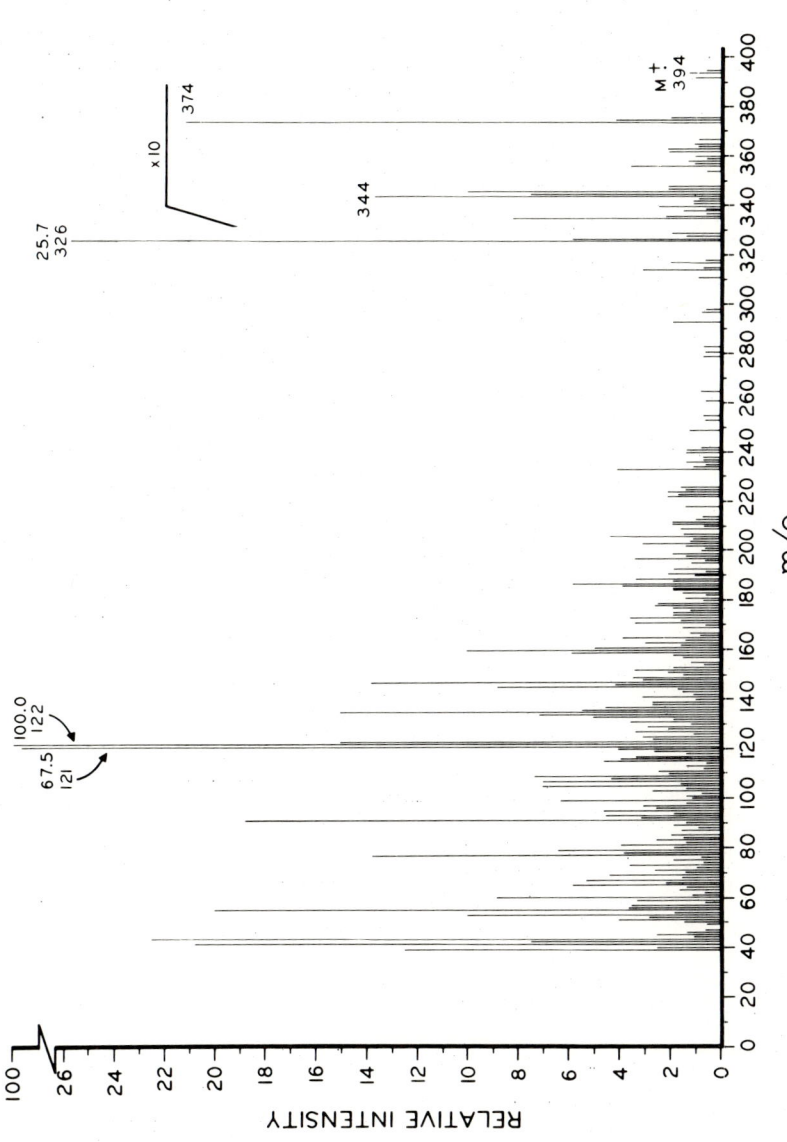

Fig.5 Low Resolution Mass Spectra of Triamcinolone Squibb. House Standard #30194-503

TABLE II

High Resolution Mass Spectrum of Triamcinolone[a]

Found Mass	Calcd. Mass	Unsat.[b]	O/E[c]	C	H	O	F
374.1747	374.1729	9.0	O	21	26	6	0
326.1528	326.1518	10.0	O	20	22	4	0
122.0741	122.0732	4.0	O	8	10	1	0
121.0659	121.0653	4.5	E	8	9	1	0

[a] Only those peaks considered to be significant to the discussion are listed. A complete element map can be obtained from Dr. A. I. Cohen, The Squibb Institute on request.
[b] Number of double bonds and rings.
[c] O - odd electron ion; E - even electron ion.

2.6 Melting Range

Like many steroids, triamcinolone does not melt sharply. The melting temperature range is wide and depends on the rate of heating and probably also the polymorphic form.

The resulting melting point temperatures (°C) have been reported without specifying the polymorphic form:

      260-262.5[4,7]
      269-271[4]
      262-264[3]
      248-250[5]
      264.2-267.9 (USP Method)[9]

2.7 Differential Thermal Analysis

Triamcinolone polymorphs "A" and "B" (see also 2.1; 2.9 and 2.10) gave different DTA patterns when heated at a rate of 15°/min. (Instrument: DuPont 900)[10].

  a. **Polymorph A**, recrystallized from 60% aqueous isopropanol (Sample #4789-35-1; Dr. G. Michel).
  1. Small endotherm at 262°
  2. Small exotherm at 266°
  3. Endotherm at 286°
  4. Exotherm at 297°

Programming the sample to 270° with subsequent cooling and reheating gave endotherms with double peaks at 277° and 281° and an exotherm at 290°. The pattern appears to be shifting to the B type configuration (see below). This indicates that the "B" polymorph is the slightly favored form[10].

  b. **Polymorph B**, recrystallized from dimethylacetamide-water (Dr. Michel, Sample #4789-33-1):
  1. Very small exotherm at 237°
  2. Endotherm (relatively broad at 278°
  3. Exotherm at 287°

Programming this sample to 250° with subsequent cooling and reheating gave an endotherm at 273° and an exotherm at 282° - i.e. very similar to initial run.

2.8 Thermogravimetric Analysis

When Squibb House Standard #30194-503 was heated at a rate of 15°/min. under nitrogen sweep no weight loss was observed to 250°. (Instrument: DuPont 900)[10].

2.9 Solubility

The following solubility data[9] (w/v) were obtained at room temperature:

≥10% in dimethylformamide
≥1% in methanol
0.7 in ethanol
2.4 in tetrahydrofuran
0.5% in dioxane
0.3% in acetone
0.09% in ethyl acetate
0.05% in chloroform
0.03% in methyl-isopropyl ketone

In water at 25° the solubility is 0.008% for both types of polymorphs (A and B)[11].

2.10 Crystal properties

Triamcinolone may exist in two polymorphs, as infrared data (see section 2.1) show. The infrared data are supported by powder x-ray diffraction data which show 2 distinct patterns (see Tables III and IV) which correspond to the A and B infrared spectra.

The solvent used for crystalline determines which polymorph is formed. Polymorph A (I) is obtained from pyridine, aqueous pyridine[3], 60% aqueous isopropanol and potassium tetraborate-water[12], polymorph B (II) from methanol, aqueous methanol[3] and dimethylacetamide-water[12].

The optical crystallographic properties of triamcinolone have been reported[13] as follows without stating which polymorph was used:

> System: Orthorhombic; Crystal Habit: Columnar Optical sign +: Axial angle: 52°; Optic orientation (assigned acc. to crystal habit). XX11a; YY11c; ZZ11b Dispersion: None observed; Refractive Indexes: $\alpha(\omega) = 1.561$; $\beta(\epsilon) = 1.582$; $\gamma = 1.680$; Density = 1.426; Refraction Experimental = 95.50; Calculated = 95.50.

TABLE III
Powder x-ray diffraction pattern of triamcinolone polymorph "A" sample #4789/35-1 (Dr. Michel) recrystallized from 60% IPA

d(A°)*	Relative Intensity**
12.5	0.09
8.37	0.12
7.8	0.20
6.3	0.18
5.92	0.34
5.69	0.57
5.37	1.00
5.05	0.52
4.78	0.25
4.69	0.60
4.19	0.24
4.10	0.51
3.90	0.28
3.81	0.22
3.72	0.32
3.63	0.21
3.37	0.27
3.27	0.30
3.22	0.29
3.06	0.21
2.94	0.29
2.75	0.23
2.44	0.06
2.36	0.10
2.30	0.05
2.22	0.09
2.12	0.03

* d = (interplanar distance) $\frac{n\lambda}{2 \sin \theta}$
λ = 1.539 A°
Radiation: $K\alpha_1$ and $K\alpha_2$ Copper
**Based on highest intensity of 1.00

TABLE IV

Powder x-ray diffraction pattern of triamcinolone polymorph "B" sample #4789/33-1 Dr. Michel recrystallized from DMA, H_2O

d(A°)*	Relative Intensity**
11.7	0.18
10.5	0.34
8.4	0.22
6.57	0.24
5.90	1.00
5.65	0.87
5.45	0.30
5.22	0.52
4.87	0.39
4.67	0.53
4.43	0.30
4.26	0.32
4.17	0.53
3.78	0.37
3.67	0.48
3.48	0.29
3.37	0.30
3.26	0.20
3.09	0.29
2.97	0.39
2.85	0.21
2.76	0.23
2.72	0.18
2.58	0.25
2.44	0.18
2.36	0.09
2.33	0.05
2.30	0.07
2.20	0.09
2.13	0.14
2.10	0.07
2.06	0.05
2.00	0.04

* d = (interplanar distance) $\frac{n \lambda}{2 \sin \theta}$

λ = 1.539A°

Radiation: $K\alpha_1$ and $K\alpha_2$ Copper

** Based on highest intensity of 1.00

3. Synthesis, Purification

Triamcinolone (I, Figure 6) is most commonly synthesized by microbiological dehydrogenation at C-1,2 of 16α-hydroxy-9α-fluorohydrocortisone (II) directly[5] or with intermediate formation of the 16,17-borate ester[14,15] or 16,21-diacetate[4,7] (V). A variety of microbiological organisms, capable of 1-dehydrogenation have been reported[16] or patented[17].

Dehydrogenation with isolated bacterial cell powders have also been reported[19,20,21]. Triamcinolone has also been obtained from 9α-fluoroprednisolone (VII)[18] by 16-hydroxylation and from triamcinolone acetonide (VI) by hydrolysis with organic acids[22,23]. Methods of purification have been patented[24].

4. Stability, Isomerization, Degradation

Triamcinolone is very stable as a solid. In aqueous and alcoholic solutions the α-ketol side chain, as in all such corticosteroids, is prone to oxidative rearrangement and degradation at alkaline pH's.

It has been reported[25] that hydrocortisone and prednisolone, when exposed to ultraviolet light or ordinary fluorescent laboratory lighting in alcoholic solution, undergo photolytic degradation of the A-ring. Since triamcinolone has the same A-ring as prednisolone it probably also is labile under these conditions. In solution triamcinolone readily isomerizes to a D-homo-analog[22] 9α-fluoro-11β,16α,17aα-trihydroxy-17aβ-hydroxymethyl-1,2-D-homo-androstadiene-3,17-dione(VIII, Figure 6) under a variety of conditions, particularly in the presence of traces of metal cations[26]. When triamcinolone is fermented anaerobically with a number of microorganisms which normally dehydrogenate at C1-2 under aerobic conditions, the 20-ketone is reduced to the 20β-alcohol (IV, Figure 6) and/or the 1,2-double bond is hydrogenated (III)[27]. Both of these reactions are individually reversible[28].

5. Drug Metabolic Products

When tritium-labeled triamcinolone was injected intravenously into beagles 20% of the dose was recovered from the urine as unchanged triamcinolone, 25% as 6β-hydroxy-triamcinolone (IX, Figure 6) and 5% as a third unidentified component which probably is not a glucuronide or sulfate conjugate. Examination of human urine after oral administration of non-radioactive triamcinolone gave similar results[29].

When the plasma protein binding properties of triamcinolone were studied it was found that tritium-labeled triamcinolone was present in the unbound state to a much greater extent than hydrocortisone[30].

6. Methods of Analysis

6.1 Elemental Analysis

Element	% Theory	Reported Ref.[4,7]	Ref.[9]
C	63.94	64.19	63.68
H	6.90	7.17	6.89
F	4.82	4.90	4.76

While in these laboratories the oxygen flask combustion method has been used without any difficulty for the determination of fluorine in triamcinolone and other fluorinated steroids[31] use of the Pregl-Roth distillation method has also been recommended[32].

FIGURE 6

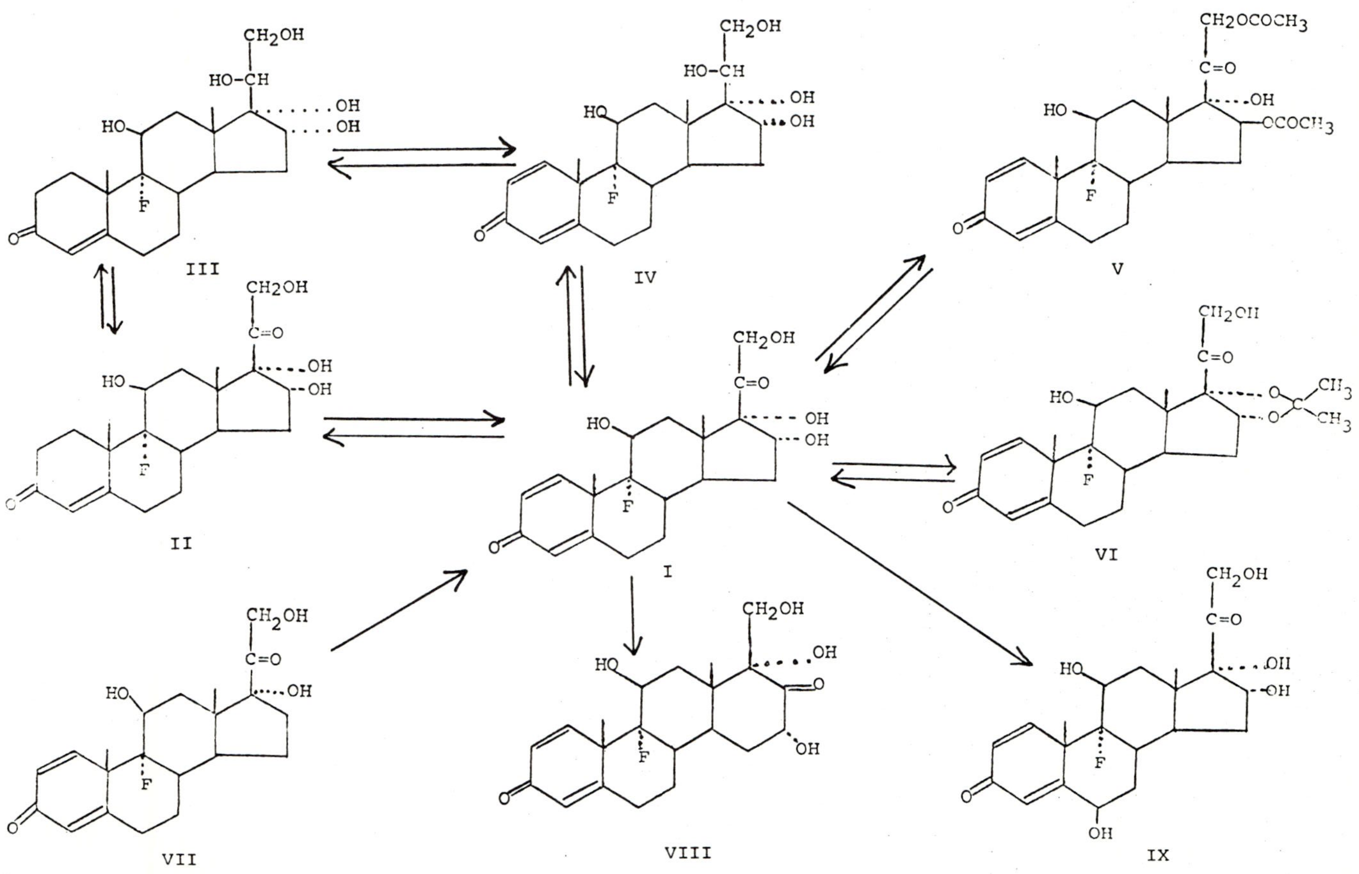

6.2 Direct Spectrophotometric Analysis

The ultraviolet absorption band at 239 mµ of triamcinolone (see 2.3) is due to the $\Delta^{1,4}$-diene-3-keto system of the A-ring (see also Section 4).

The absorbance is useful as a measure of purity from extraneous materials and can serve as a formulation batching assay. It can be used for chromatographic detection and quantitation[33].

6.3 Colorimetric Analysis

A variety of colorimetric methods have been used to detect and determine triamcinolone.

6.31 Tetrazolium Blue, in modifications of the original method for steroids with an α-ketol side chain[34] is perhaps the most widely used[3,35]. Reaction of triamcinolone with tetrazolium blue in alkaline medium measures the reducing power of the α-ketol side chain by producing a blue color (520 mµ) which can be quantitated. Color production can be stabilized by addition of chloroform[36]. The response is about 70% greater than that for non-16α-hydroxylated analogs[3]. The method is useful for general formulation assays. It also is used as a spray reagent in paper and thin layer chromatograms (Section 6.5). It is noteworthy that an increased reducing power of triamcinolone towards alkaline ferricyanide as compared with non-16α-hydroxylated steroids has also been observed[37].

6.32 Reaction of 1,4-diene-3-ketosteroids with isonicotinic acid hydrazide (isoniazid) produces a yellow hydrazone with an absorption maximum of 404 mµ[38]. This reaction can be adapted to triamcinolone for formulation assays. It also has been used as a spray reagent and for quantitation in paper and thin layer chromatography

(Section 6.5). It should be noted that 1,4-diene-3-ketosteroids react much less readily with isonicotinic acid hydrazides than do 4-ene-3-keto steroids[38].

6.33 For identification and differentiation from other steroids in formulations triamcinolone can be reacted with phenol and hydroquinone in a phosphoric-sulfuric acid mixture producing a pink color[56].

6.34 The absorption spectra of triamcinolone in concentrated sulfuric acid with maxima at 260, 310 and 390 mµ (15 min. reaction time) have been reported[3,40].

6.35 Since triamcinolone does not produce chromogens when reacted with a sulfuric acid, fructose, cysteine mixture this method can be used to determine small amounts of non-16α-hydroxylated steroids such as 9α-fluoro-hydrocortisone in the presence of triamcinolone[39].

6.36 Reacting triamcinolone and other 16α, 17α, 21-trihydroxy-20 ketones with phenylhydrazine in the presence of sulfuric acid (Porter-Silber test) gives a response of less than 10% of the chromogen formation of non-16α-hydroxylated analogs[3,41] and therefore renders useless this assay so popular for the determination of 17α-corticosteroids in biological systems.

6.4 **Polarographic Analysis**

The half wave potential (E1/2 versus saturated calomel electrode) was determined as -0.98 to -1.03 volts in tetra-n-butylammonium hydroxide-phosphate-methanol buffer pH 2[9]. Since 1,2-Dihydrotriamcinolone exhibits a half wave potential of ca -1.20 volts, this difference can be used to determine the percentages of each in mixtures of the two steroids[3].

6.5 Chromatographic Analysis

Qualitative chromatographic methods can be used for identification; quantitative methods for assessment of purity and stability of triamcinolone acetonide.

6.51 Paper Chromatographic Analysis

Paper chromatographic R_f values of triamcinolone and related steroids are reported in Table V.

A method has been developed[33] to use formation of 16α,17α-ketals and acetals *in situ* on papergrams for the early recognition of the 16α,17α-diol feature of triamcinolone and related 16α-hydroxylated steroids.

Applying a general method[44] the purity of triamcinolone can be quantitated[45] in the presence of 1,2-dihydrotriamcinolone, triamcinolone isomer, 9α-fluorohydrocortisone and 9α-fluoroprednisolone. Whatman #1 filter paper was impregnated with ethylene glycol as the stationary phase. Triamcinolone was spotted at the 100 microgram level and the chromatogram was developed with chloroform-ethyl acetate (3:2) for 20 hours. The front end of the chromatogram was allowed to run off the paper. The triamcinolone zones were located by their absorbance in the ultraviolet, cut out, eluted with an acidified methanolic solution of isonicotinic acid hydrazide and read against a standard at 415 millimicrons.

The quantitation of triamcinolone using solvent system A (Table V) has also been reported[33].

6.52 Thin Layer Chromatographic Analysis

Experience with thin layer chromato-

TABLE V

Paper Chromatographic R$_f$ Values

Solvent System* Compounds:	A(I)[33]	B(II)[33]	C(III)[33]	D(IV)[33]	E[42]	F[43]
Triamcinolone	0.18	0.42	0.14	0.13	0.22	0.93
Triamcinolone acetonide	–	–	–	0.88	–	–
Triamcinolone 16α,21-diacetate	–	–	–	0.89	–	–
Triamcinolone isomer**	0.10	0.22	0.08	0.06	–	–
1,2-Dihydrotriamcinolone	0.24	0.52	0.22	0.19	–	–
1,2-Dihydrotriamcinolone isomer	0.12	0.30	0.13	0.10	–	–
9α-Fluoroprednisolone	0.37	0.67	0.37	–	–	–
9α-Fluorohydrocortisone	0.45	0.80	0.47	0.59	–	–

* The roman numerals refer to the solvent system numbers of ref.33
** 9α-Fluoro-11β,16α,17aα-trihydroxy-17aβ-hydroxy-methyl 1,4-D-homoandrostadiene-3,17-dione[22].

The solvent systems used in Table V are the following:

		Developing Time
A(I)[33]	benzene/ethanol/water 2:1:1	4 hrs.
B(II)[33]	benzene/acetone/water 2:1:2	4 hrs.
C(III)[33]	benzene/dioxane/water/acetic acid 4:1:2:1	5 hrs.
D(IV)[33]	benzene/ethanol/water 2:1:2	5 hrs.
E[42]	chloroform/methanol/water 2:1:2	–
F[43]	Methyl ethyl ketone – 0.1$\underline{N}$ NH_4OH	4 hrs.(180 sec*)

*When run by centrifugal chromatography.

The following detection systems were used:

1. The modified Haines, Drake ultraviolet scanner[33]
2. Isonicotinic acid hydrazide[32, 38]
3. Alkaline tetrazolium blue spray[33, 42, 43]

TABLE VI

R_f or "Running distance" values (for explanation of individual values see below)

System:	1	2	3	4	5	6	7	8
Triamcinolone	0.63	0.59	0.08	0.14	0.28	0.04	0.01	0.03
Triamcinolone isomer	0.44	—	—	—	—	—	—	—
Triamcinolone acetonide	—	1.39	0.59	0.73	0.74	2.6	0.50	—
1,2-Dihydrotriamcinolone	0.72	—	0.14	0.18	0.31	0.19	—	—
9α-Fluoroprednisolone	—	—	—	—	—	—	—	—
9α-Fluorohydrocortisone	—	—	0.24	0.33	0.63	0.70	—	—

System 1[46]: Heat activated Brinkmann 200 micron silica gel plastic plate ether/N,N-dimethylformamide/acetone/methanol/water (81:12:3:3:1) 2 hours development at 5°C. Detection by U.V. light. Values given are R_f values.

System 2[47]: Kieselguhr G plate; methylene chloride/dioxane; water 2:1:1 Spray reagent: Alkaline 2,4-diphenyl-3(4-styrylphenyl)tetrazolium salt "Running distance" values related to cortisone = 1.00.

Systems 3-6[48]:Kieselguhr GF 254 plates: Spray reagent: Tetrazolium blue Solvent systems: 3 - 1,2-Dichloroethane/methanol/water 95:5:0.2 4 - 1,2-dichloroethane/2-methoxyethyl acetate/water 80:20:1 5 - Cyclohexane/ethyl acetate/water 25:75:1 6 - Stationary phase: 20% v/v formamide in acetone. Mobile phase: Chloroform/ether/water 80:20:0.5

Systems 7,8[49]: Kieselguhr G Plates Spray reagent: Tetrazolium blue Stationary phase: 10% Formamide in acetone. Mobile phase: 7 - methylene chloride/toluene (60:40). 8 - Chloroform. Values given are R_f values.

graphy of triamcinoline and related steroids is summarized in Table VI[46,47,48,49]. In addition several methods[50,51,52,53] have been reported to identify triamcinolone in mixtures of other corticosteroids, among them a method for the detection in horse urine[54].

6.53 Column Chromatographic Analysis

A column partition chromatographic procedure for triamcinolone and related steroids has been reported[33]. The sample is fractionated on diatomaceous earth (Celite) using dioxane/cyclohexane/water mixtures in the ratios 5:2:1, 5:3:1 or 5:4:1. The eluate is monitored by ultraviolet absorption. The order of elution is 9α-fluorohydrocortisone, 9α-fluoroprednisolone, 1,2-dihydrotriamcinolone, 1,2-dihydrotriamcinolone isomer, triamcinolone, triamcinolone isomer, 20β-hydroxytriamcinolone. For quantitation of individual fractions reaction with isonicotinic acid hydrazide (see Section 6.32) has been used[55]. As little as 0.05% of a comparison steroid can be detected.

7. Determination in Body Fluids and Tissues

A method for the detection and identification of triamcinolone and other corticosteroids in horse urine has been described[54].

REFERENCES

1. B. Keeler, Squibb Institute, personal communication.
2. R. J. Mesley, Spectrochemica Acta 22, 889, (1966).
3. L. L. Smith and M. Halwer, J. Am. Pharm. Assoc., Sci. Ed. 48, 348 (1959).

References Cont'd.

(4) S. Bernstein, R. H. Lenhard, W. S. Allen, M. Heller, R. Littell, S. M. Stolar, L. I. Feldman, R. H. Blank, J. Am. Chem. Soc. <u>78</u>, 5693 (1956).

(5) R. W. Thoma, J. Fried, S. Bonanno, P. Grabowich, J. Am. Chem. Soc. <u>79</u>, 4818 (1957).

(6) A. I. Cohen, Squibb Institute, personal communication.

(7) S. Bernstein, R. H. Lenhard, W. S. Allen, M. Heller, R. Littell, S. M. Stolar, L. I. Feldman and R. H. Blank, J. Am. Chem. Soc. <u>81</u>, 1689 (1959).

(8) J. Dunham, Squibb Institute, personal communication.

(9) H. Cords, Squibb Institute, personal communication.

(10) H. Jacobson, Squibb Institute, personal communication.

(11) H. Jacobson, Squibb Institute, personal communication.

(12) G. Michel, Squibb Institute, personal communication.

(13) J. A. Biles, J. Pharm. Sci. <u>50</u>, 464 (1961).

(14) R. W. Thoma and J. W. Ross, U.S. Patent 3,119,749 (1964); C. A. <u>60</u>, 9872h (1964).

(15) E. Ivashkiv, U.S. Patent 3,316,282 (1967); C. A. <u>67</u>, 11711m (1967).

(16) R. W. Thoma, A. I. Laskin, W. H. Trejo, H. Kroll, G. E. Peterson and G. P. Stickle, Sc. Rept. Inst. Super Sanita <u>1</u>, 326 (1961); C. A. <u>56</u>, 15960h (1962).

(17) L. I. Feldman, A. J. Shay and N. E. Rigler, U. S. Patent 3,037,913 (1962); C. A. <u>57</u>, 7746h (1963).

(18) J. Martinkova and J. Dyr, Coll. Czech. Chem. Commun. <u>30</u>, 2994 (1965).

(19) J. W. Ross, U.S. Patent 3,022,226 (1962); C.A. <u>57</u>, 1388a (1963).

References Cont'd.

(20) C. H. Sih, J. Pharm. Sci. $\underline{50}$, 712 (1961).

(21) R. C. Erickson, W. E. Brown, and R. W. Thoma, U. S. Patent 3,360,439 (1967); C. A. $\underline{67}$, 11711m (1967).

(22) L. L. Smith, M. Marx, J. J. Garbarini, T. Foell, V. E. Origoni and J. J. Goodman, J. Am. Chem. Soc. $\underline{82}$, 4616 (1960).

(23) J. Fried, U. S. Patent 3,177,231 (1965); C.A. $\underline{62}$, 19342a (1965).

(24) L. J. Leeson, S. A. Muller and G. M. Sieger, U. S. Patent 3,005,839 (1961); C. A. $\underline{56}$, 7396a (1962).

(25) W. E. Hamlin, T. Chulski, R. H. Johnson and J. G. Wagner, J. Am. Pharm. Assoc., Sci. Ed. $\underline{49}$, 253 (1963) and D. R. Barton and W. C. Taylor, J. Am. Chem. Soc. $\underline{80}$, 244 (1958); J. Chem. Soc. $\underline{1958}$, 2500.

(26) J. J. Goodman and L. L. Smith, Appl. Microbiol. $\underline{8}$, 363 (1960).

(27) L. L. Smith, J. J. Garbarini, J. J. Goodman, M. Marx and H. Mendelsohn, J. Am. Chem. Soc. $\underline{82}$, 1437 (1960) and J. Schmidt-Thome, G. Nesemann, H. J. Huebner and I. Alester, Biochem. Z. $\underline{336}$, 322 (1962).

(28) J. J. Goodman, M. May and L. L. Smith, J. Biol. Chem. $\underline{235}$, 965 (1960).

(29) J. R. Florini, L. L. Smith and D. A. Buyske, J. Biol. Chem. $\underline{236}$, 1038 (1961).

(30) J. R. Florini and D. A. Buyske, J. Biol. Chem. $\underline{236}$, 247 (1961).

(31) J. Alicino, Squibb Institute, personal communication.

(32) P. Joos and R. Ruyssen, J. Pharm. Belg. $\underline{19}$, 525 (1964); C. A. $\underline{62}$, 15995c (1965).

(33) L. L. Smith, T. Foell, R. deMaio and M. Halwer, J. Am. Pharm. Assoc. Sci. Ed. $\underline{48}$, 528 (1959) and L. L. Smith and T. Foell, J. Chromatog. $\underline{3}$, 381 (1960).

References Cont'd.

(34) W. J. Mader and R. R. Buck, Anal. Chem. $\underline{24}$, 666 (1952).
(35) See also USP and NF.
(36) P. Ascione and C. Fogelin, J. Pharm. Sci. $\underline{52}$, 709 (1963).
(37) N. R. Stephenson, Can. J. Biochem. and Physiol. $\underline{37}$, 391 (1959).
(38) L. L. Smith and T. Foell, Anal. Chem. $\underline{31}$, 102 (1959).
(39) C. J. Sih, S. C. Pan and R. E. Bennett, Anal. Chem. $\underline{32}$, 669 (1960).
(40) L. L. Smith and W. H. Muller, J. Org. Chem. $\underline{23}$, 960 (1958).
(41) A. Walser and H. P. Schlunke, Experientia $\underline{15}$, 71 (1959).
(42) S. C. Pan, J. Chromatog. $\underline{9}$, 81 (1962).
(43) J. A. Lowery and J. E. Cassidy, J. Chromatog. $\underline{13}$, 467 (1964).
(44) H. R. Roberts and K. Florey, J. Pharm. Sci. $\underline{51}$, 794 (1962).
(45) H. R. Roberts, Squibb Institute, personal communication.
(46) A. Vahidi and H. R. Roberts, Squibb Institute, personal communication.
(47) A. Hall, J. Pharm. Pharmacol. $\underline{16}$, Suppl. 9T (1964).
(48) C. J. Clifford, J. V. Wilkinson and J. S. Wragg, J. Pharm. Pharmacol. $\underline{16}$, Suppl. 11T (1964).
(49) D. Sonanini, R. Hofstetter, L. Auker and H. Mühlemann, Pharm. Acta. Helv. 40, 302 (1965).
(50) V. Schwarz and K. Sykora, Coll. Czech. Chem. Commun. $\underline{28}$, 101 (1963).
(51) W. Vlassak and G. Willems, J. Pharm. Belg. $\underline{19}$, 195 (1964); C. A. $\underline{63}$, 7288d (1965).
(52) R. Stainier, J. Pharm. Belg. $\underline{20}$, 89 (1965); C. A. $\underline{63}$, 11250b (1965).

References Cont'd.

(53) V. Rossetti, Biochim. Appl. **12**, 113 (1965); C. A. **64**, 11504g, (1966).

(54) M. S. Moss and H. J. Rylance, J. Pharm. Pharmacol. **18**, 13 (1966).

(55) R. Poet, Squibb Institute, personal communication.

(56) E. Ivashkiv, J. Pharm. Sci. **51**, 698 (1962).

(57) M. Heller and S. Bernstein, J. Org. Chem. **32**, 1264 (1967).

TRIAMCINOLONE ACETONIDE

K. Florey

Reviewed by N. E. Rigler

CONTENTS

1. Description
   1.1 Name, Formula, Molecular Weight
   1.2 Appearance, Color, Odor
2. Physical Properties
   2.1 Infrared Spectra
   2.2 Nuclear Magnetic Resonance Spectra
   2.3 Ultraviolet Spectra
   2.4 Mass Spectra
   2.5 Optical Rotation
   2.6 Melting Range
   2.7 Differential Thermal Analysis
   2.8 Thermogravimetric Analysis
   2.9 Solubility
   2.10 Crystal Properties
3. Synthesis
4. Stability - Degradation
5. Drug Metabolic Products
6. Methods of Analysis
   6.1 Elemental Analysis
   6.2 Direct Spectrophotometric Analysis
   6.3 Colorimetric Analysis
   6.4 Polarographic Analysis
   6.5 Chromatographic Analysis
      6.51 Paper
      6.52 Thin Layer
      6.53 Column
7. Determination in Body Fluids and Tissues
8. References

TRIAMCINOLONE ACETONIDE

1. ## Description

    1.1 **Name, Formula, Molecular Weight**
    Triamcinolone Acetonide is 9α-fluoro-11β,16α,17,21-tetrahydroxypregna-1,4-diene-3,20-dione cyclic 16,17-acetal with acetone. It is also known as 9α-fluoro-16α-hydroxyprednisolone acetonide; triamcinolone 16,17-acetonide; 9α-fluoro-11β,21-dihydroxy-16α,17α-isopropylidenedioxy-1,4-pregnadiene-3,20-dione; 9α-fluoro-16α,17α-isopropylidenedioxyprednisolone.

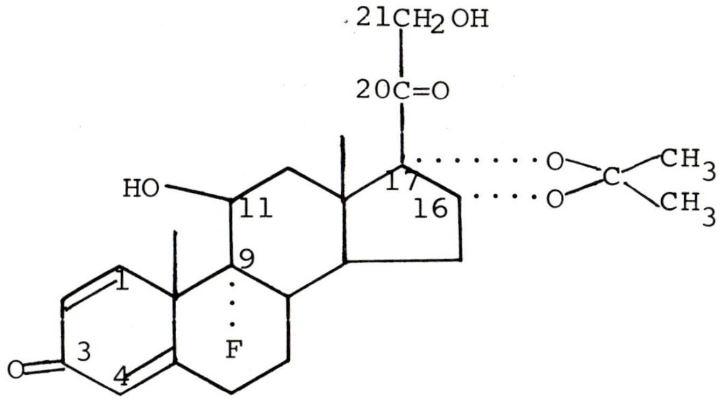

$C_{24}H_{31}FO_6$ Mol. Wt.: 434.49

    1.2 **Appearance, Color, Odor**
    White to off-white, odorless crystalline powder.

2. ## Physical Properties

    2.1 **Infrared Spectrum**
    The infrared spectrum of triamcinolone acetonide (Squibb House Standard #45885-008; IR spectrum #24226)[1] is presented in Fig. 1. The

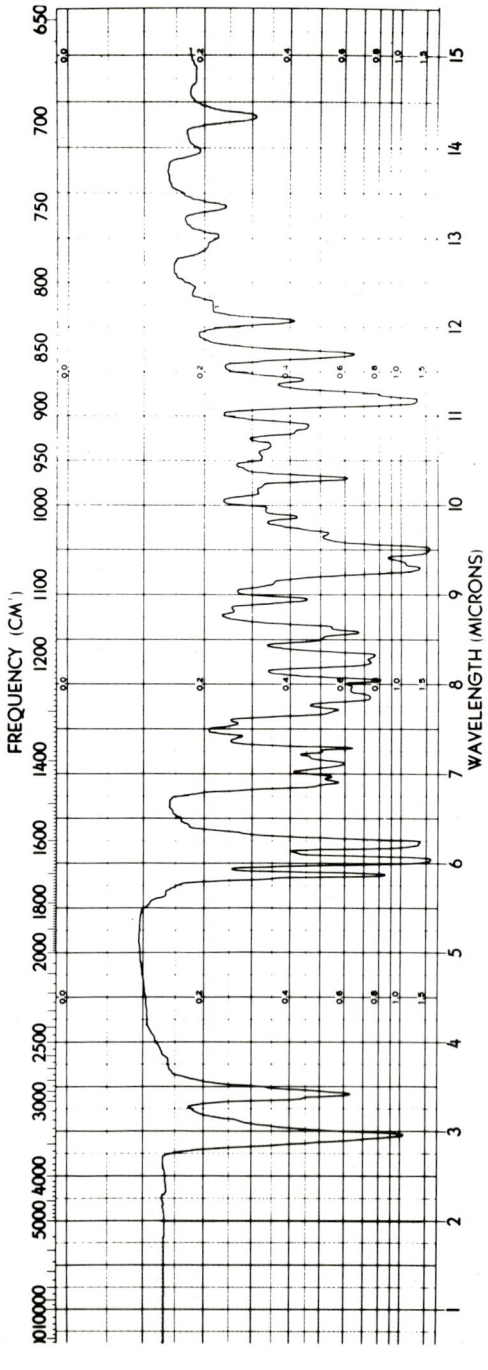

Fig.1 Triamcinolone acetonide - House Standard Batch #45885-008 - KBr pellet from MeOH solution - I.R.spectrum #24226 - Instrument:Perkin-Elmer 621

spectrum was taken in a KBr pellet from methanol solution. A spectrum of the same standard taken in a Nujol Mull is essentially identical to the one presented.

R. J. Mesley[2] in a discussion of "The infrared spectra of steroids in the solid state" assigns the following bands (cm^{-1}) to triamcinolone acetonide:
a. characteristic for 11β-hydroxy groups: 1035
b. characteristic for 21-hydroxy groups: 1100, 1057
c. characteristic for 1,4-diene-3-ones: 1408, 1300, 1247, 948, 940, 928, 892, 857, 831, 700. These assignments as well as those made by Fried[3] and Bernstein[4] essentially agree with peaks or shoulders presented in the spectrum Fig. 1.

2.2 Nuclear Magnetic Resonance Spectra

The NMR spectrum Fig. 2 was obtained by preparing a saturated solution of Squibb Standard #45885-008 in deuterochloroform containing tetramethylsilane as internal reference[32]. The spectrum is the result of nineteen individual spectra processed by a data acquisition-time averaging program[33]. The assignments shown in Table I agree with those of the previously published spectrum[10].

NMR has also been used to quantitatively analyze triamcinolone acetonide in dosage forms[34].

2.3 Ultraviolet Spectra

Bernstein[4] reported max. 238 - 239 mμ (ε 14,600) in ethanol.

Squibb House Standard #45885-008 (0.001268% in methanol) when scanned between 340 and 210 mμ exhibited a single band peaking at 238 mμ ($E_{1cm}^{1\%}$ =

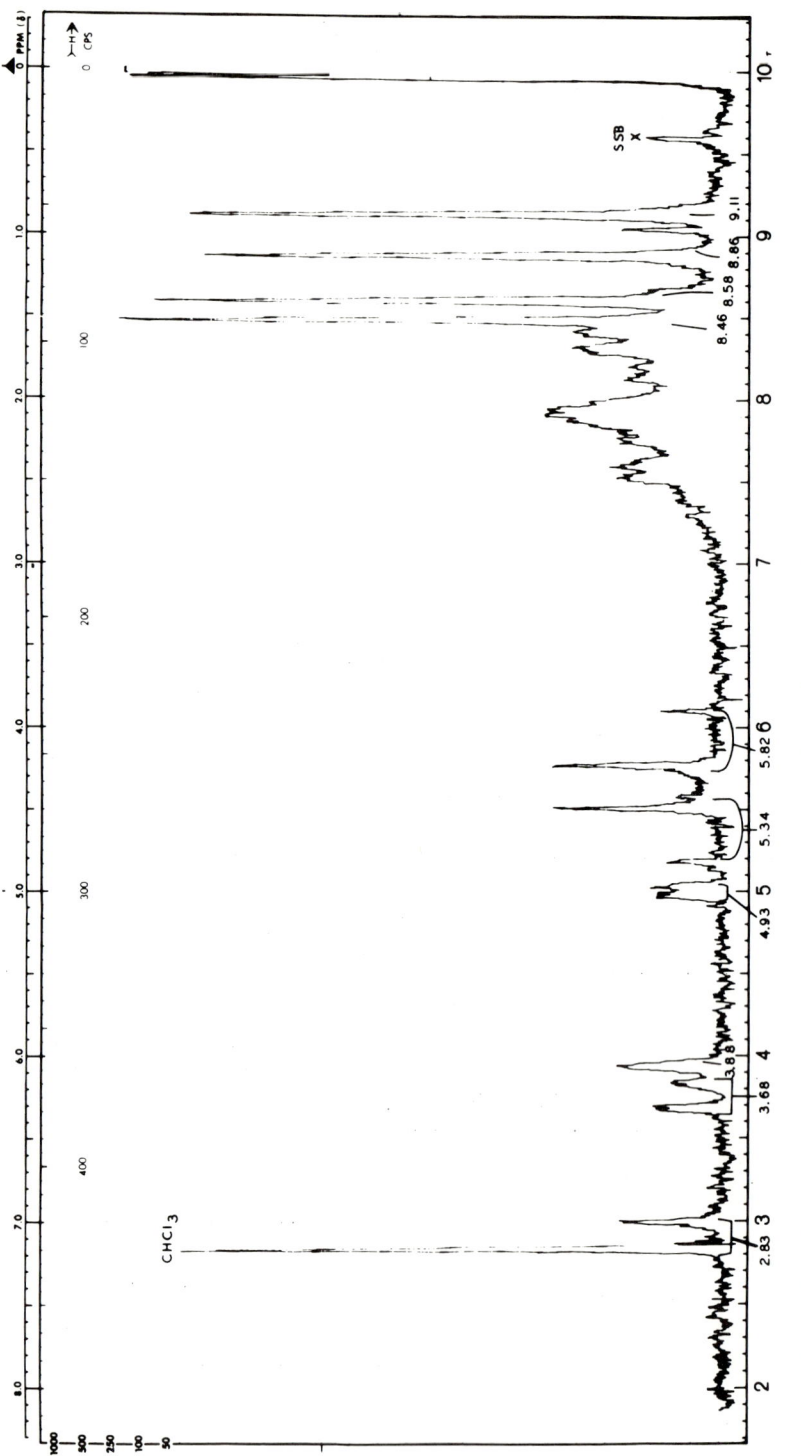

Fig. 2 Time averaged NMR spectrum of triamcinolone acetonide. Squibb House Standard #45885-008 in deuterochloroform containing tetramethyl-silane as internal reference. Instrument: Varian A-60

$3.54 \pm 1\%$; ε 15,400)[32]. This band is due to the 1,4-diene-3-one system.

TABLE I

NMR Spectral Assignments of Triamcinolone Acetonide

Protons at	Chemical Shift τ	
C-1	2.83	d; $J_{1,2} = 10.0$
C-2	3.68	q; $J_{1,2} = 10.0$; $J_{2,4} = 1.0$
C-4	3.88	m
C-11	5.6	m
C-16	4.93	d; $J = 4$
C-18	9.11	s
C-19	8.46	s
C-21	5.82	ABq; $J = 20.3$
C-21	5.34	
β-Acetonide methyl	8.86	s
α-Acetonide methyl	8.58	s

s = singlet; d = doublet; m = multiplet; ABq = AB quartet; J = coupling constant in Hz; q = quartet.

2.4 Mass Spectra

The mass spectrum of triamcinolone acetonide was obtained from Squibb House Standard #45885-008 by direct insertion of a sample into an MS-9 double focusing mass spectrometer[32]. Intensities were measured from the low resolution mass spectrum. Results are summarized as a bar graph (Fig. 3). The high resolution mass spectrum (Table II) was recorded on magnetic tape and an element map constructed with the aid of an IBM 1800-360 computer[32]. All peaks between mass 69 and 440 were evaluated. Ions of interest are listed in Table II. The molecular ion (M^+434) is quite prominent. The only fragment containing all atoms other than carbon and hydrogen is the ion at m/e 419, corresponding to the loss of one of the axial methyl groups.

There is a prominent odd-electron ion (m/e 414) with formula corresponding to the loss of hydrogen fluoride. The base peak (intensity = 100) at m/e 375 corresponds to the loss of $C_2H_3O_2$ (side chain) through cleavage between carbon atoms 17 and 20. This transition is supported by the presence of a metastable ion at m/e 324, observed in the low resolution spectrum. The base ion (m/e 375) is further fragmented by loss of acetone through cleavage of the ketal to m/e 317. This is supported by the presence of a metastable ion at m/e 268. The remaining C-16 or -17 oxygen is eliminated from ion m/e 317 as water giving rise to peak m/e 299. Ion m/e 317 also goes to m/e 279 by dehydration and dehydrofluorination again via metastable ion transition (m/e 295.5).

The last in a series of low intensity fragments with two oxygens and one fluorine is ion m/e 209 which probably represents the A- and B-ring and C-11 carbons with the full number of protons plus one re-arranged proton. The intense peak at m/e 121 represents the intact dienone A-ring with the C-6 methine and C-10 angular methyl group still attached.

TRIAMCINOLONE ACETONIDE

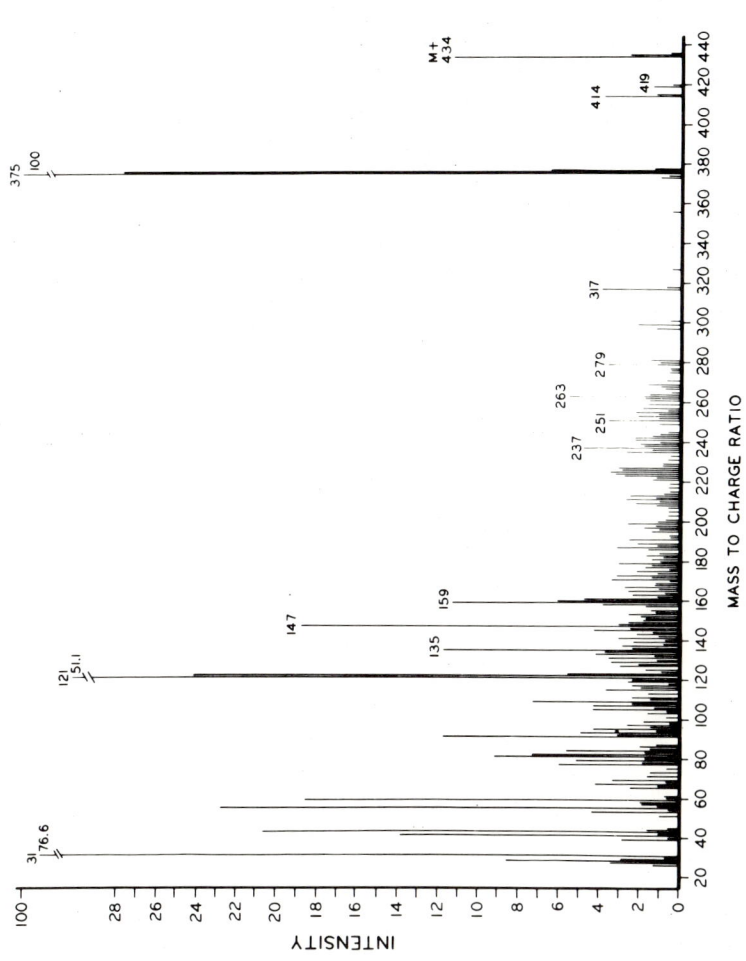

Fig. 3 Low resolution mass spectrum of triamcinolone acetonide. Squibb House Standard #45885-008. Instrument: AEI MS 902

TABLE II

High Resolution Mass Spectrum of Triamcinolone Acetonide[a]

Found Mass	Calcd. Mass	Unsat.[b]	O/E[c]	C	H	O	F
434.2114	434.2094	9.0	O	24	31	6	1
419.1884	419.1870	9.5	E	23	28	6	1
414.2052	414.2043	10.0	O	24	30	6	0
375.1970	375.1972	8.5	E	22	28	4	1
317.1542	317.1554	8.5	E	19	22	3	1
299.1416	299.1448	9.5	E	19	20	2	1
279.1365	279.1365	10.5	E	19	19	2	0
209.0950	209.0978	5.5	E	11	11	2	0
121.0655	121.0654	4.5	E	8	9	1	0

[a] Only those peaks considered to be significant to the discussion are listed. A complete element map can be obtained from Dr. A. I. Cohen, The Squibb Institute, on request.

[b] Number of double bonds and rings.

[c] O - odd electron ion; E - even electron ion.

2.5 Optical Rotation

The following rotations have been reported:

$[\alpha]_D$ + 109° (chloroform)[3]

$[\alpha]_D^{25}$ + 112° (c = 0.537 in chloroform)[4]

$[\alpha]_D^{25}$ + 124° (c = 0.5 in dimethylformamide)[5]

2.6 Melting Range

Like many steroids, triamcinolone acetonide does not exhibit a sharp melting point. The melting temperature range is wide and depends on the rate of heating.

The following melting point temperatures (°C) have been reported:
$$292 - 294^3$$
$$274 - 278^4$$
$$277 - 281^6$$
$$276 - 278 \text{ (USP method)}^5$$

2.7 Differential Thermal Analysis

A differential thermal analysis was performed on triamcinolone acetonide (Squibb House Standard #45885-008)[12].

A melting endotherm, followed by a small exotherm was observed.

The temperature of the endotherm varied with the heating rate. At a heating rate of 15°/min. the endotherm peaked at 300° and the exotherm at 305°. At a rate of 3°/min. the endotherm shifted to 281° and the exotherm to 284°.

2.8 Thermogravimetric Analysis

A thermal gravimetric analysis performed on triamcinolone acetonide (Squibb House Standard #45885-008)[12] showed a 1.0% weight loss complete at about 105°. The measurement was performed under nitrogen sweep; the heating rate was 15°/min. Additional weight was rapidly lost after the sample had melted at 300°.

2.9 Solubility

The following solubility data[5] were obtained at room temperature:

50 mg./ml. in 95% ethanol
40 mg./ml. in isopropyl alcohol
90 mg./ml. in acetone
25 mg./ml. in chloroform
250 mg./ml. in dimethylformamide

The solubilities in water as well as isotonic saline (pH 7) at 23° and 37° were determined as 0.004 ± 0.002% (40 μg./ml.).

2.10 Crystal Properties

a. The optical crystallographic properties of some glucocorticoids, among them triamcinolone acetonide, were determined by Biles[22].

The data reported for triamcinolone acetonide are as follows:
System: Trigonal; Crystal Habit: columnar; Optical Sign: -; Axial angle 0°; Optic orientation (assigned acc. to crystal habit): $\omega \parallel d$; $\varepsilon \parallel c$; Dispersion: none observed; Refractive Indices: $\alpha(\omega) = 1.595$, $\beta(\varepsilon) = 1.546$; Density: 1.323; Molar Refraction: Experimental = 109.08, Calculated = 107.39.

b. The x-ray powder diffraction pattern of triamcinolone acetonide (Squibb House Standard #45885-008) is presented in Table III[23].

c. Mesley[2], who inspected triamcinolone acetonide by infrared spectroscopy, did not discover polymorphism. (see also Section 2.1).

3. Synthesis

Triamcinolone acetonide is prepared by the reaction of acetone with triamcinolone in the presence of catalytic amounts of mineral acid. Fried[3] and Heller[6] used perchloric acid, Bernstein[4] hydrochloric acid. An alternative

TABLE III

*d (A°)	Relative Intensity**
8.72 ± 0.05	0.80
7.02	0.15
<u>5.98</u>	<u>1.00</u>
5.81	0.30
5.49	0.20
5.10	0.15
4.98	0.45
4.88	0.15
4.43	0.15
4.13	0.05
3.89	0.10
3.56	0.35
3.30	0.05
3.26	0.10
3.07	0.05
2.91	0.15
2.83	0.05
2.75	0.05
2.62	0.05

*d = (interplanar distance) $\frac{n\lambda}{2 \sin \theta}$
λ = 1.539 A°
Radiation: $K\alpha_1$ & $K\alpha_2$ Copper
**Based on highest intensity of 1.00

synthesis is the 1-dehydrogenation of the acetate of 1,2-dihydrotriamcinolone acetonide with 2,3-dibromo-5,6-dicyanoquinone[7] (see Figure 4) and subsequent saponification of triamcinolone acetonide 21-acetate.

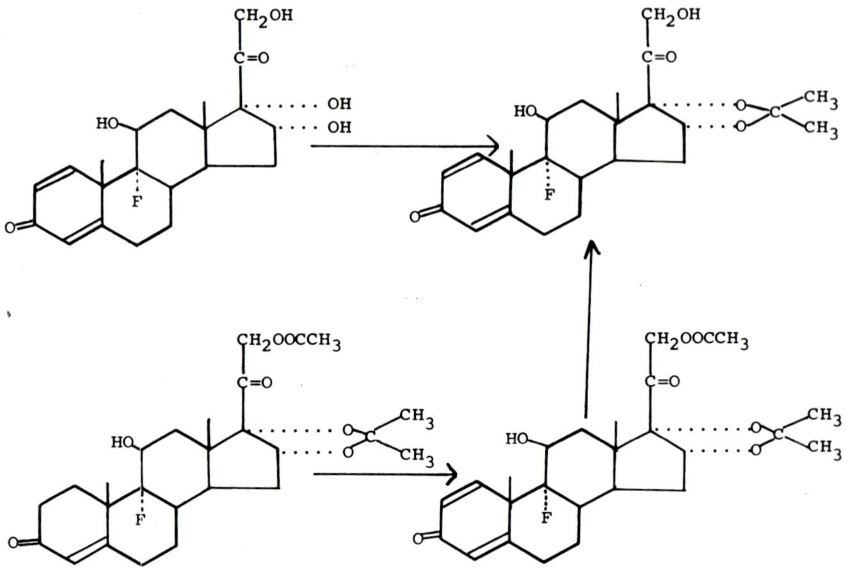

Fig. 4 Synthetic Pathways to Triamcinolone Acetonide.

4. Stability - Degradation

Triamcinolone acetonide is very stable as a solid. In aqueous and alcoholic solutions the α-ketol-side chain, as in all such corticosteroids, is prone to oxidative rearrangement and degradation at alkaline pH's. Smith, et.al.[10] oxidized triamcinolone acetonide to the corresponding etianic acid acetonide with sodium bismuthate in 50% aqueous acetic acid. It has been reported[8] that hydrocortisone and prednisolone, when exposed to ultraviolet light or ordinary fluorescent laboratory lighting in alcoholic solution, undergo photolytic degradation of the A-ring. Since triamcinolone acetonide has the same A-ring as prednisolone it probably also is labile under these conditions.

The cyclic ketal group of triamcinolone acetonide can be cleaved by a variety of organic acids[10]. For instance, subjecting the 21-acetate to the action of formic acid at slightly elevated temperatures, followed by saponification of the 16,21-diformate, yields triamcinolone[9]. While triamcinolone easily isomerizes to a D-homo-analog[10] under a variety of conditions, particularly in the presence of traces of metal cations, formation of a cyclic ketal stabilizes the molecule and no similar isomerization has been observed with triamcinolone acetonide.

5. Drug Metabolic Products

While in an experiment with tritium labeled triamcinolone in the dog Florini, et.al.[11] identified 6β-hydroxytriamcinolone as the major metabolic product, no triamcinolone acetonide metabolic products have been reported thus far.

6. Methods of Analysis

6.1 Elemental Analysis

Element	% Theory	Reported: Ref.[3]	Ref.[4]
C	66.34	66.49	66.85
H	7.19	7.31	7.23
F	4.37	–	4.71

6.2 Direct Spectrophotometric Analysis

The ultraviolet absorption band at 238 mµ of triamcinolone acetonide (see 2.3) is due to the $\Delta^{1,4}$-diene-3-one system of the A-ring (see also Section 4).

The absorbance is useful as a measure of purity from extraneous materials and can also serve as a formulation batching assay. It can be used for chromatographic detection[13] and quantitation.

6.3 Colorimetric Analysis

A variety of colorimetric methods can be used to assay triamcinolone acetonide.

6.31 Tetrazolium Blue, in modifications of the original method for α-ketol steroids by Mader and Buck[16], is perhaps the most widely used.[17,18,19] Reaction of triamcinolone acetonide with tetrazolium blue in alkaline medium gives a blue color (520 mµ) which can be quantitated. It measures the reducing power of the α-ketol-side chain and is useful for general formulation assays. It also is used as a spray reagent in paper and thin layer chromatograms (Section 6.5).

6.32 Reaction of 1,4-diene-3-one steroids with isonicotinic acid hydrazide (isoniazid) produces a yellow hydrazone with an absorption maximum at 404 mμ[20]. This reaction can be adapted to triamcinolone acetonide for formulation assays. It also has been used as spray reagent and for quantitation in paper and thin layer chromatographic systems (Section 6.5). It should be noted that 1,4-diene-3-one steroids react much less readily with isonicotinic acid hydrazides than do 4-ene-3-one steroids[20].

6.33 For identification and differentiation from other steroids in formulations triamcinolone acetonide can be reacted with phenol and hydroquinone in a phosphoric sulfuric acid mixture producing a pink color[21].

6.4 Polarographic Analysis

The half wave potential ($E_{1/2}$ versus standard calomel electrode) was determined as -1.45 volts in lithium chloride in methanol[14]. The method was not considered sufficiently accurate to be used as a quantitative assay.

Cohen[15] subjected triamcinolone acetonide to polarographic reduction in dimethylformamide and found two reducing waves:

	Wave 1	Wave 2
$E_{1/2}$ (volts vs mercury pool anode)	-1.44	-2.00
Id (diffusion current constant)	1.3	5.6
n (Apparent number of electrons transferred)	0.92	0.40

6.5 Chromatographic Analysis

Qualitative chromatographic methods can be used for identification; quantitative methods for assessment of purity and stability of triamcinolone acetonide.

6.51 **Paper Chromatographic Analysis**

Paper chromatographic R_f values of triamcinolone acetonide and related steroids in a number of solvent systems are reported[13,24,25] in Table IV.

The solvent systems used in Table IV are the following:

		Developing Time
A(IV)[13]	:benzene/ethanol/water 2:1:2	5 hrs.
B(V)[13]	:toluene/pet.ether(b.p.30-60°)/ methanol/water 12:8:3:7	2-1/2 hrs.
C(VI)[13]	:benzene/pet.ether(b.p.90-100°)/ methanol/water 5:5:7:3	3-1/2 hrs.
D[24]	:n-hexane/toluene/methanol/ water 5:5:7:3	–
E[25]	:mobile phase: toluene sat.with propylene glycol. Stationary phase: propylene glycol	20 hrs.
F[25]	:mobile phase: methylisobutyl ketone/formamide 20:1 Stationary phase: formamide	3 hrs.

The following detection systems were used:
1. A modified Haines, Drake ultraviolet scanner[13,25,26]
2. Isonicotinic acid hydrazide[13,20,25]
3. Alkaline tetrazolium blue spray[13,24]

Applying his general method[27] H. R. Roberts has worked out quantitative determinations using solvent systems E and F[25] (see above). Solvent system F was particularly well suited to quantitate residual triamcinolone. Whatman #1 paper was impregnated with the stationary phase. Triamcinolone acetonide was spot-

TABLE IV

Paper Chromatographic R_f Values of Triamcinolone Acetonide

Solvent system	A(IV)*[13]	B(V)*[13]	C(VI)*[13]	D[24]	E[25]	F[25]
Compounds:						
Triamcinolone	0.13	–	–	–	–	0.31
Triamcinolone acetonide	0.88	0.36	0.13	0.21	0.07	0.81
Triamcinolone isomer**	0.60	–	–	–	–	–
Triamcinolone isomer acetonide**	0.62	0.10	0.03	–	–	–
1,2-Dihydrotriamcinolone	0.19	–	–	–	–	–
1,2-Dihydrotriamcinolone acetonide	0.92	0.46	0.25	–	–	–
1,2-Dihydrotriamcinolone isomer	0.10	–	–	–	–	–
1,2-Dihydrotriamcinolone isomer acetonide	–	0.16	0.06	–	–	–

* The Roman numerals refer to the solvent system number of Smith[13]
**These compounds are 9α-Fluoro-11β,16α,17aα-trihydroxy-17aβ-hydroxymethyl 1,4-D-homoandrostadiene-3,17-dione and its cyclic 16aα,17aα-ketal with acetone (ref.10).

ted at the 100 microgram level in method E and at the 500 microgram level for method F. After development, spots were cut out, eluted with 0.1% isonicotinic acid hydrazide in methanol, acidified with 0.1% concentrated hydrochloric acid for one hour, and read against a standard at 415 mµ.

6.52 **Thin Layer Chromatographic Analysis**
Experience with the thin layer chromatography of triamcinolone acetonide is summarized in Table V.

6.53 **Column Chromatographic Analysis**
A column partition chromatographic procedure for triamcinolone acetonide and related steroids has been worked out by Poet[31], following essentially the procedure described by Smith et. al.[13]. The sample (40 mg) is fractionated on Celite (25 g) using a dioxane; cyclohexane:2-methoxyethanol:water 40:80:10:8 solvent system, followed by a methanol strip. The ultraviolet absorption of the eluate is monitored, and individual fractions are quantitated with isonicotinic acid hydrazide. The order of elution is 1,2-dihydrotriamcinolone acetonide, triamcinolone acetonide and finally triamcinolone (in the methanol strip).

7. **Determination in Body Fluids and Tissues**
The plasma levels of triamcinolone acetonide, after intramuscular injection, have been determined, using chloroform extraction, thin layer chromatography and determination of U.V. absorbance in the presence of fluorescein[35] or visual comparison to standards[36].

TABLE V

R_f or "Running distance" values (for explanation of individual values, see below):

System	1	2	A	B	C	D	E	a	b	c	d
Triamcinolone	–	0.59	0.08	0.14	0.27	0.04	0	0.01	–	–	0.03
Triamcinolone acetonide	0.50	1.39	0.59	0.73	0.74	2.6	0.99	0.50	0.48	0.34	–

Systems 1 and 2[28]:

System 1: Kieselguhr G plate; Dichloroethane/methylacetate/water 2:1:1; Spray reagent: Alkaline 2,5-diphenyl-3(4-styrylphenyl) tetrazolium solution; "Running distances" values related to cortisone acetate = 1.00;

System 2: Same plates; methylene chloride/dioxane/water 2:1:1; "Running distance" values related to cortisone = 1.00

Systems A–E[29]:

Kieselguhr GF 254 plates; Spray reagent: Tetrazolium blue; "Running distance" values: A,B,C,E related to hydrocortisone acetate = 1.00

D related to hydrocortisone = 1.00

TABLE V Cont'd.

Systems A-E[29] Cont'd.

Solvent systems:
A — 1,2-Dichloroethane:methanol:water 95:5:0.2
B — 1,2-Dichloroethane:2-methoxyethyl acetate:water 80:20:1
C — Cyclohexane:ethylacetate:water 25:75:1
D — Stationary phase: 20% v/v formamide in acetone
    Mobile phase: Chloroform:ether:water 80:20:0.5
E — Stationary phase: 25% v/v formamide in acetone
    Mobile phase: Cyclohexane:tetrachloroethane:water 50:50:0.1

Systems a-d[30]:

Kieselguhr G plates; Spray reagent: Tetrazolium blue
Values given are R_f values
Solvent systems:
a — methylene chloride:toluene 60:40
b — methylene chloride:toluene 50:50
c — Chloroform:toluene 25:75
d — Chloroform

References

1. B. Keeler, Squibb Institute, personal communication.
2. R. J. Mesley, Spectrochimica Acta 22, 889 (1966).
3. J. Fried, A. Borman, W. B. Kessler, P. Grabowich, and E. F. Sabo, J. Am. Chem. Soc. 80, 2338 (1958).
4. S. Bernstein, R. H. Lenhard, W. S. Allen, M. Heller, R. Littell, S. M. Stolar, L. Feldman and R. H. Blank, J. Am. Chem. Soc. 81, 1689 (1959).
5. H. Cords, Squibb Institute, personal communication.
6. M. Heller, S. Stolar and S. Bernstein, J. Org. Chem. 26, 5044 (1961).
7. A. E. Hydorn, U. S. Patent 3,035,050 (1962).
8. W. E. Hamlin, T. Chulski, R. H. Johnson and J. G. Wagner, J. Am. Pharm. Assoc. Sci. Ed. 49, 253 (1960) and D. H. R. Barton and W. C. Taylor, J. Am. Chem. Soc. 80, 244 (1958); J. Chem. Soc. 1958, 2500.
9. J. Fried, U. S. Patent 3,177,231 (1965).
10. L. L. Smith, M. Marx, J. J. Garbarini, T. Foell, V. E. Origoni and J. J. Goodman, J. Am. Chem. Soc. 82, 4616 (1960).
11. J. R. Florini, L. L. Smith and D. A. Buyske, J. Biol. Chem. 236, 1038 (1961).
12. H. Jacobson, Squibb Institute, personal communication.
13. L. L. Smith, T. Foell, R. deMaio and M. Halwer, J. Am. Pharm. Assoc. Sci. Ed. 48, 528 (1959), and L. L. Smith and T. Foell, J. Chromatog. 3, 381 (1960).
14. N. Coy, Squibb Institute, personal communication.
15. A. I. Cohen, Anal. Chem. 35, 128 (1963).

16. W. J. Mader and R. R. Buck, Anal. Chem. $\underline{24}$, 666 (1952).
17. P. Ascione and C. Fogelin, J. Pharm. Sci. $\underline{52}$, 709 (1963).
18. M. Umeda, S. Tsubota and A. Kajiita, Takamine Kenkyasho Nempo $\underline{14}$, 87 (1963; C. A. $\underline{58}$, 4379f (1963).
19. See also N. F. and U. S. P.
20. L. L. Smith and T. Foell, Anal. Chem. $\underline{31}$, 102 (1959).
21. E. Ivashkiv, J. Pharm. Sci. $\underline{51}$, 698 (1962).
22. J. A. Biles, J. Pharm. Sci. $\underline{50}$, 464 (1961).
23. N. Coy and Q. Ochs, Squibb Institute, personal communication.
24. S. C. Pan, J. Chromatog. $\underline{9}$, 81 (1962).
25. H. R. Roberts, Squibb Institute, personal communication.
26. E. von Arx and R. Weber, Helv. Chim. Acta $\underline{39}$, 1664 (1956).
27. H. R. Roberts and K. Florey, J. Pharm. Sci. $\underline{51}$, 794 (1962).
28. A. Hall, J. Pharm. Pharmacol. $\underline{16}$, Suppl. 9T (1964).
29. C. J. Clifford, J. V. Wilkinson and J. S. Wragg, J. Pharm. Pharmacol. 16, Suppl. 11T (1964).
30. D. Sonanini, R. Hofstetter, L. Anker and H. Mühlemann, Pharm. Acta Helv. $\underline{40}$, 302 (1965).
31. R. Poet, Squibb Institute, personal communication.
32. A. I. Cohen, Squibb Institute, personal communication.
33. A. I. Cohen and T. Gilmore, to be published.
34. H. W. Avdovich, P. Hanbury and B. A. Lodge, J. Pharm. Sci. $\underline{59}$, 1164 (1970).
35. R. Eberl and H. Altman, J. Rheumaforsch. $\underline{29}$, 94 (1970).
36. L. Zicha, Arzneim.-Forsch. $\underline{19}$, 340 (1969).

The author wishes to thank Mrs. Agnes LaBadie for her invaluable secretarial help in developing the format of this Profile which served as a prototype. This acknowledgement also pays tribute to all secretaries who so patiently typed these Profiles.

TRIAMCINOLONE DIACETATE

K. Florey

Reviewed by N. E. Rigler

CONTENTS

1. Description
   1.1 Name, Formula, Molecular Weight
   1.2 Appearance, Color, Odor
2. Physical Properties
   2.1 Infrared Spectra
   2.2 Nuclear Magnetic Resonance Spectrum
   2.3 Ultraviolet Spectrum
   2.4 Mass Spectrum
   2.5 Optical Rotation
   2.6 Melting Range
   2.7 Differential Thermal Analysis
   2.8 Thermogravimetric Analysis
   2.9 Solubility
   2.10 Crystal Properties
3. Synthesis
4. Stability, Degradation
5. Drug Metabolic Products
6. Methods of Analysis
   6.1 Elemental
   6.2 Direct Spectrophotometric
   6.3 Colorimetric
   6.4 Polarographic
   6.5 Chromatographic
      6.51 Paper
      6.52 Thin Layer
7. References

TRIAMCINOLONE DIACETATE

1. Description

   1.1 Name, Formula, Molecular Weight
   Triamcinolone diacetate is 16α,21-diacetoxy-9α-fluoro-11β,17α-dihydroxyl-1,4-pregnadiene-3,20-dione; also 9α-fluoro-11β,16α,17α,21-tetrahydroxy-1,4-pregnadiene-3,20-dione-16α,21-diacetate; 9α-fluoro-16α-prednisolone-16α,21-diacetate. SQ 9465.

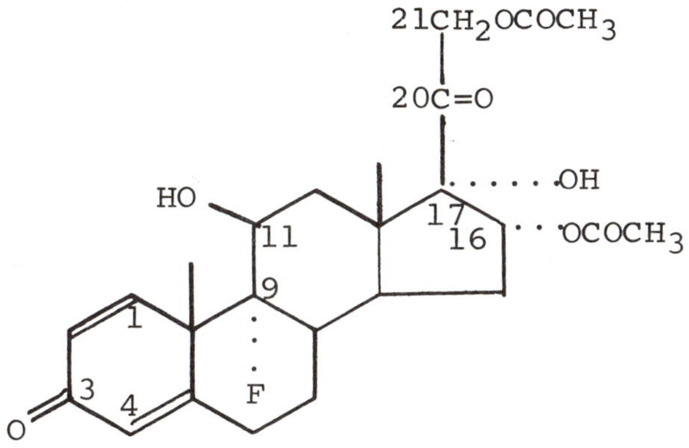

$C_{25}H_{31}FO_8$ Molecular Weight: 478.52

   1.2 **Appearance, Color, Odor**
   White to off white, odorless crystalline powder.

2. **Physical Properties**

   2.1 **Infrared Spectra**
   Triamcinolone diacetate exhibits polymorphism (see also 2.10). Smith et al[1] presented infrared spectra for two polymorphic forms. Figures 1 and 2 present infrared curves of Squibb Batch #30636-001 taken in mineral oil mull and KBr pellet from MeOH solution respectively[2]. While the latter esentially resembles the spectrum of polymorph II of Smith[1], the former has ad-

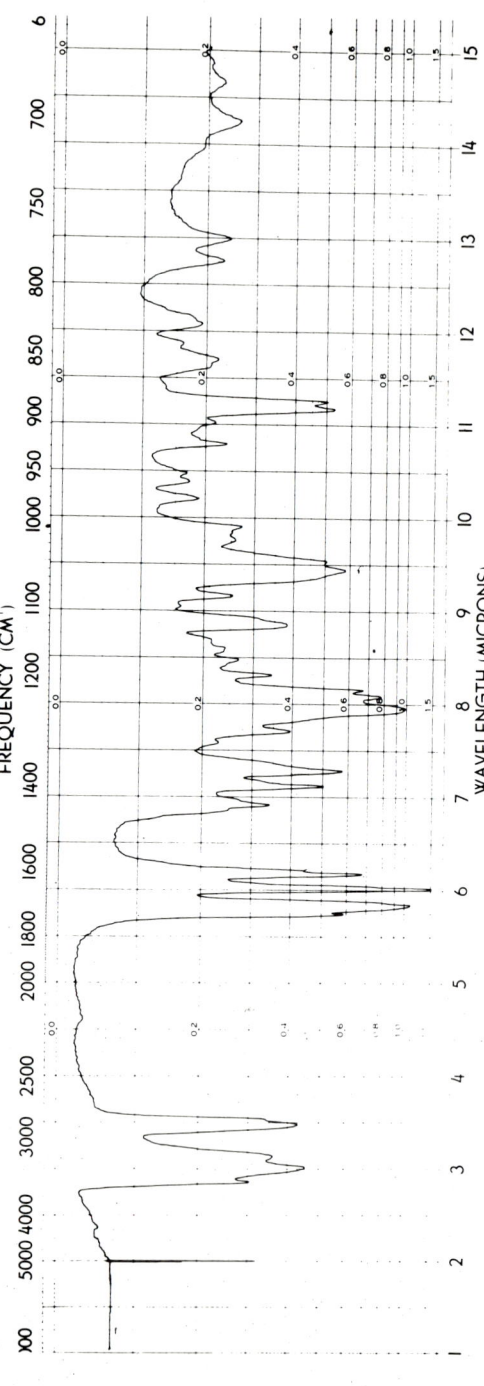

Fig. 1 Triamcinolone diacetate, batch #30636-001(4.5% moisture) I.F.curve #28604 taken in mineral oil; Instrument: Perkin Elmer 621

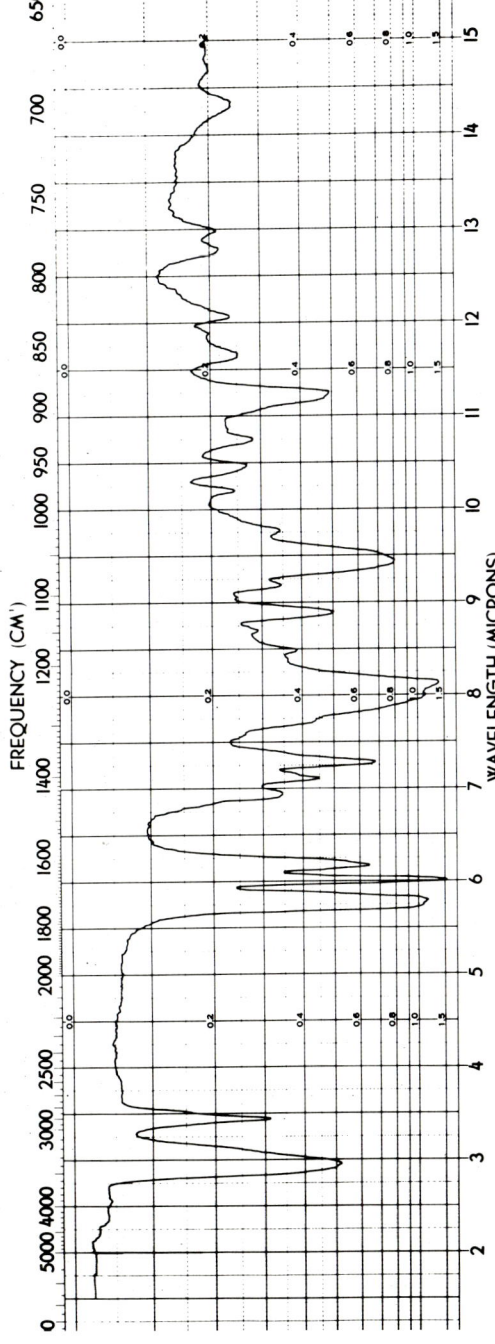

Fig. 2 Triamcinolone diacetate, batch #30636-001 (4.5% moisture) I.R.curve #28604A taken in KBr pellet from methanol solution.

ditional bands or shoulders which probably can be attributed to the presence of 4.5% moisture in the sample[2]. The significant bands (2.90; 3.40; 5.74; 6.00; 6.15 μ) agree with those previously reported[1,3,4]. The infrared spectrum in chloroform has also been recorded[5].

2.2 <u>Nuclear Magnetic Resonance Spectrum</u>
A representative spectrum and assignments of chemical shifts are presented in Figure 3 and Table I[23].

<u>TABLE I</u>

NMR Spectral Assignments
of Triamcinolone Diacetate

Protons at	Chemical Shift τ	
C-1	2.73	d, $J_{1,2}$ = 10
C-2	3.81	q; $J_{2,4}$ = 2
		$J_{1,2}$ = 10
C-4	3.99	m
C-11	5.84	m
C-16	4.56	d
C-16 (acetoxyl)	8.03	s
C-17 (OH)	4.38	s
C-18	9.12	s
C-19	8.52	s
C-21 (methylene)	5.13	m
C-21 (acetoxyl)	7.92	s

s = singlet; d = doublet; m = multiplet;
q = quartet; J = coupling constant in Hz.

TRIAMCINOLONE DIACETATE

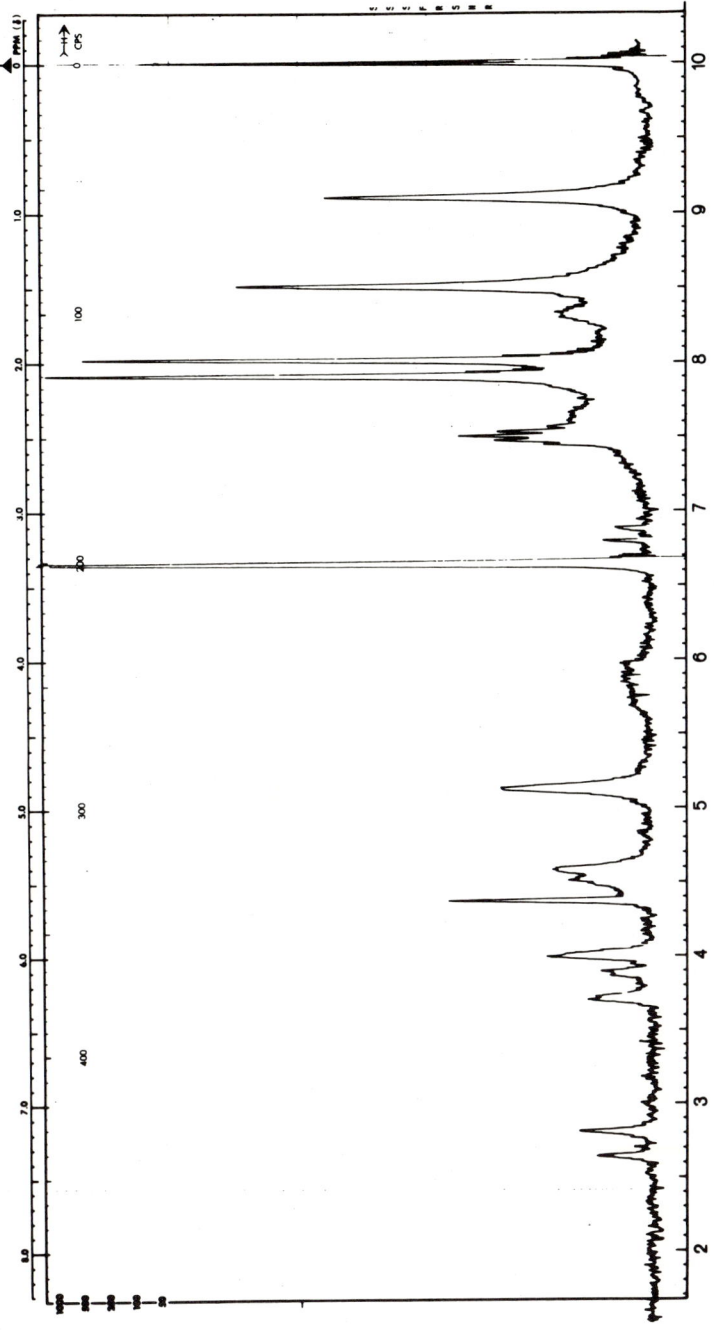

Fig. 3 NMR Spectrum of Triamcinolone Diacetate. House Standard #30636-001 in DMSO containing tetramethylsilane as internal reference. Instrument: Varian A-60

2.3 **Ultraviolet Spectrum**

Squibb batch #30630-001 when scanned between 340 and 210 mμ on a Cary 15 Spectrophotometer exhibited a single maximum at 238 mμ ($E_{1cm}^{1\%}$ = 320; ε 15,300 corrected for 4.5% moisture)[6]. This is in good agreement with values previously reported: λ max 239 mμ (ε 15,200)[3,4] and λ max 239 mμ (ε 15,270)[1].

2.4 **Mass Spectrum**

In the low resolution mass spectrum[7] summarized in Figure 4, the M^+ at m/e 478 and the M^+ -18 ion corresponding to a loss of water are very weak. However, the M^+ -20 ion, resulting from the dehydrofluorination of the molecule, is a prominent high mass ion while the ion at m/e 377, M^+ -101, arising from the loss of the side chain ($COCH_2OCOCH_3$) is the most prominent high mass ion. Characteristic of the mass spectra of fluorinated corticosteroids, the majority of fragment ions contain one oxygen, although there are several ions containing either 2, 4 or 5 oxygens. These fragments containing one oxygen extend from compositions of $C_{19}H_{19}O$ to C_7H_7O, and when steroids are acetates the mass spectrum includes the acetyl ion (C_2H_3O) at m/e 43. Of particular significance are the ions at m/e 122, 121, which are diagnostic for the presence of the A-ring dienone.

2.5 **Optical Rotation**

The following specific rotations have been reported:

$[α]_D^{25}$ + 22° (c 0.788 in $CHCl_3$)[3,4]

$[α]_D^{25}$ + 28° (c 0.38 in $CHCl_3$)[8]

$[α]_D^{22}$ + 22° (c 0.5 in $CHCl_3$)[1]

$[α]_D^{22}$ + 63° (c 0.5 in MeOH)[1]

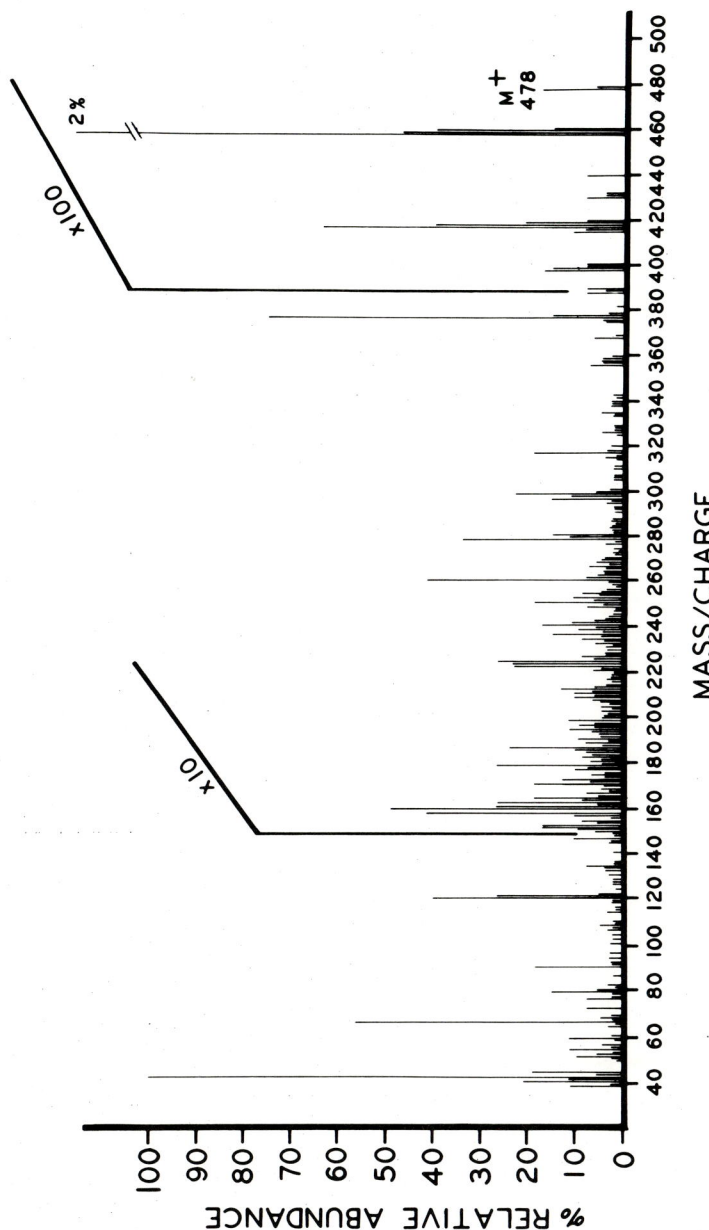

Fig. 4 Low resolution Mass Spectrum of Triamcinolone Diacetate.
Instrument: AEI MS-9

2.6 Melting Range

Like many steroids triamcinolone diacetate does not melt sharply. The melting temperature range is wide and depends on solvation[3], the rate of heating and also the polymorphic form.

The following melting point temperatures (°C) have been reported:

158-235[3,4]
186-188[4]
185-225[4]
170-180 (with gas evolution)[8]
185-232 (type I polymorph)[1]
145-236 (type II polymorph)[1]

2.7 Differential Thermal Analysis (DTA)

The erratic and wide melting range (see section 2.6) of triamcinolone diacetate is also reflected in the thermal analysis. Brancone and Ferrari[9], using a DuPont 900 analyzer at a heating rate of 20°/min, describe an endotherm at 143°, representing the hydrate, and an endotherm at 184°, representing the melting point. Using the same instrument and a heating rate of 15°/min for House Standard 30636-001 Jacobson[10] observed an endotherm at 145° (hydrate) and an exotherm at 284° (oxidative change). In vacuo (35 mm Hg) the endotherm at 145° was resolved into three endotherms occurring at 80°, 123° and 152°. With the calorimetric cell attachment, in which the sample was contained in a flat, uncovered pan, endotherms were observed at 73°, 111° and 151°. The first two endotherms are ascribed to volatile components, while the last endotherm is the melting point, which was confirmed on a Fisher-Johns block. Differences between Brancone's and Jacobson's findings can most probably be ascribed to differing crystal forms of triamcinolone diacetate and recrystallization solvents for the samples used.

2.8 **Thermogravimetric Analysis (TGA)**

At a heating rate of 15°/min under nitrogen sweep, House Standard #30636-001 gave a weight loss of 5.2%, complete by 110°. Exposure to the atmosphere of a sample dried at 120° recovered this weight loss in 20 minutes. Therefore, triamcinolone diacetate under average conditions can be expected to contain 5-6% total volatiles (water)[10].

2.9 **Solubility**

Solubility determinations were carried out on Squibb House Standard #30636-001 (Polymorph II) at 25°[10]:

Water   0.048 mg/ml
Ethanol 95%   27 mg/ml

2.10 **Crystal Properties**

According to Smith et.al.[1] triamcinolone diacetate exhibits polymorphism (see also section 2.1). The powder x-ray diffraction pattern of Squibb House Standard 30636-001 (recrystallized from methanol)[11] with 4.5% total volatile (water, see section 2.8) presented in Table II resembles polymorph II of Smith[1]. According to Smith[1] polymorph type I has been obtained from chloroform, methylene chloride, acetone, acetone/petroleum ether and benzene/petroleum ether. Type II has been obtained from acetone and from acetone/petroleum ether.

The optical crystallographic properties of triamcinolone diacetate have been reported[12] as follows without stating which polymorph was used:
System: Orthorhombic; Crystal Habit; columnar Optical Sign -; Axial Angle 69; Optic orientation (assigned acc.to crystal habit) XX c; YY a; ZZ b. Dispersion: None observed; Refractive Indexes; $\alpha(\omega) = 1.517$; $\beta(\epsilon) = 1.567$; $\gamma = 1.592$; Density 1.379; Refraction: Experimental = 111.90; Calculated = 113.99.

TABLE II
Powder x-ray diffraction pattern of triamcinolone diacetate Lot 30636-001

d($Å$)*	Relative Intensity**
14.1	0.04
9.81	0.07
9.07	0.04
6.30	0.07
5.80	0.13
5.65	1.00
5.21	0.14
5.10	0.16
4.80	0.05
4.61	0.03
3.98	0.04
3.84	0.06
3.76	0.05
3.64	0.10
3.40	0.02
3.34	0.04
3.23	0.02
3.13	0.02
3.05	0.02
2.96	0.02
2.91	0.02
2.54	0.02

* d = (interplanar distance) $\frac{n\lambda}{2 \sin \theta}$

λ = 1.539 $Å$

Radiation: $K\alpha_1$ and $K\alpha_2$ Copper

**Based on highest intensity of 1.00

3. Synthesis

Triamcinolone (I) diacetate has been prepared by microbiological[3,4] or selenium dioxide[4] dehydrogenation at C-1,2 of 16α-hydroxy-9α-fluorohydrocortisone 16α,21-diacetate (II, Fig. 5) as well as by hydrofluorination[4] of the corresponding 9,11-epoxide (III) and by acetylation of triamcinolone (V)[3,4].

4. Stability, Degradation

Triamcinolone diacetate is very stable as a solid. In mildly alkaline solution the 21-acetate group is easily split off with subsequent oxidative rearrangement and degradation of the side chain. Saponification of triamcinolone diacetate (I) to triamcinolone (V) has to be carried out under exclusion of oxygen. The D-homo-analog[14] (VI) of triamcinolone forms as a saponification byproduct. It has been reported[13] that hydrocortisone and prednisolone when exposed to ultraviolet light or ordinary fluorescent laboratory lighting in alcoholic solution undergoes photolytic degradation of the A-ring. Since triamcinolone diacetate has the same A-ring as prednisolone it probably also is labile under these conditions. When triamcinolone diacetate is fermented anaerobically with a number of microorganisms which normally dehydrogenate at C-1,2 under aerobic conditions, the 20-Ketone is reduced to the 20β alcohol[15], and/or the 1,2 double bond is hydrogenated. The reduction at carbon 20 can also be achieved with sodium borohydride and subsequent migration of the 21-acetate group to yield the 16α,20β-diacetate (IV)[15].

5. Drug Metabolic Products

While in an experiment with tritium labeled triamcinolone in the dog Florini et.al.[16] identified 6β-hydroxytriamcinolone as the major metabolic product in urine no metabolic products of triamcinolone diacetate have been reported so far.

Figure 5

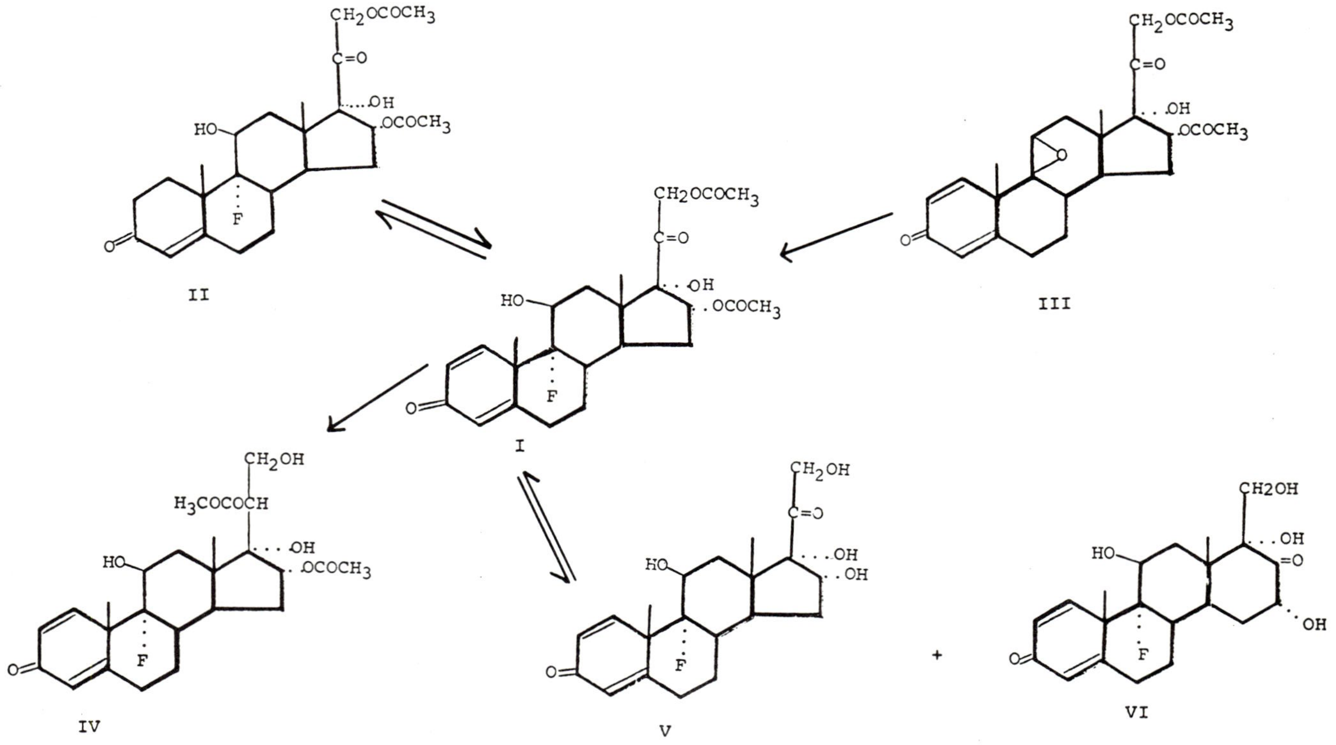

6. Methods of Analysis

6.1 Elemental Analysis

Element	% Theory	Reported Ref.[3,4]	Ref.[8]
C	62.75	63.45	62.34
H	6.53	7.44	6.74
F	3.97	4.39	-

6.2 Direct Spectrophotometric Analysis

The ultraviolet absorption band at 239 mµ (see 2.3) is due to the $\Delta^{1,4}$-3-keto system of the A-ring (see also section 4).

The absorbance is useful as a measure of purity from extraneous materials and can serve as a formulation batching assay. It can be used for chromatographic detection and quantitation[17].

6.3 Colorimetric Analysis

A variety of colorimetric methods have been used to detect and determine triamcinolone diacetate.

6.31 Reaction with tetrazolium blue in alkaline medium measures the reducing power of the α-ketol side chain by producing a blue color (520 mµ) which can be quantitated[1]. The alkalinity of the medium liberates the 21-hydroxy group prior to reaction with tetrazolium blue. Color production can be stabilized by addition of chloroform[18]. The response is about 70% greater than that for non 16α-hydroxylated analogs[1]. The method is useful for general formulation assays. It also is used as a spray reagent in paper and thin layer chromatograms. (Section 6.5).

6.32 Reaction of $\Delta^{1,4}$-3-keto steroids with isonicotinic acid hydrazide (isoniazid) produces a yellow hydrazone with an absorption maximum at 404 mµ[19]. This reaction can be adapted to triamcinolone diacetate for formulation assays.

It also has been used as a spray reagent and for quantitation in paper and thin layer chromatography (Section 6.5). It should be noted that $\Delta^{1,4}$-3-keto steroids react much less readily with isonicotinic acid hydrazide than do Δ^4-3-keto steroids[19]. The 2,4-dinitrophenylhydrazone has also been described[4].

6.33 The absorption spectra of triamcinolone diacetate in concentrated sulfuric acid with maxima at 260, 308, 375, 475-480 mµ (20 hrs. reaction time) and 100% phosphoric acid with maxima at 260, 290, 310, 375 mµ (20 hrs. reaction time) have been reported[1,20].

6.4 **Polarographic Analysis**

The polarographic half wave potential was reported as -1.01 volts, the diffusion current at 1.30 volts was 3.55 microamperes/mg/ml[1].

6.5 **Chromatographic Analysis**

Qualitative chromatographic methods can be used for identification, quantitative methods for assessment of purity and stability of triamcinolone diacetate.

6.51 **Paper Chromatographic Analysis**

Paper chromatographic R_f values of triamcinolone diacetate and related steroids are reported in Table III.

6.52 **Thin Layer Chromatographic Analysis**

Separation of triamcinolone diacetate from 1,2-dihydrotriamcinolone diacetate has been accomplished by this method[22]. Precoated Silica Gel F254 (Brinkman) was used with water-saturated ether as solvent. Development time was 60 minutes. The approximate R_f values are 0.7 for triamcinolone diacetate and 0.9 for 1,2-dihydrotriamcinolone diacetate.

TABLE III

Paper Chromatographic R_f Values

Solvent System* Compounds:	A(IV)[17]	B(V)[17]	C(VI)[17]	D[21]
Triamcinolone diacetate	0.89	0.27	0.12	—
Triamcinolone	0.13	—	—	—
Triamcinolone acetonide	0.88	0.36	0.13	—
1,2-Dihydrotriamcinolone	0.19	—	—	—
1,2-Dihydrotriamcinolone diacetate	0.92	0.40	0.22	—
Triamcinolone isomer**	0.06	—	—	—
Triamcinolone isomer diacetate	0.76	0.08	0.02	—
1,2-Dihydrotriamcinolone isomer	0.10	—	—	—
1,2-Dihydrotriamcinolone isomer diacetate	0.85	0.18	0.06	—

* The Roman numerals refer to the solvent systems of ref. 17
** 9α-Fluoro-11β,16α,17aα-trihydroxy-17aβ-hydroxymethyl-1,4-D-homoandrosta-diene14.

The solvent systems used in Table III are the following:

		Developing Time
A(IV)[17]	benzene/ethanol/water 2:1:2	5 hrs.
B(V)[17]	toluene/petroleum ether/methanol/water 12:8:13:7	2-1/2 hrs.
C(VI)[17]	benzene/petroleum ether/methanol/water 5:5:7:3	3-1/2 hrs.
D[21]	toluene saturated with propylene glycol	16 hours

The following detection systems were used:

1. The modified Haines-Drake ultraviolet scanner[17,21].
2. Isonicotinic acid hydrazide[17].
3. Alkaline tetrazolium blue spray[17].

Triamcinolone diacetate can be quantitatively determined by paper chromatography using solvent system D[21]. Whatman #1 paper was impregnated with propylene glycol in chloroform after spotting samples and standards at the 30-60 microgram level. After development of 16 hours the spots were located with a U.V. scanner, cut out and eluted with 95% ethanol. Then the absorbance of the samples was read at 239 millimicrons against a standard.

REFERENCES

(1) L. L. Smith and M. Halwer, J. Am. Pharm. Assoc., Sci. Ed. $\underline{48}$, 348 (1959).
(2) B. Keeler, Squibb Institute, personal communication.
(3) S. Bernstein, R. H. Lenhard, W. S. Allen, M. Heller, R. Littell, S. M. Stolar, Louis I. Feldman, R. H. Blank, J. Am. Chem. Soc. $\underline{78}$, 5693 (1956).
(4) S. Bernstein, R. H. Lenhard, W. S. Allen, M. Heller, R. Littell, S. M. Stolar, L. I. Feldman and R. H. Blank, J. Am. Chem. Soc. $\underline{81}$, 1689 (1959).
(5) G. Roberts, B. S. Gallagher and R. N. Jones, "Infrared Absorption Spectra of Steroids An Atlas" Vol. II Interscience Publishers, Inc. New York 1958, Plate 625.
(6) J. Dunham. Squibb Institute, personal communication.
(7) A. I. Cohen, Squibb Institute, personal communication.
(8) R. W. Thoma, J. Fried, S. Bonanno, P. Grabowich, J. Am. Chem. Soc. $\underline{79}$, 4818 (1957).
(9) L. M. Brancone and H. J. Ferrari, Microchem. J. $\underline{10}$, 380 (1966).
(10) H. Jacobson, Squibb Institute, personal communication.
(11) N. Coy and Q. Ochs, Squibb Institute, personal communication.
(12) J. A. Biles, J. Pharm. Sci. $\underline{50}$, 464 (1961).
(13) W. E. Hamlin, T. Chulski, R. H. Johnson and J. G. Wagner, J. Am. Pharm. Assoc. $\underline{49}$, 253 (1963) and D. R. Barton and W. C. Taylor, J. Am. Chem. Soc. $\underline{80}$, 244 (1958); J. Chem. Soc. $\underline{1958}$, 2500.
(14) L. L. Smith, M Marx, J. J. J. Garbarini, T. Foell, V. E. Origoni and J. J. Goodman, J.

References Cont'd.

Am. Chem. Soc. 82, 4616 (1960).

(15) L. L. Smith, J. J. Garbarini, J. J. Goodman, M. Marx and H. Mendelsohn, J. Am. Chem. Soc. 82, 1437 (1960) and J. Schmidt-Thome, G. Nesemann, H. J. Huebner and I. Alester, Biochem. Z. 336, 322 (1962).

(16) J. R. Florini, L. L. Smith and D. A. Buyske, J. Biol. Chem. 236, 1038 (1961).

(17) L. L. Smith, Th. Foell, R. de Maio and M. Halwer, J. Am. Pharm. Assoc. Sci. Ed. 48, 528 (1959) and L. L. Smith and Th. Foell, J. Chromatog. 3, 381 (1960).

(18) P. Ascione and C. Fogelin, J. Pharm. Sci. 52, 709 (1963).

(19) L. L. Smith and Th. Foell, Anal. Chem. 31, 102 (1959).

(20) L. L. Smith and W. H. Muller, J. Org. Chem. 23, 960 (1958).

(21) H. R. Roberts, Squibb Institute, personal communication.

(22) F. Dursch, Squibb Institute, personal communication.

(23) M. Puar, Squibb Institute, personal communication.

VINBLASTINE SULFATE

J. H. Burns

Reviewed by N. Neuss

J. H. BURNS

CONTENTS

1. Description
   1.1 Name, Formula, Molecular Weight
   1.2 Appearance, Color, Odor
2. Physical Properties
   2.1 Melting Range
   2.2 Optical Rotation
   2.3 Solubility
   2.4 Crystal Properties
   2.5 Ultraviolet Spectrum
   2.6 Infrared Spectrum
   2.7 Nuclear Magnetic Resonance Spectrum
   2.8 Mass Spectrum
   2.9 pK Values
   2.10 Thermogravimetric Analysis
   2.11 Differential Thermal Analysis
3. Methods of Preparation
4. Methods of Analysis
   4.1 Colorimetric Analysis
   4.2 Direct Spectrophotometric Analysis
   4.3 Stability Assay
   4.4 Thin Layer Chromatographic Analysis
   4.5 Bioassay
5. Stability - Degradation
   5.1 Dry Thermal Degradation
   5.2 Hydrolysis
   5.3 Stability in Organic Solvents
6. Metabolism
7. References

1. Description

1.1 Name, Formula, Molecular Weight

Vinblastine sulfate is the nonproprietary name assigned by the United States Adopted Names Council to the compound originally named vincaleukoblastine sulfate[1,2,3]. It is the 1:1 sulfate salt of an alkaloid obtained from the plant <u>Vinca rosea</u> Linn. (<u>Catharanthus roseus</u> G. Don) of the family Apocynaceae, better known as Madagascar periwinkle. Frequently the name is abbreviated to VLB sulfate. It is also identified by the code numbers NSC-49842 and 29060-LE[4]. The elucidation of the molecular structure, stereochemistry, and absolute configuration of this interesting compound is fully documented in the literature[5,6,7,8].

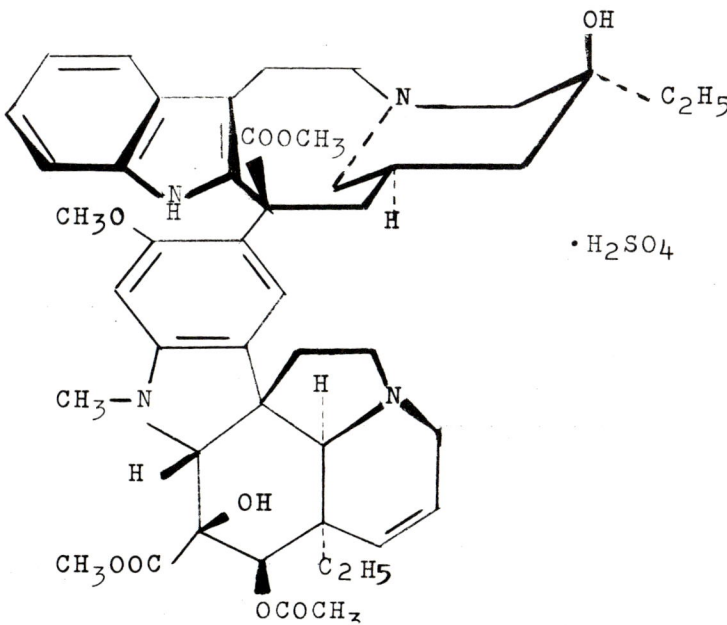

$C_{46}H_{58}N_4O_9 \cdot H_2SO_4$ Mol. Wt.: 909.07

1.2 Appearance, Color, Odor

Vinblastine sulfate is a white to slightly yellow crystalline or amorphous, odorless powder. It is very hygroscopic, relatively unstable, and quite toxic.

2. Physical Properties

2.1 Melting Range

Vinblastine sulfate melts at 284-285°C. with decomposition[2,9]. This melt was determined on the monohydrate[2,9].

2.2 Optical Rotation

The following specific rotations have been reported:

$[\alpha]_D^{22} = -28°$ (c = 1.01 in methanol)[9],

$[\alpha]_D^{26} = -28°$ (methanol)[2],

$[\alpha]_D^{22} = -36.9°$ (c = 1.01 in methanol)[1].

The last value above was the first reported in the literature. Since references 2 and 9 were co-authored by one of the authors of reference 1, it seems probable that the last value was obtained on impure material. The U.S.P. XVIII[15] specifies that acceptable vinblastine sulfate must have a specific rotation in the range -28° to -35° ($[\alpha]_D^{25}$, c = 2.00 in methanol).

2.3 Solubility

Vinblastine sulfate is soluble in water and in methanol, but only slightly soluble in ethanol[9].

2.4 Crystal Properties

The X-ray powder diffraction pattern of vinblastine sulfate has been reported by Beer et al.[9] using vanadium-filtered chromium radiation and a wavelength value of 2.2896 Å in the calculations:

VINBLASTINE SULFATE

d	I/I_1	d	I/I_1
12.9	1.00	4.13	0.08
11.7	1.00	3.95	0.04
10.7	0.02	3.73	0.02
9.35 B	0.40	3.57	0.02
8.45	0.08	3.41	0.12
7.82	0.08	3.28	0.04
7.10	0.80	3.21	0.04
6.23	0.08	3.07	0.04
5.43	0.08	3.03	0.04
5.09	0.30	2.83	0.02
4.83	0.04	2.76	0.04
4.53	0.40	2.71	0.04
4.34	0.08	2.62	0.02

2.5 Ultraviolet Spectrum

The UV spectrum[37] of vinblastine sulfate in 95% ethanol is shown in Figure 1. The following characteristic points of inflection are noted:

maximum at 214 nm, ϵ = 53,800
minimum at 246 nm, ϵ = 11,900
maximum at 262 nm, ϵ = 16,000
shoulder at 287 nm, ϵ = 13,000
shoulder at 296 nm, ϵ = 11,500

Changes in pH of the solution result in spectral change. In acidified 95% ethanol the maximum near 262 nm shifts to a slightly longer wavelength. In sufficiently alkaline 95% ethanol solutions this peak is shifted to a slightly lower wavelength and the remainder of the spectrum is altered to give that of vinblastine free base.

2.6 Infrared Spectrum

The IR spectrum[28] of vinblastine sulfate is presented in Figure 2. It is in good agreement with a previously published spectrum[1]. Interpretation of infrared spectra must often be supplemented by other known chemical properties of the compound in question. Vinblastine is no

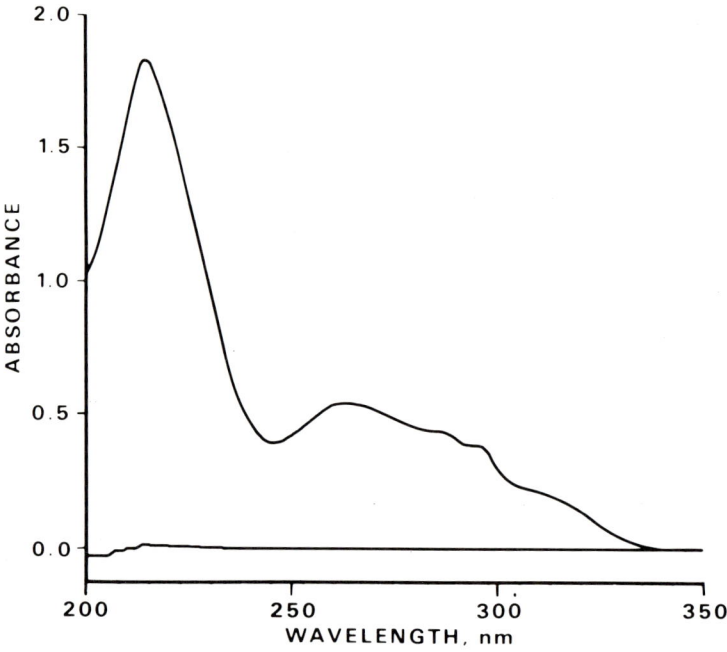

Fig. 1. UV spectrum of vinblastine sulfate in 95% ethanol; instrument: Cary model 15

VINBLASTINE SULFATE

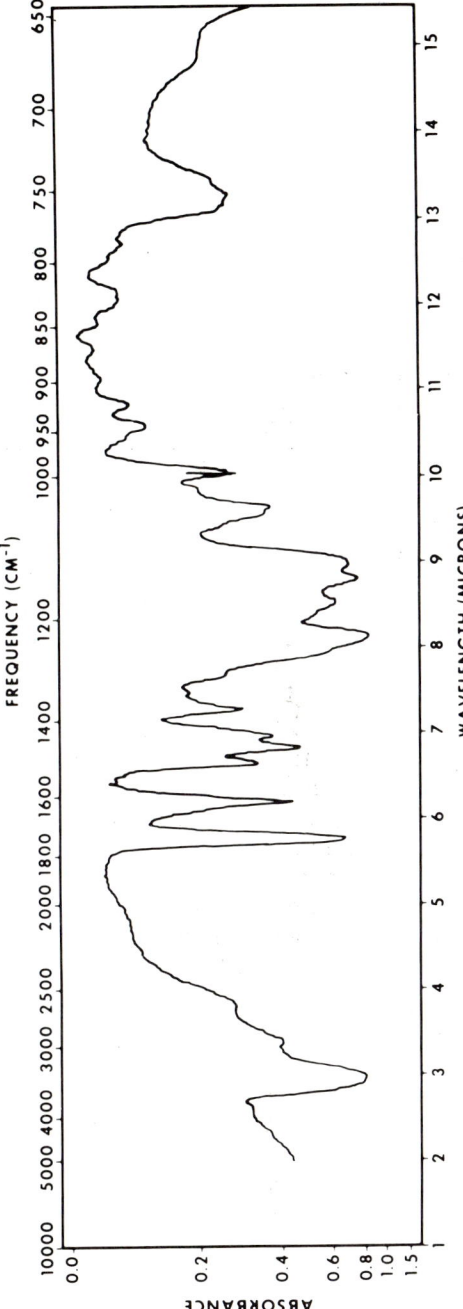

Fig. 2. IR spectrum of vinblastine sulfate in KBr pellet; instrument: Perkin-Elmer 221

exception. Its IR spectrum is very similar to that of several other alkaloids obtained from the same plant. Several articles[29,30,31,32] discuss the small differences in the spectra of these compounds.

2.7 Nuclear Magnetic Resonance Spectrum

A low resolution NMR spectrum[22] of vinblastine sulfate is reproduced in Figure 3. NMR has played an important part in the structure determinations of the Vinca alkaloids. Several review articles[29,30,31] summarize the use of NMR in the structure elucidation of VLB.

2.8 Mass Spectrum

Bommer, McMurray, and Biemann[6] have published some of the characteristic peaks of the high resolution spectrum of vinblastine free base. Their article discusses the presence of peaks at m/e = 824 and m/e = 838 shown to result from transmethylation of the parent compound. (See also ref. 35). It has recently been observed[36] that running the spectrum on the sulfate salt nearly eliminates the transmethylation reactions and gives a characteristic peak at m/e = M^+ - 18, although the molecular peak (m/e = 810) is negligible. Occolowitz suggests that obtaining spectra of both the free base and the sulfate salt may therefore be useful in interpreting mass spectral data of the dimeric Vinca alkaloids.

2.9 pK Values

Vinblastine has two titratable groups. The following pK_a values have been reported:

    5.0 and 7.0 by electrometric titration in DMF-H_2O (2:1)[9], and
    5.4 and 7.4 by electrometric titration in H_2O[2].

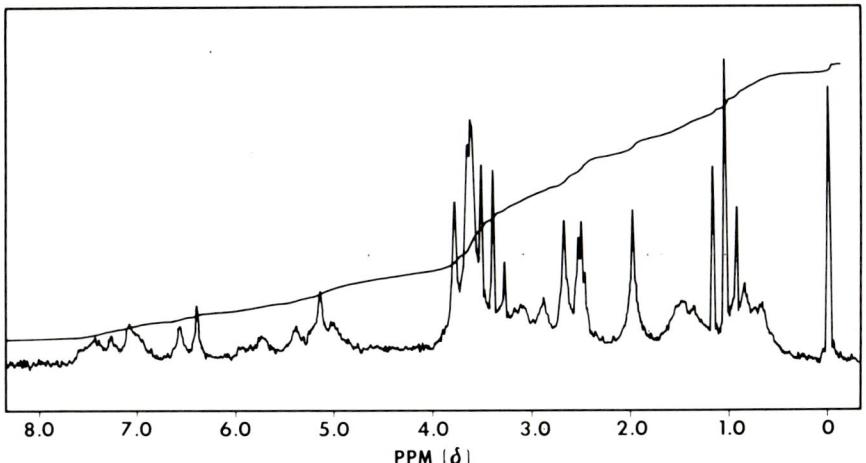

Fig. 3. NMR spectrum of vinblastine sulfate in deuterated dimethyl sulfoxide; instrument: Varian HA-60

2.10 __Thermogravimetric Analysis__

VLB sulfate is difficult to obtain free from water of hydration and/or solvents of crystallization. A thermogravimetric analysis[28] of reference standard material (Lilly lot number P-89481) showed rapid loss of weight from room temperature through 131°C. The analysis was performed using the Du Pont 950 Thermogravimetric Analyzer at a heating rate of 5°C./minute under nitrogen flowing at 44 cc./minute. Five percent weight loss was observed by 96°C., 10% by 122°C., and 12% by 131°C. The weight remained constant from 131°C. to 177°C.

2.11 __Differential Thermal Analysis__

A differential thermal analysis[28] of VLB sulfate (Lilly reference standard lot P-89481) was performed on a DuPont 900 Differential Thermal Analyzer at a heating rate of 20°C./minute. A broad endothermic phase transition peaking at 181°C. was observed.

3. __Methods of Preparation__

A total synthesis of vinblastine has not as yet been achieved. Methods of preparation involve making initial crude extracts from the periwinkle plant, followed by extraction at selected pH into organic solvents, and final separation of the complex mixture of alkaloids by column chromatography. Several methods have been devised since Noble, Beer, and Cutts[1] first reported the isolation of vinblastine as the sulfate salt. A few are briefly described here.

Beer __et al.__[9] extracted the plant leaves with aqueous-alcoholic acetic acid solution. After evaporation the residue was extracted with 2% hydrochloric acid. The HCl extract was adjusted to pH 4 with NaOH and extracted with benzene, the pH raised to 7, and further extracted with benzene. The pH 7 benzene extracts were evaporated to dryness, dissolved in benzene-methylene chloride (65:35), and passed over a neutral, partially deactivated aluminum oxide column. The column was eluted with benzene-methylene chloride using a gradient elution technique whereby the concentration of methylene chloride was linearly varied from 35% at the start to 97.5% at the completion of elution. Vinblastine rich fractions were evaporated to dryness, the residue suspended in water, the pH adjusted to 3.8 with sulfuric acid, and evaporated to dryness. The resulting vinblastine sulfate was crystallized from ethanol.

Svoboda[10] separated vinblastine from the other alkaloids of Vinca rosea by utilizing the various solubilities of their tartrates in organic solvents. Slurries of the plant or plant parts in 2% tartaric acid (pH 2) were extracted with large quantities of benzene, the benzene extract was concentrated, and extracted with 2% tartaric acid. This left the neutral alkaloids in the benzene while the weakly basic alkaloids passed into the aqueous phase. The aqueous phase was adjusted to pH 8.5 - 9.5 with NH_4OH, extracted with benzene or ethylene dichloride, the extracts evaporated, the residue dissolved in benzene, and chromatographed over an alumina column previously deactivated by 10% acetic acid. The column was eluted first with benzene, then with benzene-chloroform mixtures, and finally with chloroform. Fractions containing vinblastine were evaporated, the residue dissolved in ethanol, and ethanolic sulfuric acid added to pH 4. Upon chilling, vinblastine sulfate precipitated. It was further purified by recrystallization from absolute ethanol.

Other similar procedures are reported. One reported method[11] consisted of the initial extraction from a suspension of aerial plant parts in 12% aqueous NH_4OH into toluene and use of benzene-petroleum ether (9:1) saturated with formamide to elute the alumina column. Jovanovics, Szasz et al.[12], in a procedure similar to Svoboda's[10], extracted the plant parts first with 60% aqueous methanol containing 2% tartaric acid, and then into ethylene dichloride.

4. Methods of Analysis

4.1 Colorimetric Analysis

Formation of a deep rose color when vinblastine sulfate is heated in a solution consisting of 35 ml. of pyridine, 1 ml. of concentrated sulfuric acid, and 35 ml. of acetic anhydride containing 0.05% acetyl chloride, is the basis of an assay method reported[13] for relatively pure vinblastine sulfate. The reaction is carried out at 80°C. for 20 minutes. The absorbances of the color produced, and of a reference standard similarly treated, are measured in 1 cm. cells at 574 nm and 538 nm against water as a reference. The ratio of A_{574nm}/A_{538nm} must be in the range of 1.20-1.25 for the assay to be considered valid. The absorbance values are stable for about 20 minutes, but then begin to increase slowly. The absorbance at 574 nm is used for quantitative determination since its value changes less with time than the peak at 538 nm. The reaction is reported to obey Beer's law from 5-70 mcg. vinblastine sulfate per milliliter. Vincristine sulfate gives the same color curve[14] and must be absent when this method is employed.

4.2 Direct Spectrophotometric Analysis

Purified vinblastine sulfate may be determined by measurement of its UV absorbance in methanol[15]. Sample and reference standard are diluted to approximately 20 micrograms per milliliter in anhydrous methanol. The absorbances of the solutions are measured and compared in 1 cm. cells at the maximum at about 267 nm against anhydrous methanol in a reference cell.

Masoud et al.[16] have assayed <u>Vinca rosea</u> leaf for vinblastine content by UV measurement following extraction of the leaf, isolation of the alkaloid by TLC, removal of the TLC spot, and extraction into methanol. Measurement was made at 214 nm (free base) and compared to a standard curve reported to be linear over the range 0.8 - 8.0 mcg. vinblastine base per milliliter.

4.3 Stability Assay

The following method[17] permits measurement of the amount of vinblastine sulfate remaining intact after "dry" thermal degradation or mild acidic hydrolysis (pH 2, 50°C.). Five to ten mg. of sample dissolved in 25 ml. of pH 3.2 sodium citrate buffer are extracted with three 25-ml. portions of chloroform and the chloroform evaporated to dryness. The residue, dissolved in 5 mL of S.D. No. 3A absolute ethanol-chloroform (1:25), is quantitatively transferred to a chromatographic column prepared with 3 Gm. of alumina (Woelm, neutral, activity grade No. 1) in the same solvent. The column is eluted with the above solvent into a 50-ml. volumetric flask until about 40 ml. of eluate are collected, and the flask brought to volume with eluting solvent. An aliquot equivalent to 0.5 - 0.6 mg. of vinblastine sulfate is transferred to a 25-ml. volumetric flask, evaporated, the residue dissolved in S.D. No. 3A absolute ethanol, one drop of hydrochloric acid added, and the flask brought to volume with S.D. No. 3A absolute ethanol. The absorbance is measured in 1-cm. cells at the maximum at about 267 nm. The sample absorbance is compared to that of a known vinblastine sulfate reference standard simultaneously carried through the same procedure.

4.4 Thin Layer Chromatographic Analysis

Thin layer chromatography has been used extensively in identifying and monitoring the separation of the Vinca alkaloids. Vinblastine sulfate, vincristine sulfate, leurosine sulfate, and leurosidine sulfate can be adequately separated by the use of silica gel GF plates and development with benzene-chloroform-diethylamine (50:50:5)[17]. Ten microliters of a solution of 1 mg. of a mixture of the alkaloidal salts in 0.1 ml. of 25% water in methanol is spotted on the plate. The separation is greatly improved by drying the plate at 105°C. just prior to spotting. Thorough drying of the applied spots before chromatographing is essential. All of the separated compounds can be detected by spraying the warmed plate with a 1% solution of ceric ammonium sulfate in 85% phosphoric acid (CAS). Cone et al.[20] have reported separation of vinblastine sulfate, vincristine sulfate, and leurosidine sulfate on silica gel plates prepared using 0.5 N KOH (rather than water) and development with ethyl acetate-absolute ethanol (1:1).

Table I gives the reported R_f values of vinblastine base in several TLC systems. In general the systems in Table I are reported as giving satisfactory separations of VLB from a few specific alkaloids in each case.

The separation and identity of VLB in crude alkaloidal mixtures is often accomplished by two dimensional TLC. Thus, Masoud et al.[16] were able to separate and identify VLB in a complex mixture by TLC on silica gel G containing Radelin phosphor by eluting twice in the first direction with chloroform-methanol (95:5) and twice in the second direction with ethyl acetate-absolute ethanol (3:1). Detection was achieved by a combination of UV quenching or fluorescence and CAS reagent. VLB was the only compound near that position that caused quenching under short wavelength UV light. Farnsworth and Hilinski[19] who were able to separate the very similar alkaloids VLB, vincristine, leurosine, and leurosidine by their system in Table I, found it necessary to develop in a second direction with methanol in order to isolate the same compounds from somewhat less pure mixtures. Cone et al.[20] also demonstrated the usefulness of two dimensional TLC for separation of Vinca alkaloids. Svoboda[21] reports good separation of VLB, leurosidine, and vincristine (R_f's 0.73, 0.37, 0.54, respectively) on alumina plates by first developing the plate in ethyl acetate followed by development in the same direction with ethyl acetate-absolute ethanol (3:1).

4.5 Bioassay

The effectiveness of certain Vinca rosea extracts in prolonging the life of DBA/2 mice

Table I*

Ref.		Eluent	R_f
18	silica gel G	ethyl acetate-absolute ethanol (3:1)	0.21
18	silica gel G	n-butanol-glacial acetic acid-H_2O (4:1:1)	0.19
18	silica gel G	methanol	0.46
19	silica gel G	chloroform-methanol (95:5)	0.24
14	0.5 N LiOH/alumina	acetonitrile-benzene (30:70)	0.36
20	silica gel	ethyl acetate-absolute ethanol (3:1)	0.24
20	silica gel	chloroform	0.00
20	silica gel	ethyl acetate-absolute ethanol (1:1)	0.33
20	silica gel	ethyl acetate	0.04
20	alumina	chloroform-ethyl acetate (1:1)	0.25
20	alumina	ethyl acetate-absolute ethanol (3:1)	0.66
20	alumina	benzene	0.00
20	alumina	chloroform	0.17
20	alumina	benzene-chloroform (3:1)	0.00

* Detection by 1% ceric ammonium sulfate in 85% phosphoric acid.

implanted with P-1534 leukemia prompted an extensive investigation for the purpose of isolating the active component(s) of the plant.[38,39]

5. Stability - Degradation

    5.1 Dry Thermal Degradation[17]
        Vinblastine sulfate sealed from the atmosphere is relatively stable to heat. When lyophilized VLB sulfate contained in sealed glass ampoules is subjected to a temperature of 100°C. for 16 hours, subsequent assay by the method of Section 4.3 indicates only about 2% degradation. However, when the material is exposed to normal atmosphere and heated for 16 hours at 100°C., the assay shows that approximately 50% has degraded. Upon TLC examination [silica gel GF, C_6H_6-$CHCl_3$ - $(C_2H_5)_2NH$, 50:50:5] of the degraded mixture the major degradation product is observed as an immobile spot (unidentified) at the point of application. A second spot having a slightly lower R_f than VLB, approximating 2% of the original material, is identified as desacetylvinblastine. The IR and UV spectra of the immobile material are very similar to those of pure vinblastine.

    5.2 Hydrolysis[17]
        Aqueous solutions of vinblastine sulfate at about pH 4.5 are stable for up to 3 hours at 90°C. Heating at 50°C. for 16 hours at pH 2 results in degradation of 80 - 90% of the material. The primary product of hydrolysis under these conditions is desacetylvinblastine.

    5.3 Stability in Organic Solvents
        Vinblastine as the free base is rather unstable. It is reported[19] that vinblastine free base in benzene is stable for several weeks if kept frozen. Jakovljevic et al.[14] report that 1% solutions of vinblastine base in chloroform are stable for 24 hours under refrigeration with respect to TLC detection. Greenius, et al.[23] report

6. Metabolism

The high toxicity[24,33,34] and rather complex structure of vinblastine have limited the study of its metabolism. No metabolic studies in humans have been reported. All investigations to date have been made in rats using tritiated vinblastine. Except for trace amounts of desacetylvinblastine found[27] in the blood of rats two hours after intraperitoneal injection, no metabolites have been identified. However, other unidentified metabolites have been isolated in small amounts[26]. No pharmacokinetic constants have been observed in the literature.

Beer et al.[24] found only about 5% of the dose radioactivity in the 26 hour urine collection after intravenous injection of tritiated vinblastine (prepared by Wilzbach method) in rats. Most of this amount was excreted during the first 12 hours and consisted primarily of metabolites. Further investigation in rats revealed that 24 hours after i.v. injection about 25% of the dose radioactivity was present in the intestinal contents[25]. This was attributed mainly to the fact that 20 - 25% of the dose radioactivity was excreted in the bile during the same period. Less than 2% of the dose was found as unchanged vinblastine in this 24 hour bile collection. It is reported[26] that at least six radioactive compounds were present in the early portion of the bile collection. Twenty-four hours after i.v. injection radioactivity was found rather evenly distributed at relatively low levels among several organs with somewhat higher levels in the liver[25]. Two hours after either i.v. or i.p. injection the radioactivity was much less evenly distributed among the various tissues with up take following the i.p. injection being only one-half to two-thirds as high as that following i.v. injection[26]. Relative absorption by the various tissues, however, was found to be nearly independent of administrative route. The same authors[26] report that radioactivity was barely detectable in the brains of rats two hours after injection and suggest that possibly this explains the ineffectiveness of this drug in treating malignancies of the brain.

Grenius et al.[23] using vinblastine-4-acetyl-t (prepared by acetylation of desacetylvinblastine with tritium labeled acetic anhydride) in rats found that at the time of highest blood levels 70% of the total blood radioactivity was present in the components of the buffy coat, i.e., the interface region between the plasma layer and the packed red cells occurring when blood is centrifuged. Hebden et al.[27] continued this investigation of the distribution of VLB among blood components in the rat using vinblastine tritiated in the aromatic rings (prepared by proton exchange with tritiotrifluoroacetic acid). They reported that after i.p. doses of about 0.25 mg. kg. maximum blood radioactivity was attained in 1.5 hours at which time approximately 2% of the dose radioactivity was present in the blood. Two hours after injection they found the blood radioactivity to be distributed 60% in the platelets, 15% in the leukocytes, 15% in the plasma, and 10% in the red cells. They reported that nearly all of the platelet radioactivity was due to unchanged vinblastine while only about 50% of the plasma activity was from unchanged vinblastine. It was also observed that the decrease in total blood radioactivity with time is primarily due to loss of radioactivity from the buffy coat region.

References

1. R. L. Noble, C. T. Beer, and J. H. Cutts, Ann. N.Y. Acad. Sci. 76, 882-894 (1958).
2. N. Neuss, M. Gorman, G. H. Svoboda, G. Maciak, and C. T. Beer, J. Am. Chem. Soc. 81, 4754-4755 (1959).
3. G. H. Svoboda, N. Neuss, M. Gorman, J. Amer. Pharm. Assoc., Sci. Ed. 48, 659-666 (1959).
4. United States Adopted Names (USAN) No. 5, p. 91 (1967). United States Pharmacopeial Convention, Inc., 4630 Montgomery Ave., Bethesda, Maryland 20014.
5. N. Neuss, M. Gorman, W. Hargrove, N. J. Cone, K. Biemann, G. Büchi, and R. E. Manning, J. Am. Chem. Soc. 86, 1440-1442 (1964).
6. P. Bommer, W. McMurray, K. Biemann, ibid. 86, 1439-1440 (1964).
7. J. W. Moncrief and W. N. Lipscomb, ibid. 87, 4963-4964 (1965); Acta Crystallogr. 21, 322-331 (1966).
8. N. Neuss, M. Gorman, H. E. Boaz, and N. J. Cone, J. Am. Chem. Soc. 84, 1509-1510 (1962).
9. C. T. Beer, J. H. Cutts, and R. L. Noble, U. S. Patent 3,097,137, July 9, 1963.
10. G. H. Svoboda, U. S. Patent 3,225,030, Dec. 21, 1965.
11. K. Jovankovics and K. Szasz, Hungarian Patent 153,200, Oct. 22, 1966. C. A. 66: 118854x.
12. K. Jovanovics, K. Szasz, C. Lorincz, L. Horompo, and J. Farkas, Hung. Patent 154,715, April 30, 1968. C. A. 69: 38732c.
13. I. M. Jakovljevic, J. Pharm. Sci. 51, 187-188 (1962).
14. I. M. Jakovljevic, L. D. Seay, and R. W. Shaffer, ibid. 53, 553-557 (1964).
15. United States Pharmacopeia XVIII, p. 772, Mack Publishing Company, Easton, Pa., 1970.
16. A. N. Masoud, N. R. Farnsworth, L. A. Sciuchetti, R. N. Blomster, and W. A. Meer, Lloydia 31, 202-207 (1968).
17. R. L. Hussey, Eli Lilly and Company, personal communication.

18. N. R. Farnsworth, R. N. Blomster, D. Damratoski, W. A. Meer, and L. V. Cammarato, Lloydia 27, 302-314 (1964).
19. N. R. Farnsworth and I. M. Hilinski, J. Chromatog. 18, 184-188 (1965).
20. N. J. Cone, R. Miller, and N. Neuss, J. Pharm. Sci. 52, 688-692 (1963).
21. G. H. Svoboda, Lloydia 24, 173-178 (1961).
22. R. Laughlin, Eli Lilly and Company, personal communication.
23. H. F. Greenius, R. W. McIntyre, and C. T. Beer, J. Med. Chem. 11, 254-257 (1968).
24. C. T. Beer, M. L. Wilson, and J. Bell, Can. J. Physiol. Pharmacol. 42, 1-11 (1964).
25. Ibid. 42, 368-373 (1964).
26. C. T. Beer and J. F. Richards, Lloydia 27, 352-360 (1964).
27. H. F. Hebden, J. R. Hadfield, and C. T. Beer, Cancer Res. 30, 1417-1424 (1970).
28. A. D. Kossoy and C. D. Underbrink, Eli Lilly and Company, personal communication.
29. G. H. Svoboda, I. S. Johnson, M. Gorman, and N. Neuss, J. Pharm. Sci. 51, 707-720 (1962).
30. I. S. Johnson, J. G. Armstrong, M. Gorman, and J. P. Burnett, Jr., Cancer Res. 23, 1390-1427 (1963).
31. N. Neuss, I. S. Johnson, J. G. Armstrong, and C. J. Jansen, Advances in Chemotherapy 1, 133-174 (1964).
32. G. H. Svoboda, M. Gorman, A. J. Barnes, Jr., and A. T. Oliver, J. Pharm. Sci. 51, 518-523 (1962).
33. J. H. Cutts, C. T. Beer, and R. L. Noble, Cancer Res. 20, 1023-1031 (1960).
34. E. Frei, III, Lloydia 27, 364-367 (1964).
35. K. Biemann, Lloydia 27, 397-405 (1964).
36. J. L. Occolowitz, Eli Lilly and Company, personal communication.
37. R. C. Tiemeier, Eli Lilly and Company, personal communication.
38. I. S. Johnson, H. F. Wright, and G. H. Svoboda, J. Lab. Clin. Med. 54, 830 (1959).
39. I. S. Johnson, H. F. Wright, G. H. Svoboda, and J. Vlantis, Cancer Res. 20, 1016-1022 (1960).

VINCRISTINE SULFATE

J. H. Burns

Reviewed by N. Neuss

CONTENTS

1. Description
   1.1 Name, Formula, Molecular Weight
   1.2 Appearance, Color, Odor
2. Physical Properties
   2.1 Melting Range
   2.2 Optical Rotation
   2.3 Solubility
   2.4 Crystal Properties
   2.5 Ultraviolet Spectrum
   2.6 Infrared Spectrum
   2.7 Nuclear Magnetic Resonance Spectrum
   2.8 Mass Spectrum
   2.9 pK Values
   2.10 Thermogravimetric Analysis
   2.11 Differential Thermal Analysis
3. Methods of Preparation
4. Methods of Analysis
   4.1 Direct Spectrophotometric Analysis
   4.2 Colorimetric Analysis
   4.3 Bioassay Methods
   4.4 Colorimetric Identification
   4.5 Thin Layer Chromatographic Analysis
5. Stability-Degradation
   5.1 Dry Thermal Degradation
   5.2 Hydrolysis
   5.3 As Free Base
6. Metabolism
7. References

VINCRISTINE SULFATE

1. Description

   1.1 Name, Formula, Molecular Weight

   Vincristine is the nonproprietary name for the alkaloid originally called leurocristine[1]. It is obtained from the plant Vinca rosea Linn. (Catharanthus roseus G. Don) of the family Apocynaceae (Madagascar periwinkle). Vincristine sulfate is the 1:1 sulfate salt of vincristine. The structure of vincristine differs from that of vinblastine only by the substituent at the anilino-nitrogen in the vindoline portion of the molecule. Thus it is des-N_a-methyl-N_a-formyl vinblastine[2]. It is also known by the code numbers NSC-67574 and 37231[3] and is abbreviated to VCR and VCR sulfate. Elucidation of the molecular structure, stereochemistry, and absolute configuration of this compound is found in the literature[2,4,5,6].

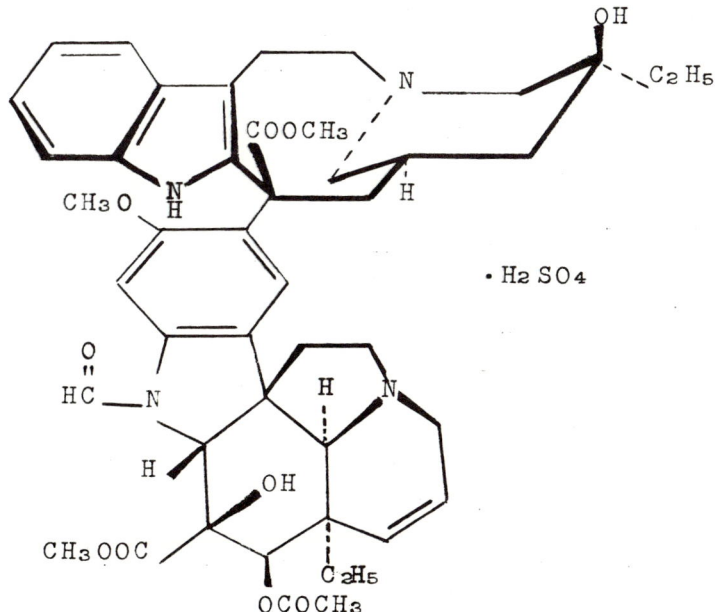

$C_{46}H_{56}N_4O_{10} \cdot H_2SO_4$ Mol. Wt.: 923.06

1.2 Appearance, Color, Odor

Vincristine sulfate is a white to slightly yellow, odorless, amorphous or crystalline powder. It is hygroscopic and very toxic[7].

2. Physical Properties

2.1 Melting Range

Vincristine sulfate after recrystallization from absolute ethanol is reported to have a melting range of 273 - 281° C. with loss of solvent occurring from 210 - 232° C.[13]

2.2 Optical Rotation

The specific rotation of vincristine sulfate in methanol has been determined[20]: $[\alpha]_D^{25} = +8.5$ (c = 0.8).

2.3 Solubility

VCR sulfate is soluble in methanol, freely soluble in water, but only slightly soluble in 95% ethanol[7].

2.4 Crystal Properties

A definitive X-ray diffraction pattern suitable for the identification of vincristine sulfate crystals has not become available to the author. Vincristine may be readily isolated as the free base by making an aqueous solution of the sulfate alkaline with NH_4OH and extracting into an organic solvent such as ethylene dichloride, chloroform, or benzene. Therefore, the X-ray diffraction pattern of vincristine free base recrystallized from methanol is presented here[13,31]. The pattern was determined at a wavelength of 2.2896 Å. using chromium radiation and a vanadium filter:

d	I/I$_1$	d	I/I$_1$
10.86	0.12	4.19	0.04
10.27	0.04	3.97B	0.20
9.73	1.00	3.83	0.02
9.26	0.30	3.77	0.02
8.82	0.30	3.62	0.08
8.59	0.60	3.57	0.08
7.44	0.60	3.42	0.12
7.10	0.30	3.35	0.04
5.89	0.20	3.24	0.04
5.66B	0.30	3.20	0.04
5.45	0.80	3.08	0.08
5.17	0.08	2.96	0.08
5.09	0.08	2.85	0.04
4.76B	0.16	2.78	0.04
4.55	0.16	2.63	0.04
4.41	0.04	2.47	0.04
4.28	0.08	2.43	0.04

2.5 Ultraviolet Spectrum

The UV spectrum[14] of vincristine sulfate is reproduced in Figure 1. The spectrum has the following characteristics:

maximum at 221 nm, ϵ = 47,100
maximum at 255 nm, ϵ = 15,400
minimum at 275 nm, ϵ = 11,400
inflection at 290 nm, ϵ = 14,000
maximum at 296 nm, ϵ = 15,600

Slight points of inflection also occur at about 260 nm and at about 305 nm.

2.6 Infrared Spectrum

The IR spectrum[15] of vincristine sulfate run in a KBr pellet is presented in Figure 2. The infrared spectra of several of the Vinca rosea alkaloids are quite similar. For example, the IR spectra of neoleurocristine, neoleurosidine, leurosidine, vinblastine, leurosine and isoleurosine (as free bases in chloroform) exhibit only small differences from that of vincristine (in chloroform).

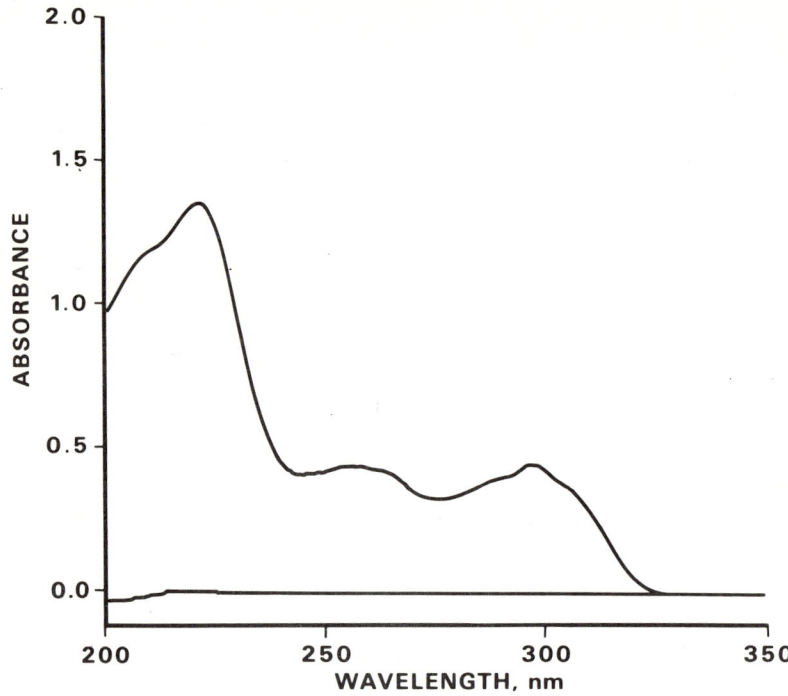

Fig. 1. UV spectrum of vincristine sulfate in 95% ethanol; instrument: Cary model 15

VINCRISTINE SULFATE

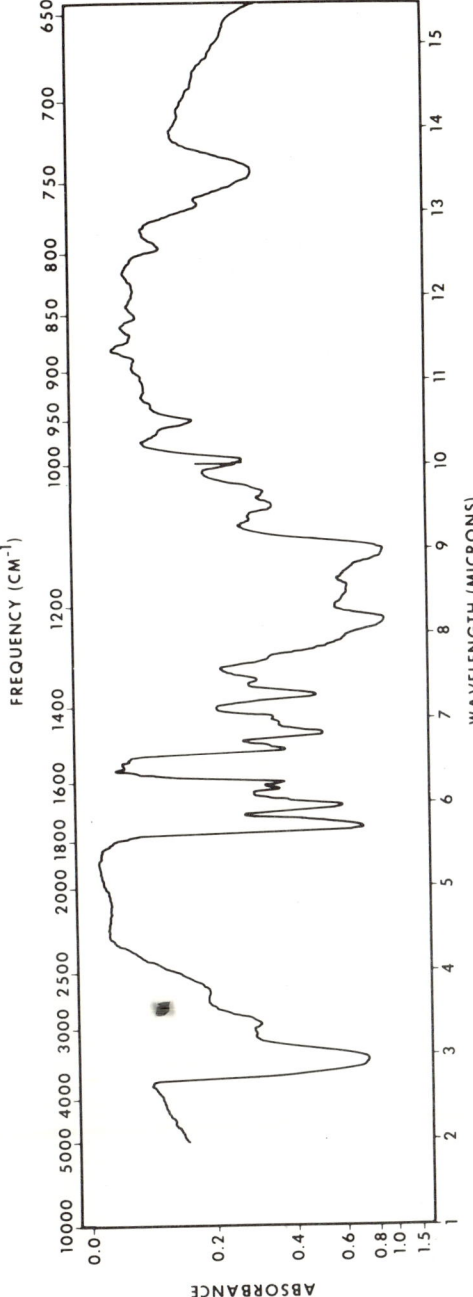

Fig. 2. IR spectrum of vincristine sulfate in KBr pellet; instrument: Perkin-Elmer 221

The similarities and differences in the IR spectra of compounds related to vincristine are discussed in several papers [25-29].

2.7 Nuclear Magnetic Resonance Spectrum

The low resolution NMR spectrum[16] of vincristine sulfate is shown in Figure 3. NMR was very useful in elucidation of the structure of vincristine[2]. The application of nuclear magnetic resonance spectroscopy to help determine the structures of vinblastine, vincristine, and other Vinca rosea alkaloids is reviewed in several papers [27, 28, 29].

2.8 Mass Spectrum

Mass spectral data for vincristine were not found in the literature. However, the type of transmethylation reactions observed by Bommer et al.[6] in recording the mass spectrum of vinblastine free base (see Vinblastine Sulfate profile, this publication) also occur with vincristine[30]. Occolowitz[30] has recently observed that the mass spectrum of vincristine sulfate is essentially free of peaks due to transmethylation and that the intensity of the molecular ion peak (824) is negligible, while a characteristic peak at m/e = M+ - 18 occurs. He suggests that if this is a general phenomenon of the dimeric Vinca alkaloids, then possibly comparison of the mass spectra of the compounds as the free base and as the sulfate salt may aid in interpretation of the mass spectral data.

2.9 pK Values

VCR free base has two titratable basic groups having pK_a values of 5.0 and 7.4 as determined by electrometric titration in 33% dimethylformamide[1].

2.10 Thermogravimetric Analysis

A thermogravimetric analysis[15] of VCR sulfate (Lilly reference standard lot P-89453) was run using the Du Pont 950 Thermogravimetric Analyzer at a heating rate of 5° C.

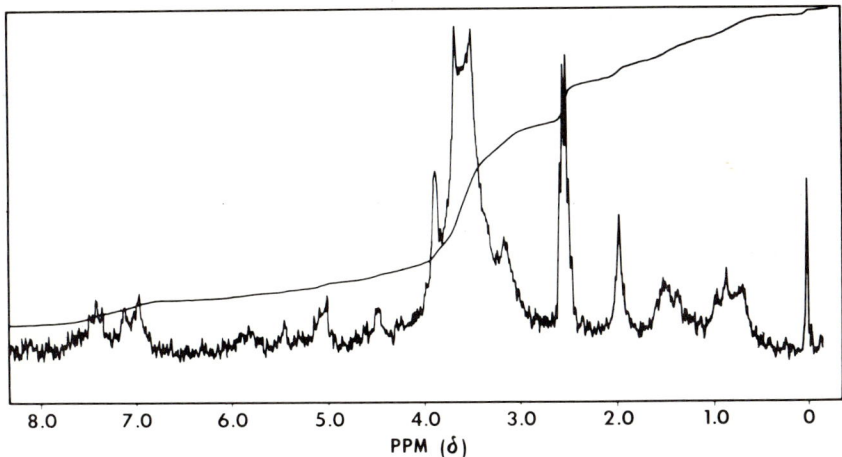

Fig. 3. NMR spectrum of vincristine sulfate in deuterated dimethyl sulfoxide; instrument: Varian HA-60

per minute under nitrogen flowing at 40 cc. per minute. Moderately rapid weight loss was observed from the start of the run at 33° C. through 110° C., 7.6% weight loss occurring over this temperature range. Weight loss from 110° C. to 177° C. was very slow amounting to a total loss of 7.9% at 177° C. Above 177° C. the rate of weight loss gradually increased at first and then became quite rapid. At 207° C. 8.8% loss had occurred, at 217° C. 10% loss, at 237° C. 15% loss, and at 276° C. a total loss of 25% was observed.

2.11 <u>Differential Thermal Analysis</u>
A differential thermal analysis of vincristine sulfate (Lilly reference standard lot P-89453) has been run[15] using the Du Pont 900 Differential Thermal Analyzer at a heating rate of 20° C. per minute. An extremely broad endotherm peaking at 127° C. was observed. This appears to correlate well with the loss of water of hydration and/or solvent of crystallization observed in the TGA analysis (Section 2.10).

3. <u>Methods of Preparation</u>
Vincristine sulfate has been obtained by addition of aqueous or ethanolic H_2SO_4 to solutions of vincristine base in methanol, ethanol, or acetone[13]. The acidified mixture was evaporated in vacuo, and the vincristine sulfate was recrystallized from absolute ethanol. It has also been obtained directly as a precipitate from certain mixtures of alkaloids in ethanol solution acidified with H_2SO_4 after removal of a preliminary precipitate which formed upon chilling for 24 hours[13]. Vincristine sulfate separated from these mother liquours after about five days of standing at room temperature.

Svoboda first obtained vincristine by rechromatography of the post-vinblastine eluates collected during the isolation of vinblastine from <u>Vinca rosea</u>[21] (also see Vinblastine Sulfate profile, this publication) and subsequently subjecting certain fractions of these new eluates to a gradient pH extraction process[1,13]. Vincristine was extracted from these fractions at pH levels 4.90, 5.40, and 5.90 and crystallized from methanol. By a somewhat shortened method[13] he combined the post-vinblastine eluates and mother liquours from the isolation of vinblastine, and extracted these into citric acid solution, adjusted the pH to about 4.4 with NH4OH and extracted with benzene, and then raised the pH to about 7.0 and again extracted with benzene. The latter extract was chromatographed and vincristine obtained from appropriate fractions as described above.

4. Methods of Analysis

4.1 Direct Spectrophotometric Analysis

Purified VCR sulfate may be assayed by comparison of its UV absorbance to that of reference material[7]. Sample and reference standard are diluted in methanol to give a concentration of about 20 mcg. VCR sulfate per milliliter. The absorbances are measured and compared in 1 cm. cells at the maximum at about 297 nm using methanol in a reference cell.

4.2 Colorimetric Analysis

The colorimetric method of Jakovljevic which was first used to assay vinblastine sulfate[8] may also be employed to assay vincristine sulfate[9]. The method is summarized in the Vinblastine Sulfate profile in this publication. The absorptivity of vincristine sulfate at the maximum absorption peak is only 75% of that exhibited by vinblastine sulfate at the same peak[9]. The method is known to be useful in measuring the stability of vinblastine sulfate and may possibly be useful in the measurement of vincristine sulfate stability[10].

4.3 Bioassay Methods

A bioassay using the acute lymphocytic P-1534 leukemia transplanted in DBA/2 mice was very useful in screening alkaloidal fractions of Vinca rosea for anti-tumor activity[32]. Activity was measured in terms of the percent increase in average survival time of mice treated with these fractions over that of untreated control mice. The observation of "indefinite" survivors among DBA/2 mice implanted with the P-1534 leukemia and treated with certain crude fractions which were chemically free of leurosine and vinblastine[33] led to the isolation of vincristine[1,28].

Dixon et al.[17] have utilized a KB cell culture system for the quantitative determination of VCR (or a cytotoxic product thereof) in sera from mice, rats, dogs, and monkeys. Standard inhibition curves were constructed by adding known amounts of vincristine to normal control serum, serially diluting, and adding to KB cell cultures. Activity levels were measured by determining the protein content of the cultures 72 hours after addition of serum from drug treated animals or control animals. The limit of detection was 0.1 mcg. of VCR per milliliter. They have proposed that if blood levels in humans are similar to that of monkeys, rats, or dogs this method may possibly be applicable to the study of the physiological disposition of vincristine in man.

Hirshaut et al.[18] have reported using long-term human leukocyte cultures for the bioassay of vincristine in the serum of patients with acute leukemia. Standard dose-response curves were constructed using known amounts of vincristine and the patient's serum prior to drug administration. Cell counts were made on day 0 and day 7. Circulating blood levels were estimated by comparing cell kills obtained after drug administration to cell kills obtained prior to drug administration using the standard curve prepared with the same patient's serum.

4.4 **Colorimetric Identification**
Jakovljevic et al.[9] have noted distinctive color reactions of various Vinca rosea alkaloids. Three reagents were employed: (a) a 1% solution of ceric ammonium sulfate in 85% phosphoric acid, (b) a 1% solution of ferric ammonium sulfate in 85% phosphoric acid, and (c) a 1% solution of ferric ammonium sulfate in 75% sulfuric acid. The test is run using 200-300 mcg. of the free base in one ml. of test reagent. Reagents (a) and (c) give a color at room temperature while reagent (b) must be heated 10 minutes in a water bath. Thus, vincristine gives with reagent (a) a blue violet color, with reagent (b) after 10 minutes heating a pink color, and with reagent (c) a blue color which changes to gray-blue. This article gives the colors formed by the three reagents with thirteen different Vinca rosea alkaloids including the four oncolytic alkaloids vincristine, vinblastine, leurosine, and leurosidine.

4.5 **Thin Layer Chromatographic Analysis**
Thin layer chromatography has proven to be a useful and efficient means of identification, separation, and purity examination of vincristine sulfate or the free base. The reader is referred to Section 4.4 of the Vinblastine Sulfate profile in this publication which contains several TLC methods and references applicable to vincristine sulfate or the free base. Cone et al.[19] have reported the separation of vincristine, leurosidine, and either vinblastine or leurosine as free bases on alumina by developing first in 100% ethyl acetate followed by a second development in ethyl acetate-absolute ethanol (3:1). Svoboda also used this system[1] and reported R_f values as follows: 0.54 for vincristine, 0.37 for leurosidine, and 0.73 for vinblastine. Detection was by means of Dragendorff's reagent. Two dimensional TLC may be useful in separation of more complex mixtures as was noted by

Farnsworth and Hilinski[11].

The R_f values of vincristine free base in a few TLC systems are given in Table I. Detection in each case was by spraying with 1% ceric ammonium sulfate in 85% phosphoric acid.

Table I

Reference	Adsorbent	Eluent	R_f
9	0.5 N LiOH/alumina	absolute ethanol-acetonitrile (5:95)	0.51
11	silica gel G	chloroform-methanol (95:5)	0.16
12	silica gel G	ethyl acetate-absolute alcohol (3:1)	0.18
12	silica gel G	n-butanol-glacial acetic acid-H_2O (4:1:1)	0.16
12	silica gel G	methanol	0.39

5. <u>Stability - Degradation</u>

   5.1 <u>Dry Thermal Degradation</u>[22]

Vincristine sulfate is quite similar to vinblastine sulfate in regard to its stability when exposed to "dry" heat. Very slight degradation is observed when the dried material is heated in sealed containers for up to 16 hours at 100° C., but when exposed to normal atmosphere at 100° C. about 50% is degraded in 16 hours. The major degradation product may be isolated from the degraded mixture by thin layer chromatography [silica gel GF, C_6H_6-$CHCl_3$-$(C_2H_5)_2NH$, 50:50:5] of the degraded mixture. It appears as an unidentified immobile spot at the point of application. In contrast to the degradation of vinblastine sulfate the desacetyl derivative of vincristine does not form under these conditions (see Vinblastine Sulfate profile, this publication).

5.2 Hydrolysis[22]

Aqueous solutions of vincristine sulfate at about pH 4.5 are essentially stable for up to three hours at 90° C. Lowering the pH of the solution substantially decreases the stability of the compound. Aqueous solutions of the compound at pH 2 maintained at 50° C. for 16 hours were found to contain only 10-20% of unchanged vincristine sulfate. Thin layer chromatography [silica gel GF, C_6H_6-$CHCl_3$-$(C_2H_5)_2NH$, 50:50:5] of an alkaline extract of the hydrolyzed mixture indicates that the degradation product is primarily one substance (unidentified) exhibiting a higher R_f than vincristine.

5.3 As Free Base

Greenius et al.[23] have stated that the Vinca alkaloids, especially as the free bases, are rather unstable. Farnsworth and Hilinski[11] have reported that vincristine as the free base decomposes under ordinary storage conditions. However, they found that benzene solutions of vincristine base were stable for several weeks if kept frozen. Similarly, Jakovljevic et al.[9] have reported that 1% solutions of vincristine in chloroform are stable under refrigeration for at least 24 hours.

6. Metabolism

The extremely high toxicity of vincristine, and therefore use of very low dosages[24], has undoubtedly limited investigation of the metabolic disposition of vincristine in animals and in humans. Neither tissue distribution studies nor isolated drug metabolic products were observed in the literature. Dixon et. al.[17] have charted concentrations of circulating blood levels over a period of time following single injections of vincristine sulfate in mice, rats, dogs, and monkeys.

Using the KB cell culture assay system mentioned in Section 4.3 to measure the concentration of drug in the sera of these animals, they found that concentration fell rapidly from peak levels attained shortly after injection to very low levels. Hirshaut et al.[18] estimated circulating blood levels following injection of the drug by using long-term human leukocyte cultures for bioassay of the serum of three patients with acute leukemia. After 2 mg. doses (i.v.) they reported the vincristine half-life was 75 minutes with a peak level of 0.4 mcg. per milliter occurring five minutes after injection.

REFERENCES

1. G. H. Svoboda, Lloydia 24, 173-178 (1961).
2. N. Neuss, M. Gorman, H. E. Boaz and N. J. Cone, J. Am. Chem. Soc. 84, 1509-1510 (1962).
3. United States Adopted Names (USAN) No. 5, p. 91 (1967).
   United States Pharmacopeial Convention, Inc., 4630 Montgomery Avenue, Bethesda, Md. 20014.
4. J. W. Moncrief and W. N. Lipscomb, J. Am. Chem. Soc. 87, 4963-4964 (1965); Acta Cryst. 21, 322-331 (1966).
5. N. Neuss, M. Gorman, W. Hargrove, N. J. Cone, K. Biemann, G. Büchi and R. E. Manning, J. Am. Chem. Soc. 86, 1440-1442 (1964).
6. P. Bommer, W. McMurray and K. Biemann, ibid., 86, 1439-1440 (1964).
7. United States Pharmocopeia XVIII, p. 773, Mack Publishing Company, Easton, Pa. 18042 (1970).
8. I. M. Jakovljevic, J. Pharm. Sci. 51, 187-188 (1962).
9. I. M. Jakovljevic, L. D. Seay and R. W. Shaffer, ibid., 53, 553-557 (1964).
10. I. M. Jakovljevic, Eli Lilly and Company, personal communication.
11. N. R. Farnsworth and I. M. Hilinski, J. Chromatog. 18, 184-188 (1965).
12. N. R. Farnsworth, R. N. Blomster, D. Damratoski, W. A. Meer and L. V. Cammarato, Lloydia 27, 302-314 (1964).
13. G. H. Svoboda, A. J. Barnes, Jr., and R. J. Armstrong, U. S. Patent 3,205,220, September 7, 1965.
14. R. C. Tiemeier, Eli Lilly and Company, personal communication.
15. A. D. Kossoy and C. D. Underbrink, Eli Lilly and Company, personal communication.
16. R. Laughlin, Eli Lilly and Company, personal communication.

17. G. J. Dixon, E. A. Dulmadge, L. T. Mulligan and L. B. Mellet, Cancer Res. $\underline{29}$, 1810-1813 (1969).
18. Y. Hirshaut, G. Weiss and E. Blackham, Clin. Res. $\underline{16}$, 360 (1968).
19. N. J. Cone, R. Miller and N. Neuss, J. Pharm. Sci. $\underline{52}$, 688-692 (1963).
20. R. E. Mensel, Eli Lilly and Company, personal communication.
21. G. H. Svoboda, N. Neuss and M. Gorman, J. Am. Pharm. Assoc., Sci. Ed. $\underline{48}$, 659-666 (1959).
22. R. L. Hussey, Eli Lilly and Company, personal communication.
23. H. F. Greenius, R. W. McIntyre and C. T. Beer, J. Med. Chem. $\underline{11}$, 254-257 (1968).
24. E. Frei, III, Lloydia $\underline{27}$, 364-367 (1964).
25. G. H. Svoboda, M. Gorman, N. Neuss and A. J. Barnes, Jr., J. Pharm. Sci. $\underline{50}$, 409-413 (1961).
26. G. H. Svoboda, M. Gorman, A. J. Barnes, Jr., and A. T. Oliver, ibid., $\underline{51}$, 518-523 (1962).
27. G. H. Svoboda, I. S. Johnson, M. Gorman and N. Neuss, ibid., $\underline{51}$, 707-720 (1962).
28. I. S. Johnson, J. G. Armstrong, M. Gorman and J. P. Burnett, Jr., Cancer Res. $\underline{23}$, 1390-1427 (1963).
29. N. Neuss, I. S. Johnson, J. G. Armstrong and C. J. Jansen, Advances in Chemotherapy $\underline{1}$, 133-174 (1964).
30. J. L. Occolowitz, Eli Lilly and Company, personal communication.
31. N. Neuss, Lilly Collection of Physical Data of Indole and Dihydroindole Alkaloids, Vol. 2, Part 1, Lilly Research Laboratories, Eli Lilly and Company, Indianapolis, Indiana (1964).
32. I. S. Johnson, H. F. Wright, G. H. Svoboda and J. Vlantis, Cancer Res. $\underline{20}$, 1016-1022 (1960).
33. I. S. Johnson, G. H. Svoboda and H. F. Wright, Proc. Am. Assoc. Cancer Res. $\underline{3}$, 331 (1962).